Elements of Hypersonic Airbreather Design and Development

Ernst Heinrich Hirschel · Werner Staudacher ·
Mirko Hornung · Daniel Kliche

Elements of Hypersonic Airbreather Design and Development

Springer

Ernst Heinrich Hirschel
Zorneding, Germany

Werner Staudacher
Zorneding, Germany

Mirko Hornung
Grafing/Straußdorf, Germany

Daniel Kliche
Munich, Germany

ISBN 978-3-031-94218-1 ISBN 978-3-031-94219-8 (eBook)
https://doi.org/10.1007/978-3-031-94219-8

This Springer imprint is published by the registered company Springer Nature Switzerland AG
The registered company address is: Gewerbestrasse 11, 6330 Cham, Switzerland

Preface

Airbreathing hypersonic flight is a fascinating topic. Many attempts have been made to advance it in the direction of actual operation. Apart from military applications, which are not considered here, no study or project advanced far enough. Notable exceptions are the flights of the X-43A and the X-51A, which were pure experimental flight vehicles of rather small size.

The last two decades of the previous century have seen a number of studies of single-stage-to-orbit (SSTO) and two-stage-to-orbit (TSTO) space transportation systems. Airbreathing propulsion in form of ramjet and scramjet propulsion did play a deciding role. Another topic with much appeal was that of long-haul airbreathing hypersonic passenger flight, with flight distances ranging up to half around the globe. Such studies extended and continue into the present century. Of particular interest remained and still remains the propulsion technology. That regards the ramjet and in particular the scramjet, and also the detonation engine concept.

Basic research was and is conducted on propulsion technology and also concerning—after a, to some extent inexplicable neglect—the all deciding topic of laminar-turbulent transition of hypersonic boundary layers. The topic of real-elastic versus perfect-elastic airframe structure design technology, too, has not found the needed interest. It is of very large importance when considering the partly quite large airframes of potential hypersonic airbreathers and the necessary ground testing of their static and dynamic elastic properties, this all in view of the topic of airframe-propulsion integration.

Flight nowadays is considered to be hypersonic at flight Mach numbers above $M_\infty \approx 5$. The ramjet-propulsion domain today is seen in the interval $3 \lessapprox M_\infty \lessapprox 7$, followed by the scramjet-propulsion domain, which reaches up to $M_\infty \approx 12$. Because of the rising of dynamic pressure with M_∞, both flight domains imply flight at altitudes much higher than those of the turbojet domain. Thermal loads on the flight vehicle—depending strongly on the state of the boundary layer, either laminar or turbulent—are rising extremely with rising flight Mach number. Hence knowledge of the location, the shape and the extent of the laminar-turbulent transition zone is of utmost importance. However, this is a problem, which still is not mastered to the necessary degree.

The thermal loads topic via structural aerothermoelasticity also regards the airframe and the airframe-propulsion integration. For the latter, six airframe-propulsion integration (API-) types are defined in this book. All plays a deciding role also in the final flight-vehicle certification. The API-type plays an important role also regarding flyability and controllability of the airbreather under consideration.

All in all it must be emphasized that it is not sufficient to create the concept of an airbreathing hypersonic aircraft or that of a globe-spanning airbreathing hypersonic cruiser. The topics design, development, ground testing, verification and finally certification before commission into service begins, are of eminent importance. Hence this book is dedicated to a sketch of relevant flight vehicle design, development and engineering issues. That to a degree is incomplete, in fact even lacking important details of some phases.

In view of the urging environmental challenges one has to ask whether—civil—airbreathing hypersonic flight should still be a worthwhile topic. There are at least three answers. (1) With TSTO-systems using LH2 as propellant it could be possible to reduce the environmental impact of launch to orbit in comparison to pure rocket launch. (2) The access to space with a TSTO-system from places far in the north, which was the rationale of the German Hypersonics Technology Programme in the 1980s/1990s, is a viable reason to pursue airbreathing hypersonic flight. (3) Long-haul hypersonic passenger flight, which however, appears to be problematic in view of the said environmental impact.

Regarding the design and development of hypersonic airbreathers a large number of phenomena and problems is present, which are not existing in ordinary aircraft design. Particular problems exist with the available simulation means. The designer must be aware of the potentials and the shortcomings of both the computational and experimental tools and facilities. This mainly regards the fields of aerothermodynamics, propulsion and structural mechanics.

Since the early 1980s, when airbreathing hypersonic flight became an interesting topic, disciplinary and multidisciplinary computation and optimization made marvelous developments. The second "mathematization wave" did happen. That was due to the enormous developments of computer power and storage and to the same degree due to the very impressive progress in the field of numerical algorithms. Experimental simulation with its opto-electronic measurement techniques also flourished thanks to the computer power and storage capabilities, which became available.

Computational flow simulation allows to treat steady and unsteady flow fields past and through configurations of all kinds, including surface radiation cooling—which is the permitting factor of hypersonic flight from the structural side—and thermo-chemical phenomena. Movable surfaces as well as aerothermoelastic deformations can be handled, and in particular also heat-transfer phenomena, both by conduction and radiation.

Laminar and attached turbulent flow can well be treated, surface irregularities can be modeled to a degree. The major unsolved problem is laminar-turbulent transition. The location, the shape and the extent of the transition zone has a very high impact regarding viscous flow phenomena and thermal loads. Separation of turbulent flow can be modeled with scale-resolving methods.

Combustion modeling in ramjet and in particular in scramjet engines faces the problem of turbulent mixture and ignition due to the very short residence times. The following expansion through the exhaust nozzle with thermo-chemical freezing effects can be handled.

Experimental flow simulation is well advanced, the quiet wind tunnel technique permits extremely important research and high-speed opto-electronic diagnostic techniques allow high-resolution insight into flow phenomena. Regarding hypersonic flow the big problems are that surface radiation cooling cannot be simulated experimentally, and that the disturbance environment of laminar-turbulent transition cannot be represented to the needed degree. With high Mach numbers, small model sizes and very short flow residence times over the model, the resulting diminutive measurement times are another limitation.

Structural mechanics faces a scale of thermal loads, which is not present in low-speed aircraft. Temperatures range from the cryogenic ones in fuel tanks and parts of the ram or scram engine to those of the radiation cooled surfaces of the external flow path over the flight vehicle. They may reach significantly more than 2000 K. It must be emphasized that, in the first place, only surface radiation cooling of the external flow path makes hypersonic flight of any kind possible by keeping surface and structure temperatures in the limits of available structural material. In the internal flow path, in particular in the engine's combustion chamber, even higher temperatures are present. Active cooling is a necessity there.

As a consequence, depending on the maximum flight speed and its duration, different materials, ranging from metallic to ceramic ones, become necessary. Subcritical surface imperfections are required, as well as a coating, which permits effective surface radiation cooling along the external flow path, and in general a low surface catalycity and anti-oxidation properties. Aerothermoelasticity plays a large role. Depending on the kind of airframe-propulsion integration—we distinguish five types of them—it can be a deciding factor regarding the flyability of the vehicle.

One problem of very high importance is that the mandatory dynamic structural tests of the assembled airframe with a hot primary structure at true flight temperatures appear not to be possible. That of course depends on the overall size of the flight vehicle, which in some studies goes up to nearly 100 m in length. A computational simulation is not possible because of the joint-modeling problem, which appears far from being solved. Joint modeling brings in scale/simulation problems comparable to the scale problem of turbulent flow.

Ground testing at true flight temperatures of the fully assembled propulsion system—from the inlet through the engine as such to the nozzle—of large hypersonic airbreathers extending over up to 30 m length also appears not to be feasible with the required means and facilities.

In both cases, ways to overcome these problems must be found and have been proposed, for instance, with the Hypersonic Technology Development and Verification Concept of the former German Hypersonics Technology Programme.

Regarding flyability and controllability of hypersonic airbreathers, a range of other items play a large, even deciding role, but are not addressed.

This sketch of design and development issues of hypersonic airbreathers shows the range of topics, which are treated in this book. However, by far not all is discussed in depth. Some general design problems are identified and illustrated but solutions are not discussed in detail. The book rather aims for an introduction into the field.

Overall, very little material from design studies is available, most from the former German Hypersonics Technology Programme. Regarding winged re-entry flight the "Lessons Learned" from the Space Shuttle Orbiter's first five flights, edited by J. P. Arrington and J. J. Jones (NASA CP-2283 (1983)), give a very valuable summary of technical details. For airbreathing hypersonic flight it is more or less only the book "Road to Mach 10—Lessons Learned from the X-43A Flight Research Program" by C. Peebles (Library of Flight, AIAA, Reston, Va. (2008)) with the 29 technical papers in the annex, which conveys much of the gained knowledge.

The authors of this book worked for almost five decades in different fields of hypersonics: at the German Aerospace Research Establishment (DVL/DFVLR, now DLR), in industry (MBB/Dasa, now Airbus), Industrieanlagen-Betriebsgesellschaft (IABG)), and in research and teaching at the Technical University Aachen (RWTH), the Technical University München (TUM), the University of the Armed Forces München and the University Stuttgart. They were involved in many major technology programs and projects of winged re-entry flight and hypersonic airbreathing flight.

The book is intended for graduate students, doctoral students, design, and development engineers, and technical managers. We believe that airbreathing hypersonic flight has a bright future. We hope that our book will help the reader to familiarize oneself with a number of problems and approaches regarding the design and development of airbreathing hypersonic flight vehicles.

Zorneding, Germany — Ernst Heinrich Hirschel
Zorneding, Germany — Werner Staudacher
Grafing, Germany — Mirko Hornung
München, Germany — Daniel Kliche
February 2025

Acknowledgements The authors are much indebted to several persons, who read the book or parts of it, and provided critical and constructive comments.

First of all, we would like to thank G. Simeonides and C. Weiland, who read all of the manuscript. Their suggestions and input were very important and highly appreciated.

Factual information, data, and illustrative material was made available for the book by many colleagues. We wish to thank M. Böswald, Ch. Breitsamter, O. Brodersen, G. De Chiara, Th. Eggers, M. Franze, R. Friedrich, H.-U. Georg, S. Görtz, M. H. Gojny, J. Häberle, St. Hein, S. Karl, H. Kuczera, F. Lu, J. Martinez, L. Reimer, H. Rick, W. Schröder, J. Steelant, H. Weihs, and C. Weiland.

Special thanks go to C. Weiland, who contributed illustrations for the book.

Ernst Heinrich Hirschel
Werner Staudacher
Mirko Hornung
Daniel Kliche

Competing Interests The authors have no competing interests to declare that are relevant to the content of this manuscript.

Contents

Chapter 1
Introduction

Hypersonic airbreathers in the context of this book are winged and also non-winged hypersonic flight vehicles with either ramjet or scramjet (supersonic combustion ramjet) propulsion. This of course implies turbojet propulsion for the domains of subsonic, transonic and supersonic flight Mach numbers. The exceptions are the launch from a carrier aircraft and/or with a booster rocket used in the first stage of flight.

The intent of the book is to give the reader an outline of the major design and development issues of airbreathing hypersonic flight vehicles. It is not intended to give solutions to particular design problems, but to identify those, to show their importance and to give reference to the pertinent literature and to further reading. Very important is to realize that the design and development issues of such flight vehicles basically are different compared to those of winged or non-winged re-entry vehicles.

Predominantly we consider flight vehicles, which start from the ground. Much of the content of the chapters of course also holds for other launch modes.

In the following Sect. 1.1 we give an overview of studies and projects of airbreathing hypersonic flight. Section 1.2 then looks at three classes of hypersonic flight vehicles of which the cruise and acceleration vehicles (CAVs), and the ascent and re-entry vehicles (ARVs) are the topic of this book.

A short historical survey of ramjet and scramjet propulsion is the topic of Sect. 1.3. Four key technologies and five types of airframe-propulsion integration are considered in Sects. 1.4 and 1.5. Particular development problems, namely requirements creep, obsolescence and mass issues, which are of high significance especially for airbreathing hypersonic flight vehicles, are looked at in Sect. 1.6. Closing the introduction a short overview is given over the contents of the chapters of the book.

E. H. Hirschel et al., *Elements of Hypersonic Airbreather Design and Development*,
https://doi.org/10.1007/978-3-031-94219-8_1

1.1 Airbreathing Hypersonic Flight: Overview of Studies and Projects

Airbreathing hypersonic flight, i.e., airbreathing flight at Mach numbers $M_\infty \gtrsim 5$, see Sect. 2.1.1, became a possibility with the maturation of ramjet and later scramjet propulsion, Sect. 1.3. Ramjet propulsion first concerned mainly military applications and later space transportation systems, whereas presently military scramjet applications are emerging worldwide. This section gives a short overview of the projects and the achievements so far. A global overview and a look in particular at European efforts up to about the year 2011 is given in [1]. That mainly concerns re-usable space transportation systems with single-stage-to-orbit (SSTO) systems and the lower stages of two-stage-to-orbit (TSTO) systems.

In 1986 US President Ronald Reagan was asking for an *Orient Express*, which with a flight Mach number $M_\infty = 25$ would permit to reach low Earth orbit or to fly from Dulles Airport in Washington DC to Tokyo in about two hours. That call led to a number of proposals and work on SSTO-systems—and on airbreathing propulsion systems—such as the National Aerospace Plane (NASP), with the technology demostrator X-30. NASP was finally abandoned due to unsurmountable problems with the hydrogen-fueled ram/scramjet propulsion system, airframe-propulsion integration, and the structure and materials concept. The program was canceled in 1993.

In Europe at that time a number of proposals, system and technology studies and also projects with airbreathing first stages of TSTO-systems and airbreathing/rocket SSTO-systems was initiated and finally also terminated [1].

In the U.K. it were SSTO-concepts, first the Horizontal Take-Off and Landing concept HOTOL, which was pursued from 1982 to 1988, followed by the Interim HOTOL and finally the SKYLON concept, which was terminated in 2016.

In France, beginning around 1986, several TSTO- and SSTO-concepts were studied, for instance the SSTO-system ORIFLAMME, and the TSTO-systems RADIANCE and STAR-H. Finally all converged into the program PREPHA in the period 1992 to 1996. That basically was a research and technology program on hypersonic scramjet propulsion investigating topics like operation and flight-vehicle integration, computational tools for aerothermodynamics and combustion, and others.

Germany, from 1987 to 1995, had the Hypersonics Technology Program with the TSTO-reference concept SÄNGER [1], see also [2]. Studied also were several hypersonic technology experimental aircraft versions under the name HYTEX [1]. In the field of aerothermodynamics a cooperation with Norway and Sweden started in 1990. Three Collaborative Research Centers, funded by the Deutsche Forschungsgemeinschaft (DFG) at Aachen, Stuttgart and Munich together with the German Aerospace Center (DLR) worked on related topics, with the TSTO-system ELAC as reference concept [3].

The European Space Agency (ESA) initiated the Future European Space Transportation Investigations Program (FESTIP), which lasted from 1994 to 1998 [1]. Conceptual design work was performed for 18 system concepts, only two were

TSTO-systems with an airbreathing lower stage. The integrated international concept team had also some experts from Russia.

Since the beginning of the 1980s special efforts were spent on the advancement of scramjet technology. NASA and CIAM performed flight tests over the Soviet Union. Later also France joined in, see, e.g., [4, 5]. The four HyShot flight tests—with hydrogen as propellant—in Australia, performed by the University of Queensland with different international partners, were dedicated to supersonic combustion experiments, see, e.g., [6, 7]. HyCAUSE, i.e., Hypersonic Collaborative Australian/United States Experiment, was a flight experiment with a configuration with a three-dimensional inward-turning inlet, in contrast to the usually two-dimensional inlet [8].[1] The flight on June 15, 2007, which was to reach $M_\infty = 10$, was not successful.

The program Hyper-X of NASA was to demonstrate the performance of a hydrogen-propelled scramjet integrated into an aircraft. Three flight tests of the X-43A were attempted, the first one failed, in March 2004 with the second one $M_\infty = 6.83$ was reached, and with the third one finally $M_\infty = 9.68$ in November 2004, [9, 10], see also Sects. 5.1.1 and 6.4.

Hyper-X was the first of hypersonic flight vehicles in the Next Generation Launch Technology (NGLT) program of NASA [11]. NGLT had the goal to advance hypersonic airbreathing vehicle system technologies in view of future space launch systems in general. In the meantime military applications in the form of hypersonic cruise missiles with scramjet propulsion became an urgent topic. Worldwide efforts started to develop corresponding systems for the low hypersonic Mach number domain around $M_\infty \approx 5$ to 6.

In the years 2003 to 2010 in HyFly a dual-combustion ramjet (DCR), i.e., a ramjet with a combustion chamber for both subsonic and supersonic combustion and with a circular inlet, was tested. HyFly, funded by the U.S. Navy and DARPA, was to investigate the technology of a hypersonic tactical guided ramjet missile with jet propellant JP-10 [12]. Several flight tests were performed, the last one in the year 2010, but none led to a full success. In the year 2020 the DCR technology again became a topic of interest with the hypersonic cruise missile designated HyFly 2.

The waverider X-51A of the US Air Force on May 26, 2010 performed the, until then, longest hypersonic flight of a scramjet propelled vehicle [13]. The flight with the Mach number $M_\infty = 5$ lasted 210 s. The propellant was JP-7, which also was used to cool the scramjet engine.[2] The attempt on June 13, 2011 to reach $M_\infty = 6+$ failed due to inlet unstart.

In the Soviet Union already in the 1960s a TSTO-system, the Spiral 50–50, with an airbreathing lower stage was pursued. In the 1990s in the Oryol program the TSTO-system MIGAKS with an airbreathing lower stage was investigated. Military developments followed. In the year 2020 Russia claimed to have reached $M_\infty = 8$ with the cruise missile 3M22 Zirkon. Its NATO code for the ship-based version is SS-N-33 Zirkon.

[1] We use the notation 'inlet' instead of 'intake'.

[2] For the implications of JP-7 see Sect. 2.6.

Table 1.1 Technology readiness levels (NASA)

TRL 1	Basic principles observed and reported
TRL 2	Technology concept and/or application formulated
TRL 3	Analytical and experimental critical functions and/or characteristic proof of concept
TRL 4	Component and/or breadboard validation in laboratory environment
TRL 5	Component and/or breadboard validation in relevant environment
TRL 6	System/subsystem model or prototype demonstration in a relevant environment (ground or space)
TRL 7	System prototype demonstration in a space environment
TRL 8	Actual system completed and "flight qualified" through test and demonstration (ground or flight)
TRL 9	Actual system "flight-proven" through successful mission operations

The European Union funded the Long-Term Advanced Propulsion Concepts and Technologies (LAPCAT) study, coordinated by ESA-ESTEC. It followed ATLAS-I, ATLAS-II, FAST20XX, HEXAFLY and STRATOFLY. LAPCAT I lasted from the years 2005 to 2008 with 12 partners from industry, research institutions and universities. LAPCAT II followed from 2008 to 2012, HEXAFLY-INT from 2014 to 2019 and STRATOFLY from 2018 to 2020. The overarching objective of these studies and projects was to advance the technology of airbreathing hypersonic flight in view of possible future antipodal connections [14–16], similar to the aforementioned Orient Express concept of the U.S. It is expected that the Technology Readiness Level TRL = 6, Table 1.1, can be reached by the year 2035 for hypersonic transatmospheric flight.

In the last couple of years there was seen a worldwide turn of airbreathing hypersonic flight toward hypersonic cruise missiles, presently with ramjet propulsion only. Such missiles, together with boost-glide vehicles, the latter without propulsion unit, today pose the largest technological challenges of flight-vehicle development.

Sustained hypersonic flight capability for future military applications already was a topic at NATO in the 1990s [17]. Considered were long range immediate-reaction reconnaissance, high speed interception of air targets, long range precision strike against hardened and/or time critical targets, and also access to space. As strategic benefit was seen the inherent reduction in time-to-target and low vulnerability.

It can be expected that in the future a technological spillover from present military applications to civil hypersonic flight, also to TSTO- and SSTO-systems may happen.

1.2 Hypersonic Flight Modes and the Related Flight Vehicles

When speaking of hypersonic flight—we only treat flight in the Earth's atmosphere—one has to distinguish between at least three modes and the associated classes of flight vehicles. The different modes lead to different configurational and multidisciplinary design and development problems and also to different flow-physical challenges.

The three flight modes are re-entry flight, cruise and acceleration flight as well as ascent and re-entry flight. Accordingly we speak of re-entry vehicles (RVs), cruise and acceleration vehicles (CAVs), and ascent and re-entry vehicles (ARVs).[3] The latter two vehicle classes are the topic of this book.

Re-entry vehicles (RVs)—we consider only re-entry flight from low Earth orbit (LEO)—are either non-winged vehicles, i.e., capsules, or winged vehicles. Capsules are in operation from the 1960s on and presently dominate the field. Winged re-entry vehicles are the former US Space Shuttle Orbiter, the Russian BURAN with only one flight, the abandoned European HERMES project, the abandoned European-US X-38 project, the abandoned Japanese Hope-X project, the abandoned German HOPPER project, and presently the active Boeing X-37 and the Dream Chaser of the Sierra Nevada corporation. Their common mission is an aero-braking flight from LEO to the Earth surface. Characteristic features of this vehicle class are listed in the second column of Table 1.2.

Cruise and acceleration vehicles (CAVs), third column of Table 1.2, are flight vehicles with airbreathing propulsion, which encompasses turbojet modes, ramjet modes, and scramjet modes. These vehicles, up to recent times were mainly in the conceptual phase. Cruise vehicle means air-transportation flight, whereas acceleration flight means flight of, for instance, lower stages of two-stage-to-orbit (TSTO) space transportation systems, up to the separation of the upper orbital stage.

The third vehicle class, ascent and re-entry vehicles, (ARVs), includes single-stage-to-orbit vehicles (SSTO) with either airbreathing propulsion or mixed air-breathing/rocket propulsion, fourth column of Table 1.2. Also here airbreathing vehicle developments remained at the conceptual level. Examples are, for instance, the US American National Aerospace Plane (NASP) or X-30 and the British HOTOL.

The difference between the first two vehicle classes is evident when considering their missions. Because RVs basically fly a braking trajectory, they have a blunt shape in order to achieve a high aerodynamic drag. This is evident when looking at capsules, which however fly at a small angle of attack in order to allow for trajectory corrections on a limited scale, see, e.g., [19]. Winged re-entry vehicles have a blunt nose, and blunt leading edges, too. In addition, in the high Mach number regime they fly at a high angle of attack. With the Space Shuttle Orbiter that was around $\alpha = 40$.

[3] In [18] as well as in [19] the terms *re-entry vehicle* (RV), *cruise and acceleration vehicle* (CAV) and *ascent and re-entry vehicle* (ARV) are used, and are employed here, too.

Table 1.2 Comparative consideration of the aerothermodynamic features and multidisciplinary design features of three major classes of hypersonic vehicles

Item	Re-entry vehicles (RV's)	Cruise and acceleration vehicles (CAV's)	Ascent and re-entry vehicles (ARV's)
Mach number range	28–0	0–7 to 12	0 to 7/12–28
Configuration	Blunt	Slender	Opposing design requirements
Flight time	Short	Long	Long(?)/short
Angle of attack	High	Low	Low/high
Drag	High	Low	Low/high
Aerodynamic lift/drag ratio	Low	High	High/low
Flow field	Compressibility-effects dominated	Viscosity-effects dominated	Viscosity-effects/compressibility-effects dominated
Laminar-turbulent transition	Important	Very important	Opposing situations
Thermal surface effects: 'viscous'	Not important	Very important	Opposing situations
Thermal surface effects: 'thermo-chemical'	Very important	Important	Opposing situations
Thermal loads	High	Medium/high	Medium/high
Thermo-chemical effects	Strong	Weak/medium/strong	Medium/strong
Rarefaction effects	Initially strong	Weak	Medium/strong
Critical components	Trim and control surfaces	Inlet, nozzle/after-body, trim and control surfaces	Inlet, nozzle/afterbody, trim and control devices
Special problems	Large Mach-number span	Propulsion integration, thermal management	Propulsion integration

LEO re-entry vehicles of either kind usually have a cold primary structure and a passive heat protection system, Sect. 6.1. Cooling is via surface radiation, Sect. 3.8, which means that a surface coating besides the anti-oxidation and anti-recombination properties needs to have a high radiative emissivity, which gives the windward oriented surfaces a near-black appearance. The emissivity coefficient is around $\epsilon \approx 0.8$ to 0.9. A winged re-entry vehicle has a low emissivity lee-side coating of near white appearance in order to allow to manage in orbit the heat radiated from the sun, see the former Space Shuttle Orbiter, the X-37 and the Dream Chaser.

CAVs in contrast to RVs need to have a small aerodynamic drag to accelerate and/or cruise at high speed through the atmosphere. This leads to slender configurations with small nose and leading-edge radii, which fly at small angle of attack. They may have a hot primary structure, however, with the requirement to house cryogenic fuel tanks. The vehicle surface including the lee side is radiation cooled. Contradictive requirements exist at the vehicle's nose and leading edges, Sect. 3.5. They need to have small radii in view of the aerodynamic wave drag, but sufficiently large radii in order to allow for radiation cooling that needs large boundary-layer thicknesses. Usually this leads to ceramic materials for these vehicle surface elements, Chap. 6.

We distinguish six basic types of airframe-propulsion integration (API), Sect. 1.5. With the API-types 4 and 5, up to now the most prominent ones, the lower side of the vehicle houses the propulsion system including the inlet and the nozzle. The latter cannot be a bell nozzle, because tail striking at the high angles of attack during take-off and approach and landing must be avoided. The nozzle either can be a single expansion ramp nozzle (SERN) or a linear spike nozzle. The lower side of the flight vehicle hence is a highly coupled lift and propulsion system. Demanded is a configuration with a low-drag aerodynamic shape that ensures flyability and controllability in the entire flight envelope.

These short characterizations reveal the contradicting demands on the third vehicle class of Table 1.2, the ascent and re-entry vehicle (ARV) class. To reach orbit with purely airbreathing propulsion, as was the approach with NASP, certainly is a pipe dream. For an ARV the whole sequence of turbojet, ramjet/scramjet and rocket propulsion appears to be the only possibility to reach LEO. And that certainly is far away into the future, if it is possible at all, given the different design characteristics.

1.3 Ramjet and Scramjet Propulsion

The literature on ramjet and scramjet propulsion systems is immense. Here we give a short overview only, pointing in particular to [20]. Turbojet propulsion came into practical application during the 1940s. The concept of ramjet propulsion goes back to Ren Lorin, who in 1913 made a proposal in this regard. Whether the propulsion system of the German Fieseler Fi 103, the V-1 flying bomb, called Vengeance Weapon 1, can be considered as a ramjet, is arguable. A true ramjet propelled flight vehicle was devised by Ren Leduc in France, and supersonic speeds were achieved in 1957 with the Nord Aviation Griffon II, which had turbojet/ramjet propulsion.

Supersonic missiles with ramjet propulsion, which have to start with a rocket booster, were studied worldwide during and after the Second World War and reached practical application. In the 1950s the scramjet, a ramjet with supersonic combustion, came into consideration. In the hypersonic flight regime the pure ramjet concept leads to very high pressures and temperatures in the engine, which can no more be handled.

The supersonic combustion ramjet, the scramjet, is the way out. In the whole flow path through the engine supersonic flow is present. Thus avoiding the high pressure and temperature, which would result if the flow is decelerated to subsonic flow. This of course goes together with the tremendous challenge to achieve combustion in a supersonic flow with a very short residence time of the flow in the engine's combustion chamber.

Whereas ramjet propulsion is a well understood propulsion mode, scramjet propulsion for hypersonic flight, i.e., at flight Mach numbers $M_\infty > 5$, more or less is in its infancy. The flights of the X-43A with $M_\infty = 6.83$ in March 2004 and $M_\infty = 9.68$ in November 2004 and that of the X-51A with $M_\infty = 5$ in 2010 were milestones. These flights were more or less design-point flights, the question of off-design operation appears not to have been answered. Not much is known about the Russian Zirkon cruise missile, for which a maximum Mach number of 8 is claimed. In general it appears that presently military applications are driving the scramjet technology worldwide.

When we talk about a flight vehicle with scramjet propulsion, for instance a CAV, we tacitly mean a propulsion system, which consists of a turbojet component, a ramjet, and a scramjet component. These components can come as mixed components, for instance a turboramjet or a ram-/scramjet (Dual Mode Ramjet (DMR)). Except for Chap. 4, where we treat basics of ramjet and scramjet propulsion, we simply speak of ramjet or scramjet propulsion, without reference to the turbojet and/or mixed systems. The latter may include also rocket propulsion either for the launch regime or the high speed regime of launch into orbit.

The detonation engine as propulsion means is a topic since the 1940s. It is a possible engine concept for hypersonic flight. Despite extensive research efforts no application is known up to now. In [21–23] detailed information can be found.

1.4 Key Technologies

The key technologies treated in this book are aerothermodynamics, propulsion, structure and materials, and control and guidance.

- Aerothermodynamics provides via airframe shaping and optimization the lift and with that the flyability and controllability of the flight vehicle while aiming to keep drag within acceptable limits. Along the vehicle's external flow path it further gives the mechanical loads—surface pressure and skin friction—and in particular also the thermal loads—temperature and heat flux—on the airframe. A special topic is airframe-propulsion integration, which depends on the airframe-propulsion integration (API-) type, see Sect. 1.5. Chapter 3 is devoted to an overview of the field of aerothermodynamics as a design technology.
- Propulsion has to cover the whole flight Mach-number span of the vehicle. That regards the core engines of turbo, ramjet and, if applicable, scramjet propulsion in their different combinations. Exceptions are—primarily—military applications with rocket booster for launch. The link to aerothermodynamics is via the external and in particular the propulsion flow path with the inlet system and the isolator—the duct between inlet and combustor—the combustor and finally the exhaust nozzle. The latter, depending on the type of airframe-propulsion integration—see below—can be a single expansion ramp nozzle (SERN), a bell nozzle (not likely an option for CAV-type vehicles), a plug nozzle or a linear spike nozzle. Major

aspects of ramjet and in particular scramjet propulsion are treated in Chap. 4. Chapter 5 deals with issues of airframe-propulsion integration.

- Structure and materials provide the airframe in view of the mandatory aerodynamic lift, stabilization, trim and control surfaces, the propulsion flow path, the internal volumes, the mechanical loads, and the thermal loads. The thermal loads topic has two challenges, the high temperatures present at the external surface and along the propulsion flow path, as well as the extremely low temperatures in the fuel system, if cryogenic fuel is used. The external flow path to a high degree has surface radiation cooling, whereas active cooling is mandatory for the internal (propulsion) flow path. All that is the topic of thermal management of the flight vehicle. Structure and materials are the topic of Chap. 6.
- Control and guidance is a field which supplements the above key technologies. Flight vehicle control is not only a topic of aerodynamic stabilization, trim and control surfaces, but also of the thrust-vector change with flight Mach number, if a single expansion ramp nozzle is employed. Guidance concerns the flight path of the aircraft in view of its mission including take-off and approach and landing, but also in view of the mechanical and thermal loads and the propulsion demands. A short characterization is given for instance in [19]. Critical elements are the air-data system, either as flush ports in the vehicle nose, or as opto-electronic system, for instance LIDAR, and the notch filters, which prevent the elastic airframe deformation fluctuations to enter and adversely affect the aircraft's flight control system.

1.5 Basic Types of Airframe-Propulsion Integration

The topic 'airframe-propulsion integration' directly governs the connections between the key technologies in the sense of Cayley's design paradigm, Sect. 2.2. Different integration types have emerged so far for hypersonic airbreathers, with different merits and different challenges.

A major effect is present, if forebody pre-compression is applied. That is used in order to reduce effectively the inlet capture area, but leads to a high net-thrust sensitivity depending on the angle of attack of the flight vehicle, Sect. 5.5.

Another effect is related to the SERN-type nozzles. These are necessary in order to avoid tail striking during take-off and approach and landing. Such nozzles however lead to changes of the thrust vector direction with changing flight Mach number, Sect. 5.7. Whether linear spike nozzles are a way out is not established.

There are many studies and projects regarding hypersonic flight vehicles. We tentatively select six basic Airframe-Propulsion Integration types (API-types), which are listed with increasing complexity. For the corresponding studies and projects see Sect. 1.1 above.

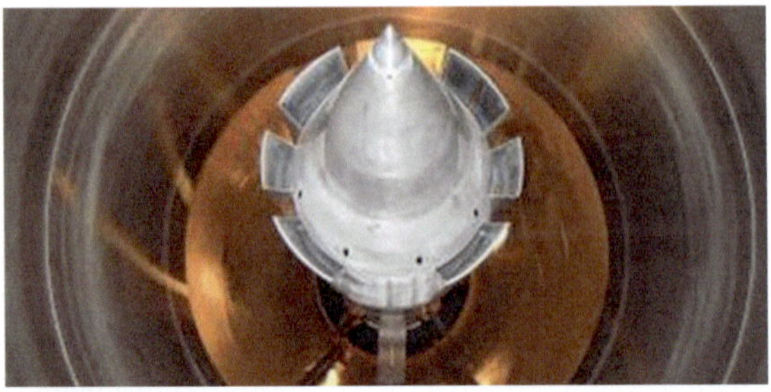

Fig. 1.1 The HyFly circular inlet [24]

Fig. 1.2 The LAPCAT-A2 configuration [14]

- API-type 0 has a circular inlet with a central cone, Fig. 1.1. The propulsion unit is also circular, the nozzle is a bell nozzle. Lift for this type is provided by wings. No direct coupling of the lift and the propulsion function is present.
  HyFly was a hypersonic cruise missile demonstrator with a dual-combustion ramjet. It was terminated. HyFly 2, the successor, also with a dual-combustion ramjet, is of API-type 3.
- API-type 1 has podded propulsion units for instance below the wing and at the wing tips, Fig. 1.2.
  This is an artist impression of the civil Mach 5 transport proposal studied in the LAPCAT program, with its origins in the SKYLON proposal. The flight vehicle with pre-cooled turbo-ramjet engines has a trapezoidal wing and a canard. There is no interaction between the fuselage plus the wing and the circular inlets of the loosely integrated propulsion units. Provided that sufficient ground clearance is given through take-off and landing, the nozzles are bell nozzles. Aerothermoelastic wing behavior will affect the engines in particular those at the wing tips.
- As API-type 2 the study STRATOFLY MR2/MR3 can be seen, Fig. 1.3.
  It has a waverider configuration with an inward-turning inlet on the upper side, like the HyCAUSE vehicle [8]. The propulsion flow path with three compression ramps, the isolator, the combustor chamber, and the bell nozzle is completely inside the airframe. Lift generation at the lower side of the aircraft hence is decoupled from the thrust generation.

Fig. 1.3 STRATOFLY MR2 with inward turning inlet [16]

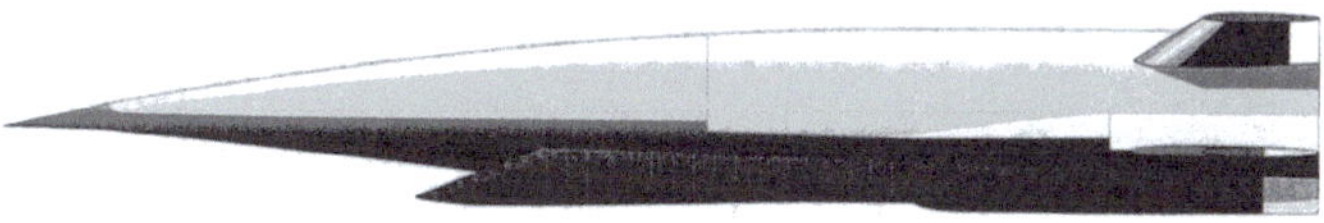

Fig. 1.4 X-51A as example of an API-type 3 flight vehicle [25]. Short pre-compression surface, $L_{pc} \approx 2\,\text{m}$, no SERN, bell nozzle

- API-type 3 appears to be typical for present-day hypersonic cruise missiles. Examples are the waverider X-51A, Fig. 1.4, HyFly 2, and the SS-N-33 Zirkon. The inlet is at the lower side of the flat forebody, which serves as short pre-compression (pc) surface, like with API-type 4 vehicles.[4] The flat forebody up to the first inlet ramp acts as pre-compression surface, reducing the inlet capture area—the inlet is a two-dimensional one—of the propulsion unit.
  The missiles, launched for instance with a rocket booster, have bell nozzles, because there is no need to approach and land horizontally. With a bell nozzle no thrust-vector change happens along the flight trajectory. If the nozzle is a swivel nozzle, used for trajectory changes/corrections, the deflections will be small. In any case the nozzle will not have the consequences, which a SERN-type nozzle has.
- API-type 4 flight vehicles also have the inlet at the lower side with a short fore-body pre-compression surface. An example is the X-43A, Fig. 1.5. The X-43A as experimental flight vehicle with scramjet propulsion was the first to reach high hypersonic flight Mach numbers. Lift generation and propulsion happens at the lower side of the flight vehicle, see the API-type 5 characterization.
  Another example is the Lockheed SR-72, where the inlets—left and right side—are at the wing roots and parts of the forebody and the wing surface act as pre-compression surfaces. Aerothermoelasticity effects seem not to play a significant role. The nozzle maybe of SERN type (X-43A) or a round or rectangular one (SR-72).

[4] Forebody: short and long, both as absolute measures, are qualitative statements related to whether major aerothermoelastic effects are to be expected.

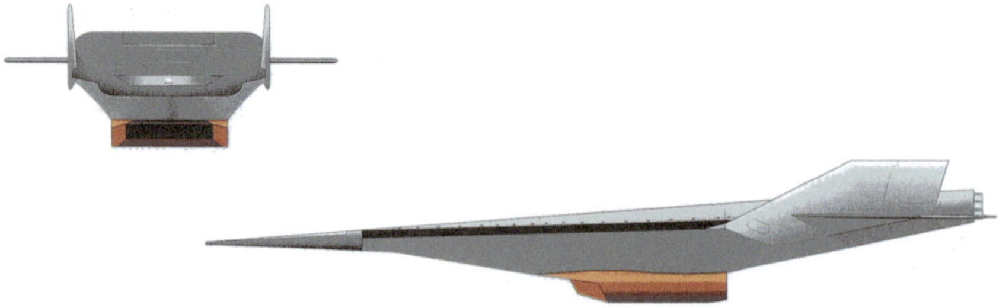

Fig. 1.5 Example of an API-type 4 flight vehicle: NASA's X-43A, short pre-compression surface, $L_{pc} \approx 0.95$ m, SERN [25]. Upper left: front view, lower middle: side view

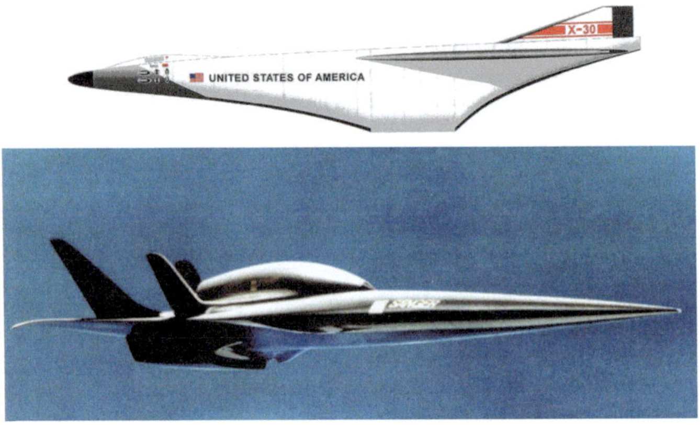

Fig. 1.6 Examples of API-type 5 flight vehicles (long pre-compression surface, SERN). Upper part: the SSTO-system NASP/X-30 [25], $L_{pc} \approx 16$ m. Lower part: the lower stage—with the upper stage mounted—of the TSTO-system SÄNGER of the German Hypersonics Technology Programme [1], $L_{pc} \approx 52$ m

- API-type 5 to a degree can be considered as a classical way of hypersonic airframe-propulsion integration with the lower side of the forebody being a long pre-compression surface. NASP/X-30, configuration 201, 1990, upper part of Fig. 1.6, as well as the lower stages of TSTO-concepts in the 1980s and the 1990s were of this type. The lower stage of the SÄNGER TSTO-system, lower part of Fig. 1.6, is also an example.

  The propulsion unit is located at the lower side of the flight vehicle. The long slender forebody minimizes the wave drag. It further provides the needed volume for fuel tanks and so on. The nozzle is of SERN type in order to permit rotation of the aircraft during take-off and approach and landing. Wings are necessary in order to provide additional lift and to assure low-speed flight qualities, see, e.g., [26].

  The lower side of the flight vehicle thus serves for both, lift generation and propulsion. Cayley's design paradigm, Sect. 2.2, hence is invalid. Problems associated

Table 1.3 Airframe-Propulsion Integration types and their characteristic complexities

Integration type	Characteristic complexities
API-type 0	Cylindrical missile body, lift is provided by wings. Circular inlet, bell nozzle, no lift-propulsion interaction
API-type 1	This type has podded engines like a transport aircraft. Circular inlets. Aerothermoelastic challenges arise with wing-tip engines, hence a coupling of lift and thrust will be a problem
API-type 2	Engine-onset flow is fully inside the body, the nozzle is of bell type. Inward turning rectangular ramp inlet. No coupling into lift and pitching moment
API-type 3	This missile type has a flat lower-side forebody, serving as pre-com-pression surface and a bell nozzle. Rectangular ramp inlet. Coupling of lift and propulsion system
API-type 4	Short forebody lower-side pre-compression and possible SERN result in coupling of lift and thrust. Rectangular ramp inlet. With large flight vehicles aerothermoelastic effects of the relatively short forebody lead to a certain net-thrust sensitivity
API-type 5	Long forebody lower-side pre-compression and SERN result in high coupling of lift and thrust. Rectangular ramp inlet. Aerothermoelastic effects of the long forebody lead to a high net-thrust sensitivity

with this type of integration are aerothermoelasticity effects, the wall-temperature difference between the lower and the upper side, the net-thrust sensitivity and the thrust-vector change of the SERN-type nozzle. Similar issues—except for the forebody aerothermoelastic effects—arise with API-type 4 flight vehicles.

These types of airframe-propulsion integration pose different problems regarding the four key technology fields sketched in the preceding Sect. 1.4. A major technological problem arises with the coupling of lift and propulsion. The coupling is zero with API-type 0, low with type 1, followed with some distance by type 2 and 3 and highest with type 4 and in particular type 5.

Of importance too is the inlet performance in terms of cross-section reduction and total-pressure recovery. This is discussed in Sect. 5.2.

In Table 1.3 the API-types are listed together with their characteristic complexities.

1.6 Particular Development Problems

Development of new and technologically highly advanced flight systems can take many years, even decades. The development of the Space Shuttle took about nine years until the first mission was flown. The fighter aircraft F-35 needed 14 years until entry into service (EIS). About ten years into EIS were needed with the passenger aircraft B787 and A350.

With long development times—but not necessarily only with them—at least three problems arise and have to be taken care of: requirements creep, obsolescence, and mass increase. They all affect the iron triangle "performance, cost, schedule".

Requirements creep means that the requirements on the product change with time, a problem more typical with military products. The tactical or the strategic setting changes with time and the customer reacts with new or modified requirements. In any case development time and cost will rise, sometimes even dramatically. The same holds for hypersonic airbreathers with their long technical maturation and development times, for the X-43A see, e.g., [10].

Obsolescence of devices, materials and general approaches is a possibility in particular with very advanced projects, which take long times to develop. Take for instance a new highly advanced material for high-temperature application which makes existing materials obsolete. Change in a late stage of development may require a revamping of the design. If accepted, costs will rise and the EIS will be shifted.

Mass increase of the airframe and/or the propulsion system during the design and the development phase of aircraft is a common fact. It often happens also during the opening of the flight envelope. A particular problem too, is the location of the center of gravity. Its position relative to the center of pressure determines stability and control requirements of the aircraft, see, e.g., [27]. Regarding the longitudinal stability, ballast in the fuselage nose or a change of the sweep angle of the wing are means of correction.[5] Tuning of the aeroelastic properties of wings with ballast can also become necessary.

For modern passenger aircraft with matured, metal-construction technology the dry mass empty increase up to EIS is usually below 2%. For composite airframes no data are yet publicly available. Commercial passenger aircraft have structural weight fractions of about 20%.

The mass increase during an aircraft development and envelope opening phase is taken into account with the so-called contingency mass. This mass margin is defined already in the flight-vehicle definition phase. In the design of fighter aircraft one today assumes a contingency mass of about 6 to 8% of dry-mass empty. From the first flight to full mission deployment another 2 to 4% mass growth are a rule.

For the TSTO system SÄNGER, Sect. 2.8, the payload/dry-mass empty ratio was 4.4% related to the lower stage, and 2.5% related to the whole TSTO system, see, e.g., [29]. The contingency mass both for the lower stage and for the whole system, was only 3.1% of dry-mass empty. This was much too optimistic. For a large air vehicle of this kind and for the propulsion and speed domain, even today only insufficient design and operation experience is available.

---

[5] Early German examples of sweep-angle correction are for instance (a) the wing of the Focke-Wulf Fw 200 Condor, where the outer wing portions had to be swept back by about 7°, and (b) the Messerschmitt Me 262, where the wing was swept back from original 4° to 18°. This still leads to the wrong impression that from the beginning the aircraft was a swept-wing aircraft. Later versions of the aircraft with truly swept wings were studied, which finally led to the pioneering P 1101 V-1 project, see, e.g., [28].

1.7 The Content of the Chapters

The following Chap. 2 discusses hypersonic airbreather design on a basic level. The differences to the design of conventional aircraft are highlighted, Trends and sensitivities are emphasized. In the next four chapters major technologies are discussed: aerothermodynamics, propulsion, airframe-propulsion integration, and structure and materials.

Chapter 3 treats the aerothermodynamic features of the external flow path, in particular high temperature real-gas effects, the thermal state of the vehicle surface and for the considered flight vehicles the extraordinarily important problem of laminar-turbulent transition. General design considerations are given and the stage integration as well as the separation process of TSTO concepts are examined.

Ramjet and scramjet propulsion are the topic of Chap. 4. The basic features of scramjet propulsion are discussed. Airframe-propulsion integration is treated in Chap. 5. Accounting of forces and moments is considered as well as forebody pre-compression and the ensuing sensitivities. Inlet and nozzle (SERN) topics are treated, also coolant flow, and the interactions of the propulsion system with the flight vehicle's trim and control.

Chapter 6 looks at structure and materials aspects, hot and cold structures, the latter with a thermal protection system, all with the respective materials, fuel-tank integration and so on. The overall problem of aerothermoelasticity is addressed.

Technology development and verification is considered in Chap. 7. Experimental ground-testing facilities, computational simulation, and also experimental flight vehicles are examined and possible Technology Development and Verification (TDV) approaches discussed. Chapter 8 sketches the German Hypersonics Technology Development and Verification Concept. Closing notes are given in Chap. 9.

In all following chapters problems are given with a solution guide and the solutions provided in Chap. 10. Properties of the Earth standard atmosphere are given in Appendix A.

Constants, functions etc. are listed in Appendix B. Symbols, abbreviations and acronyms are given in Appendix C. Copyrights of figures, a subject and a name index close the book.

References

1. Kuczera, H., Sacher, P.W.: Reusable Space Transportation Systems. Springer, Berlin, Heidelberg (2009)
2. Koelle, D.E., Sacher, P.W., Grallert, H.: Deutsche Raketenflugzeuge und Raumtransporter-Projekte. Bernard und Graefe, Bonn (2007)
3. Jacob, D., Sachs, G., Wagner, S. (eds.): Basic Research and Technologies for Two-Stage-to-Orbit Vehicles. Final Report of the Collaborative Research Centers 253, 255 and 259 of the Deutsche Forschungsgemeinschaft. Wiley-VCH, Weinheim (2005)
4. Roudakov, A.S., Semenov, V.L., Hicks, J.W.: Recent Flight Test Results of the Joint CIAM-NASA Mach 6.5 Scramjet Flight Program. NASA TM 206548 (1998)

5. Bouchez, M., Roudakov, A.S., Kopchenov, V., Semenov, V.L., Scherrer, D.: French-Russian analysis of Kholod dual-mode ramjet flight experiments. In: AIAA-Paper 2005-3320 (2005)
6. Smart, M.K., Hass, N.E., Paull, A.: Flight data analysis of the HyShot 2 scramjet flight experiment. AIAA J. **44**(10), 2366–2375 (2006)
7. Sunami, T., Itoh, K., Satoh, K., Komuro, T.: Mach 8 ground tests of the hypermixer sramjet for HyShot-IV flight experiment. In: AIAA-Paper 2006-8062 (2006)
8. Walker, S., Rodgers, F., Paull, A., Van Wie, D.M.: HyCAUSE flight test program. In: AIAA-Paper 2008-2580 (2008)
9. Marshall, L.A., Bahm, C., Corpening, G.P., Sherrill, R.: Overview with results and lessons learned of the X-43A Mach 10 flight. In: AIAA-Paper 2005-3336 (2005)
10. Peebles, C.: Road to Mach 10–Lessons Learned from the X-43A Flight Research Program. Library of Flight, AIAA, Reston, Va (2008)
11. Moses, P.L., Rausch, V.L., Nguyen, L.T., Hill, J.R.: NASA hypersonic flight demonstrators-overview, status, and future plans. Acta Astronaut. **55**, 619–630 (2004)
12. Foelsche, R.O., Leylegian, J., Betti, A., Chue, R., Marconi, F., Beckel, S., Tyll, J., Charletta, R., Bakos, R.: Progress on the development of a freeflight atmospheric scramjet test technique. In: AIAA-Paper 2005-3297 (2005)
13. Warwick, G.: Scramjet success—first X-51A waverider flight meets major objectives. Aviation Week and Space Technology (2010)
14. Steelant, J.: Sustained hypersonic flight in Europe: technology drivers for LAPCAT II. In: AIAA-Paper 2009-7240 (2009)
15. Steelant, J., Langener, T., Di Matteo, F., Hannemann, K., Riehmer, J., Kuhn, M., Dittert, C., Scheuerpflug, F., Jung, W., Marini, M., Pezzela, G., Cicala, M., Serre, L.: Conceptual design of the high-speed propelled experimental flight test vehicle HEXAFLY. In: AIAA-Paper 2015-3539 (2015)
16. Steelant, J., Langener, T.: The LAPCAT-MR2 hypersonic cruiser concept. In: ICAS paper 2014-0428 (2014)
17. N.N.: Future Aerospace Technology in the Service of the Alliance, vol. 3. AGARD-CP-600, Sustained Hypersonic Flight (1997)
18. Hirschel, E.H.: Basics of Aerothermodynamics: Second, Revised. Springer, Berlin (2015). Progress in Astronautics and Aeronautics, AIAA, Reston, Va, vol. 204. Springer, Heidelberg (2004)
19. Hirschel, E.H., Weiland, C.: Selected Aerothermodynamic Design Problems of Hypersonic Flight Vehicles. Progress in Astronautics and Aeronautics, AIAA, Reston, Va, vol. 229. Springer, Heidelberg (2009)
20. Fry, R.S.: A century of ramjet propulsion technology evolution. J. Propuls. Power **20**(1), (2004)
21. Lu, F.K., Braun, E.M.: Rotating detonation wave propulsion: experimental challenges, modeling, and engine concepts. J. Propuls. Power **30**(5), 1125–1142 (2014)
22. Shaw, I.J., Kildare, J.A.C., Evans, M.J., Chinnici, A., Sparks, C.A.M., Rubaiyat, S.N.H., Chin, R.C., Medwell, P.R.: A theoretical review of rotating detonation engines (2019). https://doi.org/10.5772/intechopen.90470
23. Jiang, Z., Teng, H.: Gaseous Detonation Physics and Its Universal Framework Theory. Springer, Singapore (2022)
24. Holden, M.S., Wadhams, T.P., MacLean, M.: Experimental studies in the LENS supersonic and hypersonic tunnels for hypervelocity vehicle performance and code validation. In: AIAA-Paper 2008-2505 (2008)
25. Gorn, M.H., De Chiara, G.: X-Planes from the X-1 to the X-60: An Illustrated History. Springer Praxis Books (2021)
26. Hirschel, E.H., Rizzi, A., Breitsamter, C., Staudacher, W.: Separated and Vortical Flow in Aircraft Wing Aerodynamics. Springer, Heidelberg (2021)

27. Chudoba, B.: Stability and Control of Conventional and Unconventional Aerospace Vehicle Configurations. Springer Aerospace Technology (2019)
28. Meier, H.U. (ed.): German Development of the Swept Wing–1935-1945. Library of Flight, AIAA, Reston, Va (2010)
29. Hirschel, E.H., Weiland, C.: Design of hypersonic flight vehicles: some lessons from the past and future challenges. CEAS Space J. **1**, 3–22 (2011)

Chapter 2
Basic Considerations of Hypersonic Airbreather Design

This chapter is devoted to a presentation of basic considerations of airbreather design for hypersonic flight, mainly for the Airframe-Propulsion Integration type 4 and 5, Sect. 1.5. Initially definitions of flight domains and domains of propulsion modes are given. After that the design procedure is discussed. Important is the observation that Cayley's design paradigm, which is the blueprint for any successful technical design, loses its validity when considering the design of airbreathing hypersonic flight vehicles of API-type 4 and 5.

It follows a consideration of the objectives of multidisciplinary design work, of the flight environment and of systems and operational constraints.

After a short look at the differences between conventional and hypersonic flight vehicle design, the implications of fuel are considered. Single-stage-to-orbit (SSTO) versus two-stage-to-orbit (TSTO) systems is the topic of the next section. The reference concept of the subsequent sections is presented. A discussion of inevitable performance trends of airbreathing hypersonic flight vehicles and a sensitivity study close the chapter.

2.1 Definitions

2.1.1 Flight Domains

To begin with we ask what is hypersonic flight. Flight speed or Mach number regimes are well defined, except for hypersonic flight. The definitions result from the occurrence of fluid-mechanical phenomena and the behavior of the lift and the drag coefficient, which we single out for our consideration. The reader must note that we consider the coefficients, not the actual forces. The latter are found after multiplication of the coefficients with the reference area A_{ref} and the dynamic pressure of the

E. H. Hirschel et al., *Elements of Hypersonic Airbreather Design and Development*,
https://doi.org/10.1007/978-3-031-94219-8_2

free-stream $q_\infty = 0.5\ \rho_\infty v_\infty^2$, where ρ_∞ is the density of the atmosphere, a function of the flight altitude—see Appendix A—and v_∞ the flight speed.

The density ρ_∞, like the temperature T_∞ and the pressure p_∞ vary with the altitude H. Density and pressure decrease monotonically with increasing altitude, the temperature decreases up to about 20 km altitude, rises then to about 50 km altitude and again decreases up to about 100 km altitude, Fig. A.1 in Appendix A.

Flight with the same flight speed at different flight altitudes hence occurs at different Mach numbers, because the speed of sound $a_\infty = \sqrt{\gamma R T_\infty}$ in the atmosphere depends on the temperature T_∞, which changes with altitude, Sect. 2.4. The ratio of specific heats γ and the specific gas constant R of atmospheric air at rest, Appendix B, can be considered as independent of the altitude up to about 100 km altitude.

The first flight domain to be considered is the subsonic domain. The flight speed is well below the speed of sound a_∞.

If the density ρ in the flow field past the flight vehicle changes only negligibly, we speak of incompressible flow with ρ = const. and note that for slender configurations at small angle of attack linear lift theory holds. In the subsonic flight domain we generally have a more or less constant zero-lift drag coefficient C_{D_0}, whereas as the lift coefficient increases with the Mach number according to the Prandtl-Glauert rule: $C_L \propto 1/\sqrt{1 - M_\infty^2}$ [1–3].

The subsonic domain extends up to the lower critical Mach number M_{crit_l} of the transonic domain. That number is reached, when supersonic flow appears over the body. Usually considered is an airfoil or a wing. The supersonic flow appears as a pocket, which is terminated by a shock wave. The transonic domain has the problem of the sudden, very strong drag rise, the drag divergence at the drag divergence Mach number $M_{dd} > M_{crit_l}$. The drag divergence is joined by a lift drop and finally by transonic buffet, which by interaction with the elastic wing structure leads to the transonic buffeting.[1] The transonic domain is one of the acceleration pinch-points of hypersonic flight vehicle design, Fig. 5.36.

The upper critical Mach number M_{crit_u} of the transonic domain is reached when the whole flow is supersonic, except for the subsonic pocket at the blunt nose of the airfoil or at the leading edge of the wing. Beyond this Mach number the supersonic domain is present. The lift coefficient drops according to the Ackeret rule $C_L \propto 1/\sqrt{M_\infty^2 - 1}$.

The upper limit of the supersonic domain, which is the lower one of the hypersonic domain, is not clearly defined. The aerodynamic force and moment coefficients as well as the shape of the bow shock of the flight vehicle and the characteristics of the supersonic part of the flow field become asymptotically independent of the Mach number at a value of it, which depends on the body shape, typically at $M_\infty \approx 6$ to 10. This is the Mach number independence principle of Oswatitsch.[2]

[1] For a compact discussion, see, e.g., Sects. 2.4 and 9.1 of [4].

[2] For a compact derivation of this principle see Sect. 6.8 of [5].

Because the body shape governs the independence, the principle is not used to define the lower bound of the hypersonic domain. Today instead one uses $M_\infty = 5$. At this Mach number vibration excitation of the air molecules is present in the front of blunt noses and leading edges, but not yet dissociation, Sect. 2.4.

2.1.2 *Turbo-, Ram and Scramjet Domains*

The propulsion flight domains do not coincide with the flight speed domains. The upper Mach number bound of turbojet propulsion today is $M_\infty \approx 4$. Beyond this Mach number ramjet propulsion is possible up to $M_\infty \approx 6$ to 7. Above these values the scramjet domain begins.

Figure 2.1 shows this graphically in terms of the specific impulse. The specific impulse I_{sp}, i.e. the impulse gained per unit mass of propellant, is considered when comparing propulsion means, in our case airbreathing propulsion and rocket propulsion.[3] The larger I_{sp} is, the less propellant is needed to obtain a given amount of momentum. Important is to note that for hydrocarbon fuels the domains are much narrower than for hydrogen, especially for scramjets.

The figure shows scramjet hydrogen propulsion possible up to $M_\infty = 24$. Behind this is the fact that hydrogen has faster reactions than hydrocarbon fuels and also a higher cooling capacity. Section 2.6 below is dedicated to a detailed discussion of the two roles of fuel.

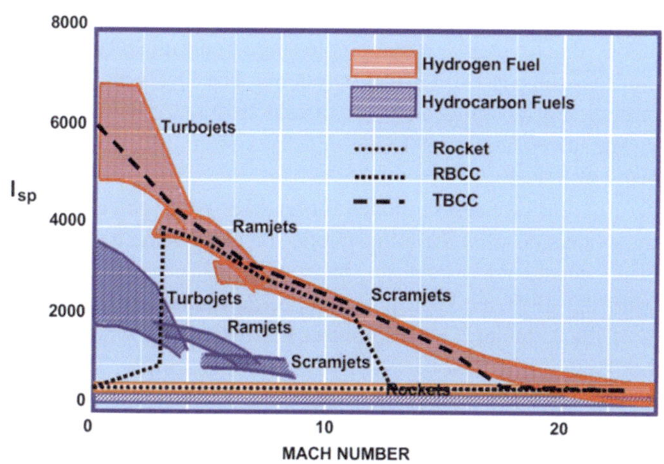

Fig. 2.1 Specific impulse of different propulsion means as function of the flight Mach number [6]. RBCC: rocket-based combined cycle propulsion, TBCC: turbine-based combined cycle propulsion

[3] The specific impulse is defined as the quotient of the nozzle exit velocity v_e and the gravitational acceleration g_0: $I_{sp} = v_e/g_0$ [s]. It can also be written in terms of the thrust force F and the exhaust mass per second $\dot{m}$: $I_{sp} = F/(g_0\,\dot{m})$ [s]. Another definition in use is $I_{sp} = F/\dot{m}$ [m/s].

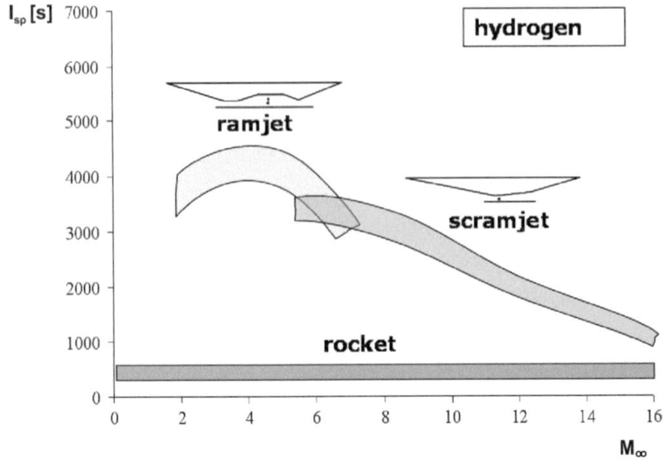

Fig. 2.2 Specific impulse I_{sp} of different (hydrogen-fuel) propulsion systems as function of the flight Mach number [7]

A different picture is given in [7]. Airbreathing hypersonic flight vehicles can be classified as (a) cruise and acceleration vehicles (CAV) or (b) as acceleration vehicles for orbital access (ARV), either civil or military ones. Both vehicle classes use turbojet propulsion for the low-speed part of flight, as long as they are not launched with a booster rocket, which is seen with military flight vehicles. The turbojet domain is followed by the ramjet and, if it applies, by the scramjet domain.

Acceleration vehicles can be either single-stage to orbit (SSTO) vehicles, or the lower stage of two-stage to orbit (TSTO) systems. The SSTO or ARV vehicle must meet the contradicting design goals of small vehicle drag for ascent, and a large one for re-entry. The lower stage of a TSTO or CAV system, like a cruise vehicle only has the demand of low drag.

In [7] the different propulsion modes for an ARV-type vehicle are investigated. Airbreathing propulsion uses the atmospheric oxygen for the combustion process. This results in a considerably large I_{sp}, because the oxygen does not enter the fuel balance. Figure 2.2 gives the typical large specific impulse of the ramjet and the smaller one of the scramjet.[4] Both impulses are decreasing with increasing flight Mach number. Both are much larger than that of the rocket engine. The latter, however, does not depend on the Mach number.

The performance of a flight vehicle's propulsion system depends not on the fuel consumption per unit time, but on the integrated consumption over the mission

[4] The data underlying the curves in Figs. 2.2, 2.3 and 2.4 are the result of a literature search made in the frame of the thesis [7]. The widths of the curves represent the data spread of airbreathing and rocket engines all with hydrogen as fuel.

duration. The acceleration capability a of a vehicle with weight W_v is directly proportional to the thrust T and inversely proportional to the mass m_v of the vehicle:

$$a \sim \frac{T - B}{m_v}, \tag{2.1}$$

where B is a function of vehicle drag, mass, and the trajectory angle γ.

The engine weight W_e (or mass) is a part of the vehicle weight (or mass). A small light engine with large thrust and little fuel consumption would be the ideal. A characteristic measure for a propulsion system hence is the thrust-to-weight ratio T/W_e, Fig. 2.3.

The figure in this respect shows the advantage of rocket propulsion over ramjet and scramjet propulsion. Rocket propulsion requires a smaller integration effort into the vehicle structure and in particular needs no inlet—more in general no airframe-propulsion integration—and therefore specifically builds much lighter.

Regarding CAV-type flight vehicles, one often finds in the literature the acceleration vehicle discussed as a vehicle for orbital access. In our diction such a vehicle would be an airbreathing ARV-type vehicle, with the contradicting design goals of small vehicle drag for ascent, and a large one for re-entry. Here we understand the acceleration vehicle mainly as the lower stage of a TSTO system, which performs an ascent to upper stage separation, and then returns to Earth.

However, many of the problems we address here are the same or similar for airbreathing CAV- and ARV-type vehicles, if we consider only the ascent mission of the latter. The CAV-type vehicle may employ ramjet and/or scramjet propulsion. The ARV-type vehicle operates beyond the ramjet propulsion domain and could employ

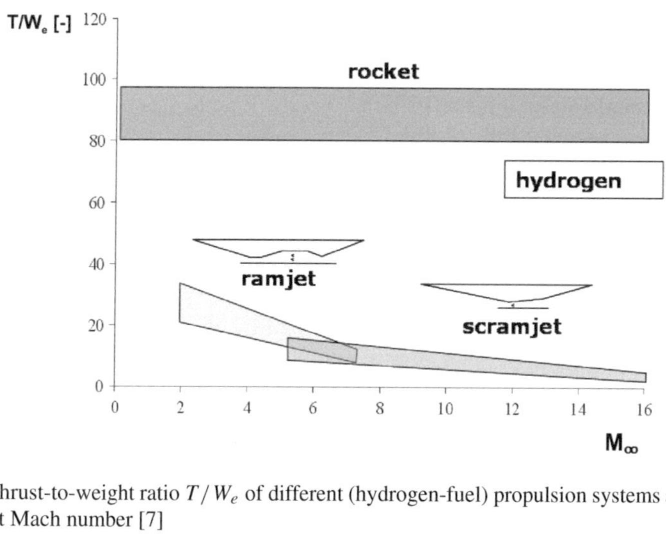

Fig. 2.3 Thrust-to-weight ratio T/W_e of different (hydrogen-fuel) propulsion systems as function of the flight Mach number [7]

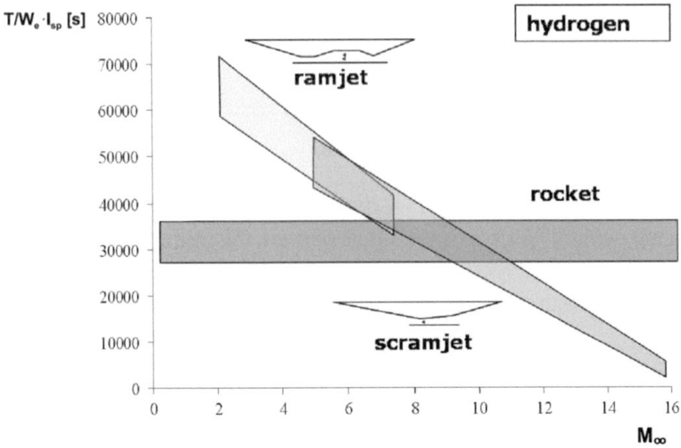

Fig. 2.4 Acceleration missions: specific performance $(T/W_e) \cdot I_{sp}$ of different (hydrogen-fuel) propulsion systems as function of the flight Mach number [7]

scramjet propulsion up to $M_\infty \approx 12$ and then rocket propulsion. Of course in any case turbojet propulsion is needed for the low-speed part of flight.

The use of the different propulsion modes for an ARV-type vehicle demands the closer look which is given in [7]. When discussing the issue of airbreathing propulsion versus rocket propulsion, usually the specific impulse I_{sp}, i.e. the impulse gained per unit mass of propellant, is considered. The larger it is, the less propellant is needed to obtain a given amount of momentum, as was noted above.

However, it is the product of specific impulse and thrust-to-weight ratio, the specific performance $(T/W_e) \cdot I_{sp}$ of a propulsion system, which shows its relevance for a flight vehicle with a pure acceleration mission, viz. a ARV-type vehicle, Fig. 2.4.

In Fig. 2.4—keep in mind that it is based on hydrogen as fuel—we see that airbreathing propulsion has, due to its large specific impulse, an advantage over rocket propulsion, but only in a limited flight Mach number range, until the thrust-to-weight issue comes into play.

Where to switch from airbreathing to rocket propulsion cannot be determined from Fig. 2.4, because the characteristics of the whole flight system have a strong influence on the performance. Today the upper limit of scramjet flight is assumed to be $M_\infty \approx 12$.

We conclude that a higher acceleration capability of a propulsion system—in this case a rocket propulsion system—at very high flight Mach numbers can more than compensate its smaller specific impulse. The achievement of a sufficiently positive thrust-minus-drag balance with the highly integrated lift and airbreathing propulsion systems of CAV-type and ARV-type flight vehicles is a major factor in this consideration.

2.2 The Design Procedure, Cayley's Design Paradigm

The design and development approach of hypersonic airbreathers at first can be seen basically as being the same as that of conventional aircraft. The background is given by Cayley's design paradigm. Sir George Cayley (1773–1857) was an early British aviation pioneer, who conceived the essentials of the aircraft as we know it today [8].

Cayley's design paradigm, [9], reads (not in his own words):

- Assign functions plainly to corresponding sub-systems, e.g.,
  - wing $\Rightarrow$ provision of lift,
  - propulsion system $\Rightarrow$ overcoming of drag,
  - horizontal stabilizer and elevator $\Rightarrow$ longitudinal trim, stabilization and control,
  - vertical stabilizer and rudder $\Rightarrow$ lateral (directional) stabilization and control,
  - fuselage $\Rightarrow$ payload accommodation,
  - et cetera.
- Have the different functions and the corresponding sub-systems only weakly and linearly coupled. Then you can treat and optimize each function and sub-system more or less independent of the others. In this way you treat and optimize the whole aircraft, which integrates all functions and sub-systems.

This paradigm has been proven to be decisive in the design of aircraft for the subsonic, the transonic, even for the supersonic flight domain. However, the quest for more performance and efficiency, in particular the opening of the hypersonic flight domain, has led to higher and higher integrated functions and sub-systems, and with that to a weakening of Cayley's paradigm. This is different for different kinds of flight vehicles, for different API-types and in each case does not necessarily encompass all major functions and sub-systems.

In the context of our considerations it is important to note that also with the engineering disciplines, which are involved in the design and development of flight vehicles, a differentiation similar to that of sub-systems and functions—according to Cayley's paradigm—has taken place.[5] This, however, was and is also a strong technology driving factor.

The differentiation of the engineering disciplines, though, had also adverse effects. It led, for instance, to the presently still established sequential and iterative design cycles with a weak interaction of the disciplines. It further led in some cases to autonomy drives of disciplines by duplicating—under the euphemism "adaptation"—skills and tools of other disciplines. These skills and tools then often do not participate in the sub-sequential developments of the mother disciplines.

In view of these developments, the meaning of Cayley's design paradigm is expanded to cover also the disciplines. Besides the differentiation of functions/sub-systems (first aspect) we have that of the disciplines (second aspect). The significant

---

[5] This differentiation also is present in university education and in research at universities and research establishments.

observation now is, that both aspects of Cayley's design paradigm are persistently weakening in modern aircraft design. This definitely holds the more for the design and development of airbreathing hypersonic flight vehicles of high API types.

An important issue for such vehicles, in general for CAV/ARV-type flight vehicles, is that with the present design and development approach the actual quantification of the aerothermo/servo-elastic properties of the vehicle's airframe would be made only very late in the development process. Then the first—already completely defined and developed—airframe has been assembled. All this holds for the propulsion system, too, and in particular also for the fully integrated flight vehicle.

Partly this quantification process can extend deep into the flight envelope opening process. Changes—actually "repair solutions"—which must be made of the airframe, if the said properties don't meet the requirements, become necessary. They can be very costly and in any case will increase the structural mass of the vehicle, but in the extreme can lead to an ending of the project.

Similar problems exist with regard to the structures and materials layout of airbreathing hypersonic vehicles in view of the thermal loads associated with their high speed flight. Basically two structural concepts are to be considered: (1) hot primary structure, (2) cold primary structure with a thermal protection system (TPS).

For CAV-type vehicles the first concept seems to be that of choice. However, at least the tanks of cryogenic fuel with their insulation needs must be integrated. This could ask for a TPS at major parts of the structure.

For ARV-type vehicles the second concept is necessary for the windward side during re-entry. Hot primary structure elements are a possibility for the remainder of such a vehicle.

Structural concepts of either kind cost weight, not to mention maintenance and repair efforts during vehicle lifetime. The real performance of a concept will become apparent only after the first flight(s). Again "repair solutions" to either improve a concept or to shed unnecessary mass will be costly and time consuming. Important in this regard are the ground-test issues, Chap. 7, see also the discussion in [10].

All this must be seen in view of the payload fraction (payload mass/take-off mass), which, for instance, a two-stage-to-orbit (TSTO) space transportation system like SÄNGER has [11]. For the SÄNGER lower stage this was estimated to be approximately 4.4% (for a conventional rocket launcher with similar payload capacity: approximately 2–4 per cent). Compared to payload fractions of modern transport aircraft of 40–50%, these are very small numbers. This means that the design is much more critical for an hypersonic airbreather than for a transonic aircraft. Hence the development risk is very much higher, which must be seen in the context of the also very much larger development time and costs. In this regard see also Sect. 1.6.

A good chance to overcome the weakening of Cayley's design paradigm (first aspect), lies in the developments of computational simulation, see, e.g., [9]. In general it appears that a strong and non-linear coupling of functions/sub-systems asks also for a strong discipline coupling. But there are also cases, where anyway a strong disciplines coupling is mandatory, even if functions/sub-systems obey Cayley's design paradigm.

The treatment of design problems in the classical way is partly possible, but often it leads to increasingly large time and cost increments. Design risks can become high, and, especially with very strong couplings of functions/sub-systems, they can become untenable. The discussion shows, that the weakening of Cayley's design paradigm in both aspects asks for new approaches, which are presently in development everywhere. This holds for aeronautics in general, but especially for hypersonic airbreathers.

The availability of enormous computer power and the capabilities of information technologies make a post-Cayley design paradigm (integrated design) a viable option in the sense of the Virtual Product [9]. That is understood when speaking of true multidisciplinary design and optimization, Chap. 7. Of course present-day design and development is multidisciplinary. That, however, usually only in the sense, that the parent discipline brings in the other disciplines in terms of simple and partly very approximate methods.

True multidisciplinary design and optimization must overcome the second aspect of Cayley's design paradigm. It must bring together the best suited tools in a strong coupling of the disciplines. The methods of numerical aerodynamics are one of the key elements of such approaches. Large advancements, however, are necessary in flow-physics and thermodynamic modeling [5].

Large advancements–and a new thinking–are necessary, too, in structural mechanics. New structure-physics models are necessary in order to permit to quantify the influence of joints of all kind (non-linearities, damping)—the joint problem—, of non-linear deformations (buckling), et cetera. This is true in particular when static and dynamic aeroelastic properties of the hot airframe are to be described and have to be optimized with high accuracy and reliability, Chap. 7.

Actually a shift from the *perfect-elastic* to the high-fidelity *real-elastic* airframe consideration and modeling is necessary already in the early design phases, and not late in the development process. This holds analogously also for thermal protection, propulsion integration, guidance and control, etc.

Figure 2.5 illustrates how aircraft shapes vary with flight domains. In the subsonic domain the wings generally are not swept. This is in contrast to the transonic domain, where depending on the cruise Mach number leading-edge sweep up to thirty degrees can be observed. The supersonic domain sees the slender delta wing with canard, often a double delta wing. Fighter shapes may feature a hybrid wing consisting of a strake and a trapezoidal wing, see, e.g., [4]. In the hypersonic domain the configuration is even more slender, with flyability and controllability as the limits.

When considering the propulsion means in the various flight domains, Fig. 2.5 in addition shows how Cayley's design paradigm loses its validity. In the subsonic and the transonic domain aircraft obey the paradigm well. Particularly in the transonic domain the high-lift configuration needs an extra consideration. With growing engine bypass ratio the engine cowl is more closely coupled to the wing and in that configuration locally a strong interaction of cowl and wing flow exists [4].

The fighter aircraft in Fig. 2.5 with $M = 2.35$ has its engine inlet at the lower side of the fuselage, which points to a weak coupling of lift and propulsion. In the hypersonic domain, with ramjet and scramjet propulsion at the core of airbreathing hypersonic

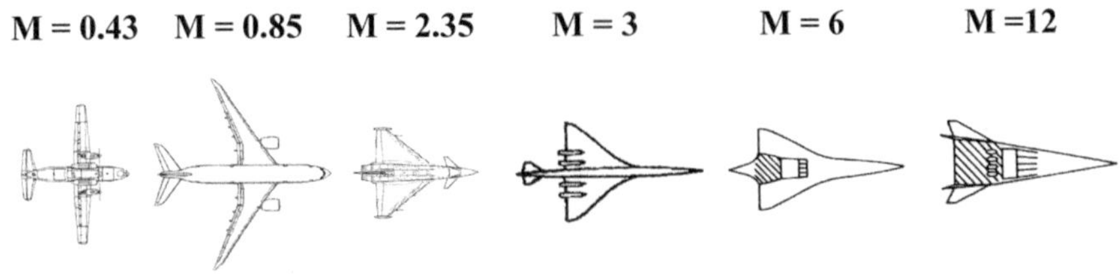

Fig. 2.5 Aircraft shapes as function of the nominal flight Mach number. From the left: subsonic domain, transonic domain, supersonic domain, high supersonic domain, low hypersonic domain, high hypersonic domain. The last three sketches, [12], show the aircraft seen from below

flight, the engines for instance are located below and close to the fuselage/wing and now a close coupling of lift and propulsion exists, API-type 4 and 5, Sect. 1.4. In this case Cayley's design paradigm is completely violated.

2.3 Challenges of Multidisciplinary Design Work

It is a challenging task to conduct a multidisciplinary design process for CAVs and ARVs with the goal to optimize their performances and missions. This demands a strong linkage of the various disciplines (second aspect of Cayley's design paradigm). The global process encompasses a technological and an economical part.[6] The technological part mainly regards the

- aerodynamic shape,
- aerothermodynamic performance including stability, controllability and maneuverability,
- aerodynamic and thermal loads, external and internal flow path,
- aeroelastic and aerothermoelastic properties of the airframe and the propulsion system and their components,
- ascent, descent, and contingency trajectories,
- guidance and control,
- masses: vehicle, propellant, payload,
- internal lay-out including propulsion system, fuel tanks, payload compartment, crew cabin and the definition of the center-of-gravity and the moments of inertia,
- structures and materials layout including thermal protection system,
- trim, control and lift enhancement surfaces,
- undercarriage,
- propulsion system,

[6] Regarding the latter customarily the main items considered are costs, operations, life cycle issues and the rate of return of investment.

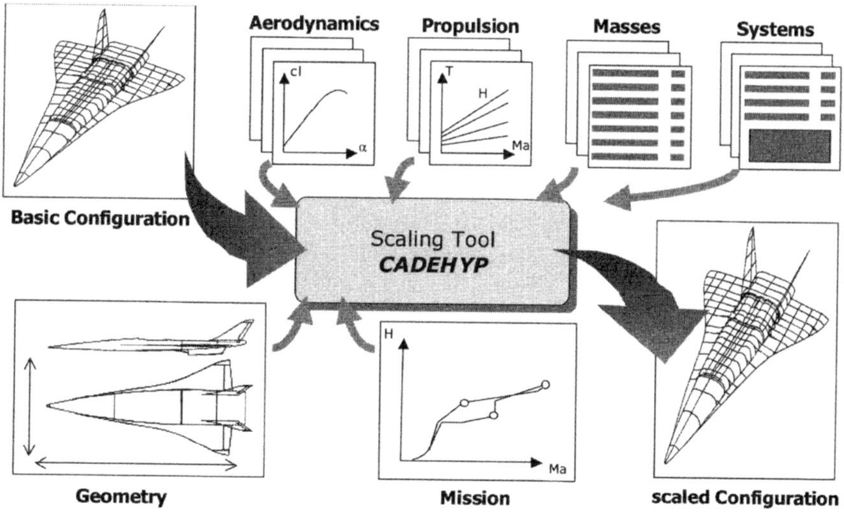

Fig. 2.6 Schematic of the CADEHYP scaling tool [7]

- reaction control system,
- thermal management (airframe, propulsion system, tanks, ...),
- ... .

The challenge is to deal with a large number of design variables and constraints due to the demands of the various disciplines, which contribute to the definition and synthesis of the system. The disciplines in general are not independent of each other. Some of the interactions between them are so strong that only a closely coupled design approach can lead to the desired results.

Reference [13] gives an overview of the objectives of multidisciplinary design approaches. In particular also the different methods are discussed, which are to be considered when dealing with an optimization problem. Here we give with Fig. 2.6 the schematic of the scaling tool CADEHYP [7], see also [14].[7] In the background again is the TSTO space transportation system SÄNGER.

Beginning with a basic configuration the input from the contributing technical disciplines is iterated in view of the intended mission of the flight vehicle. The result then is a 'scaled', i.e., with the intended performance compatible configuration. The figure holds for CAV as well as ARV type flight vehicles.

---

[7] CADEHYP (Computer Aided Design Hypersonics) is an industrial tool for the design and scaling of re-usable space transportation systems.

2.4 Hypersonic Airbreathing Systems: The Flight Environment and Systems and Operational Constraints

The Earth atmosphere with its properties poses the flight environment of airbreathing hypersonic aircraft. Here this concerns cruise and acceleration vehicles (CAV) with assumed flight Mach numbers up to $M \approx 12$, but also the airbreathing ascent of possible ascent and re-entry vehicles (ARV), also up to that Mach number. Note that the crucial part of ARV flight is the re-entry from the orbital altitude.

Short overviews over the main flow phenomena and similarity parameters are given in Figs. 2.7, 2.8 and 2.9. They predominantly concern a flight vehicle's external flow path, but also strongly determine its propulsion system flow path. Both paths together represent the flight vehicle as such. Not considered is the situation in the combustion chambers of the turbo-/ram-/scram engines of the aircraft as well as their downstream flow path including either a single expansion ramp nozzle (SERN) or a circular nozzle, see the different integration types in Sect. 1.5.

The three figures, going back to [5], cover the altitude domain $0\,\text{km} \leqq H \leqq 65\,\text{km}$, and the flight-speed domains $0\,\text{km/s} \leqq v_\infty \leqq 5.5\,\text{km/s}$ and $0\,\text{km/s} \leqq v_\infty \leqq 7\,\text{km/s}$. Nominal design points are indicated for the supersonic passenger aircraft CONCORDE (circle at the lower left), for the reference concepts 1 to 4 of a German hypersonic technology study [15] (black dots), for the SÄNGER TSTO space transportation system (staging condition) [16] (circle in the middle), and for the NASP/X-30 (cruise) [17] (open ended rectangular bar).

Red dots indicate the flight points of the X-43A in the year 2004 and the X-51A in the year 2010. For all these flight vehicles see the overview in Sect. 1.1. Also included are typical trajectory data of re-entry vehicles, the former Space Shuttle Orbiter and the former European space plane project HERMES.

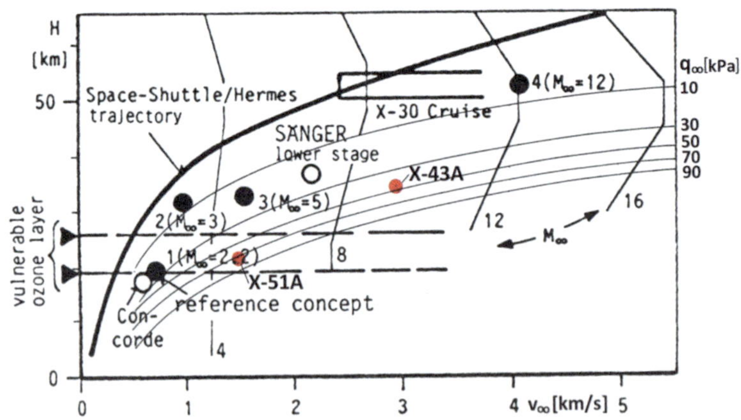

Fig. 2.7 Hypersonic flight-vehicle concepts in the velocity-altitude map with flight Mach numbers M_∞ and iso-dynamic-pressure lines

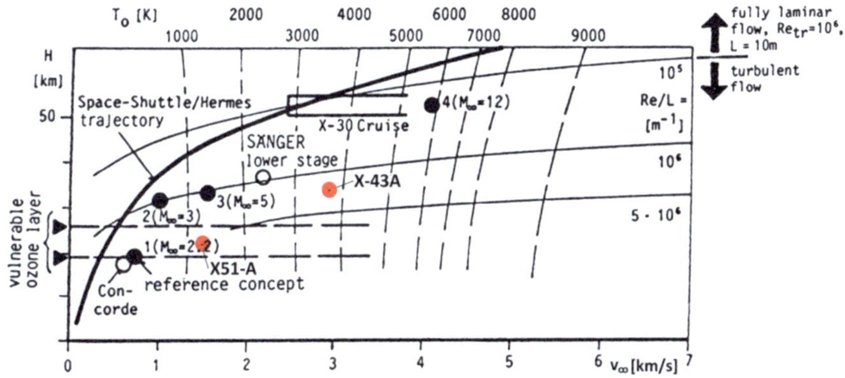

Fig. 2.8 Hypersonic flight-vehicle concepts in the velocity-altitude map with flight unit Reynolds numbers ($Re/L \equiv Re^u_\infty$), and equilibrium real-gas total temperatures T_o ($\equiv T_t$)

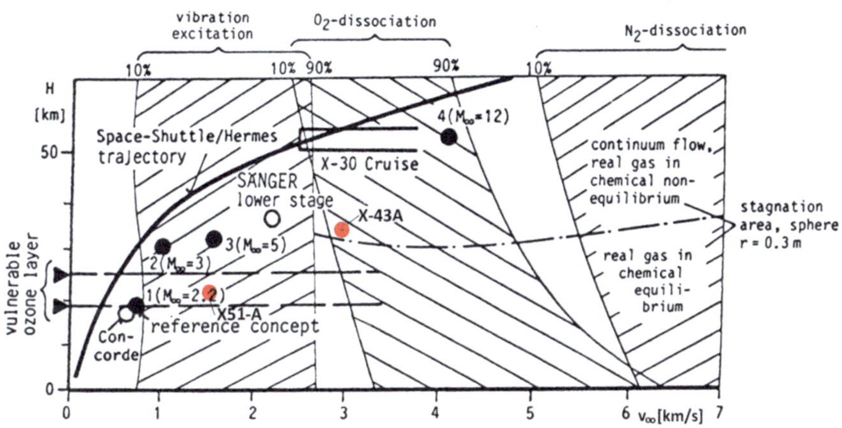

Fig. 2.9 Hypersonic flight-vehicle concepts in the velocity-altitude map with high-temperature real-gas effects

Iso Mach-number lines are given in Fig. 2.7. It is important to observe that depending on the flight altitude different flight speeds v belong to a given iso-Mach-number line. The reason for this effect is that the temperature T in the atmosphere is varying to a considerable degree with the altitude H, see Appendix A.

The dynamic pressure $q_\infty = 0.5\ \rho_\infty\ v_\infty^2$—inserted as iso-lines with values $q_\infty =$ 10 to 90 kPa—is an important quantity. It governs the aerodynamic loads on the flight vehicle structure, the performance of the airbreathing propulsion system as well as the aerodynamic performance of the vehicle shape and its trim, stabilization and control surfaces.

Hence the trajectories and the nominal design points in the figure reflect flight with—besides others—a dynamic pressure constraint [13]. Airbreathing vehicles with ram- and/or scram-jet propulsion fly at q_∞ = 30–90 kPa. The X-43A flew for

12 s at $M_\infty = 9.68$ with $q_\infty \approx 50$ kPa and the X-51A for 210 s at $M_\infty = 5$ with $q_\infty \approx 70$ kPa. For winged re-entry vehicles, like the former Space Shuttle Orbiter, the constraint is $q_\infty \lessapprox 15$ kPa. The nominal trajectory in the figures is below 10 kPa.

Figure 2.8 gives for the considered flight environment the distribution of two other very important quantities, the equilibrium real-gas total temperatures T_0 and the unit Reynolds number $Re^u = Re/L = \rho_\infty v_\infty/\mu_\infty$.

The lines of constant total temperature (equilibrium real gas) reflect the major problem of any hypersonic flight, the thermal loads on the external and the internal flow-path structure. Surface radiation cooling, Sect. 3.7, permits that even with increasing flight speed most of the vehicle's structure does not need active cooling. This does not hold for the internal propulsion flow path—inlet, propulsion unit, throat region of the nozzle—where active cooling is mandatory. In any case, the thermal household of the flight vehicle, with a cold or even with a hot primary structure—the cooling problem of internal systems of course remains—together with the needs of the propulsion system, poses a major challenge.

It is to be noted already here that surface radiation cooling of the external flow path—more in general the *thermal state of the surface*, Sect. 3.7—strongly influences also wall shear stress, boundary-layer thickness, separation inclination and so on—the *thermal-surface effects*, Sect. 3.10. All this depends very strongly on the location and the stream-wise extent of the laminar-turbulent boundary-layer transition region, Sect. 3.11.

The major similarity parameter regarding laminar-turbulent transition is the Reynolds number.[8] The iso-unit Reynolds-number lines in Fig. 2.8 at higher altitudes are reasonably parallel to iso-altitude lines. They indicate that at flight below approximately 50 km altitude boundary layers will become turbulent, if the unit Reynolds number $Re^u_\infty = \rho_\infty u_\infty/\mu_\infty \approx 10^5\ \mathrm{m}^{-1}$ is taken—with due reservations—as zero-order criterion.

For the design of airbreathing flight vehicles hence laminar-turbulent transition and turbulence, Sect. 3.11, are a fundamental problem, more than for winged re-entry vehicles. The laminar portion at the front part of the flight vehicle will be small at small altitudes. At higher altitudes it may extend over larger portions of the vehicle surface. Still today the accurate prediction of transition is a tremendous challenge, see, e.g., [5].

The situation regarding high-temperature real-gas effects—see, e.g., [5]—in the considered flight domain is shown in Fig. 2.9. The data are related to the equilibrium real-gas total temperatures T_0 shown in Fig. 2.8. At flight speeds below $v_\infty \approx 0.8$ km/s air can be considered as a calorically and thermally perfect gas. Vibration excitation must be taken into account for $0.8\,\mathrm{km/s} \lessapprox v_\infty \lessapprox 2.6\,\mathrm{km/s}$. Above $v_\infty \approx 2$ km/s dissociation of oxygen occurs. Nitrogen is not affected at the Mach numbers in the considered flight domain.

[8] Regarding the role of similarity parameters in the hypersonic flight domain see, e.g., [5, 18].

The dissociation of oxygen below $H \approx 25$ km generally is in equilibrium, above that altitude non-equilibrium effects may play a role. Ionization effects first occur at higher flight speeds than those considered here and perhaps can be ignored.[9]

The reader must be aware that these statements are of approximate validity. In reality the shape of the vehicle, its size, and its flight attitude play a major role.

Finally a look at possible low-density effects, i.e., slip flow and temperature jump at the vehicle surface, is necessary. We give a very short summary, based on the considerations in [5].

As shown in Appendix A, density and pressure decrease rapidly with increasing altitude. The temperature is not constant, up to 100 km altitude it changes between $T \approx 200$ and 300 K. The mean free path λ, not considered in that appendix, increases fast with altitude due to the behavior of the density:

$$\lambda = \frac{16}{5} \frac{\mu}{\rho} \frac{1}{\sqrt{2\pi RT}}. \tag{2.2}$$

The mean free path is the average distance that a gas particle travels between successive collisions with other particles. It is used for a qualitative characterization of the flow regimes, which are encountered in hypersonic flight.

The ratio of the mean free path and a characteristic length L of a flow field is called the "overall" Knudsen number Kn:

$$Kn = \frac{\lambda}{L}. \tag{2.3}$$

The characteristic length L can be a body length, a nose radius, a boundary-layer thickness, or the thickness of a shock layer, depending on the problem considered.

Regarding the Knudsen number three approximate flow regimes are of interest:

- continuum flow with no-slip wall-condition $u_w = 0$, and no wall-temperature jump $T_{gw} = T_w$:

$$Kn \lessapprox 0.01, \tag{2.4}$$

- continuum flow with slip effects (slip flow $u_w > 0$ and temperature jump $T_{gw} > T_w$ at the body surface):

$$0.01 \lessapprox Kn \lessapprox 0.1, \tag{2.5}$$

- disturbed free molecular flow:

$$0.1 \lessapprox Kn \lessapprox 10. \tag{2.6}$$

No sharp limits exist between the three flow regimes, but we can say that here only the first two regimes are of interest. Figure 2.10 shows the Knudsen number

[9] Ionization happens at temperatures above 5,000 K to 6,000 K [19]. That means above flight speeds of 3.5 to 4 km/s.

Fig. 2.10 Knudsen numbers Kn_∞ as function of the altitude H [5]: **a** SÄNGER lower stage, $L \approx 80$ m; **b** Space Shuttle Orbiter, $L \approx 30$ m; **c** X-38, $L \approx 8$ m; **d** sharp nose cone, (D =) $L \approx 0.3$ m; **e** pitot probe, (D=) $L \approx 0.01$ m; **f** measurement orifice, (D =) $L \approx 0.001$ m

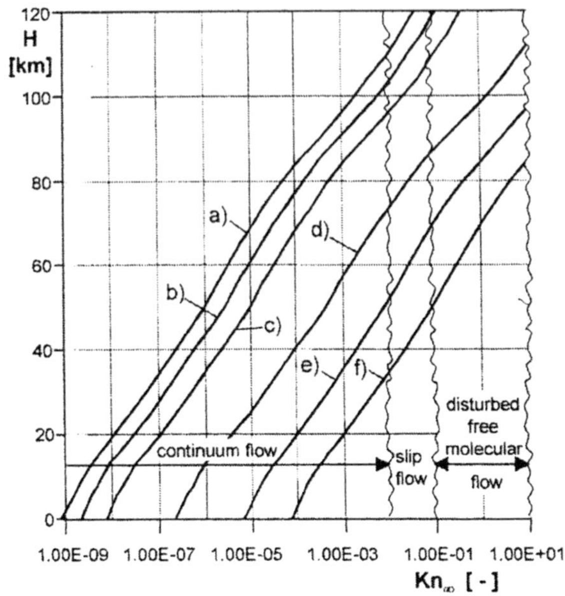

$Kn_\infty = \lambda_\infty / L$ of some hypersonic flight vehicles and some vehicle components, with the characteristic length L taken as vehicle length or component diameter.

Globally the lower stage of SÄNGER (case a), with a maximum flight altitude of approximately 35 km, the Space Shuttle Orbiter (case b), and also the X-38 (case c, look also at Fig. 6.46 in [18]), remain fully in the continuum regime in the interesting altitude range below approximately 100 km. For the sharp nose cone (case d) slip-flow effects can be expected above 75 km altitude. The pitot probe (case e) is in the continuum regime only up to approximately 50 km altitude because of its small diameter, and likewise the measurement orifice (case f) only up to 35 km altitude.

We conclude that low density effects regarding hypersonic airbreathers at most are of interest for measurement devices, but not for the external and propulsion flow path. The situation is different, if the structure of shock waves plays a role, for instance in combustion chambers. This concerns discrete numerical computation methods because shock waves are only a few mean-free paths thick [5].

The flight environment is one item to be considered. Other items directly concern the flight vehicle. For guidance objectives and trajectory control variables see the short remarks in [13]. Here some statements are made regarding systems and operational constraints.

Basic systems and operational constraints of CAV and ARV vehicles are given in the following listing:

- The dynamic pressure q_∞ is a measure of both mechanical loads on the vehicle's structure and the demands of the airbreathing propulsion system. A typical ascent trajectory of a CAV can be a 50 kPa trajectory for both the turbojet and the ramjet mode. If both the ramjet and the scramjet mode are to be employed, dynamic-pressure ranges of 25 kPa $\lessapprox q_\infty \lessapprox$ 95 kPa have been considered, see, e.g., [20].

- Flight of CAV-type vehicles below 40 to 60 km altitude means that the viscous flow predominantly is turbulent, Fig. 2.4. The determination of location, shape and extent of the laminar-turbulent transition zone is the big problem, Sect. 3.11. The forward stagnation point usually is the primary reference location where thermal loads are constrained. In addition flat portions downstream of the nose region with turbulent flow may be chosen as reference locations.
- Besides the normal load factor with $n_{z,max} = q_\infty C_Z A_{ref}/(m\, g) = 2$ to 2.5, which generally is assumed for the stage-separation process of a TSTO space transportation system, also the axial load factor with $n_{x,max} = q_\infty C_X A_{ref}/(m\, g) = 3$ to 3.5 is a constraint, see, e.g., [13].[10] Related to this issue is the degradation of passenger comfort with hypersonic passenger transport aircraft. In [13], too, for ordinary, untrained passengers acceptable acceleration and deceleration data are given. Not to be forgotten is the g-reduction, which also is an issue in passenger transport.
- With API-type 4 and 5 flight vehicles the influence of the thrust vector on the longitudinal forces and the pitching moment, Sect. 5.7, implies that trimmability as well as stability and controllability of the flight vehicle are critical operational constraints. If the thrust-vector angle of a SERN-type nozzle cannot be restricted or controlled in one or the other way, the only available degree of freedom to trim the vehicle is the—symmetrical—elevon deflection. If the elevon angle is large, a substantial trim drag D_{trim} drag will result.[11]

  At the same time, depending on the overall forces and moments balance, a decrease of the lift L may happen, and a strong decrease of the lift-to-drag ratio L/D. Besides that, the longitudinal stability characteristics of the vehicle are influenced. Also important in this regard are the resulting hinge moments (actuator performance) together with the mechanical and thermal loads on the elevons.

In addition to these basic systems and operational constraints others need to be respected. This holds in particular for airbreathing propulsion systems. The hypothetical flight corridor of CAV-type flight vehicles in the velocity-altitude map is given in Fig. 2.11 [20]. There AREA 1 concerns, for instance, the lower stages of TSTO space transportation systems and hypersonic aircraft, whereas AREA 2 and 3 concern SSTO (ARV-type) systems.

The AREA 1 flight corridor in Fig. 2.11 is bound on the lower altitude side by structural pressure limits and thermal loads on the airframe, the propulsion system and the control surfaces with their hinge-moment limits.

At the upper side of the flight corridor lift and combustion limits play a role and also the limit of the aerodynamic control authority. Control aspects possibly make the flight corridor much narrower in reality. Even larger regions must be excluded due to propulsion issues [20].

---

[10] C_X and C_Z are the axial and the normal force coefficients. For the axis definition see Sect. 2.9.2.1 and also [21].

[11] A body flap, like with the Space Shuttle Orbiter, is not a way out.

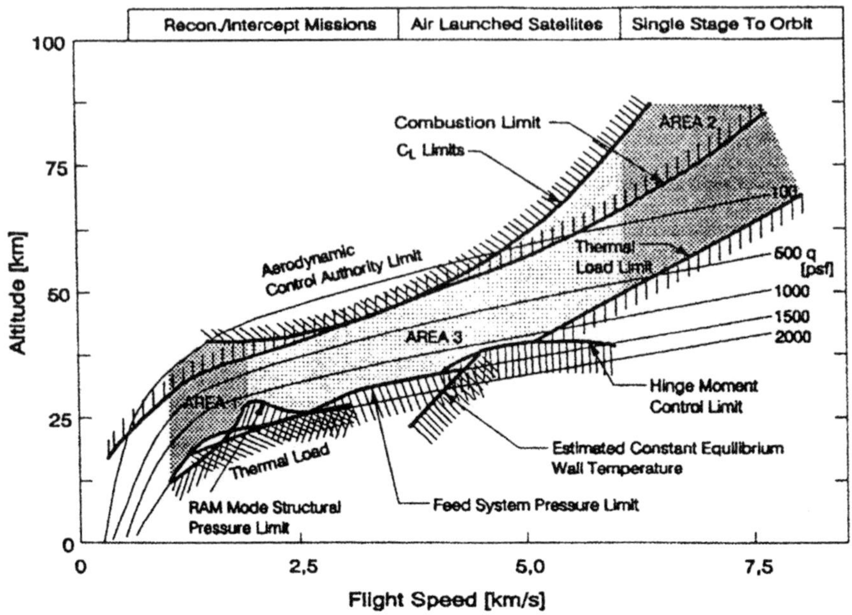

Fig. 2.11 Flight corridors and system and operational constraints of CAV- and ARV-type flight vehicles in the altitude-velocity map [20]

In conclusion it can be stated that flight trajectory design and optimization for future airbreathing CAV- and ARV-type flight vehicles poses extremely large challenges. Cayley's design paradigm is invalid for API-type 3 and in particular API-type 4 and 5 vehicles. Non-linear aerodynamics/structure dynamics/propulsion/flight dynamics/flight control coupling issues must be identified. New vehicle design and test approaches must be developed, but also new trajectory design and optimization approaches.

2.5 Differences Between Conventional and Hypersonic Flight Vehicle Design

In conventional subsonic, transonic and supersonic—up to $M_\infty \approx 3$—aircraft design a number of technical issues exists, which today can be considered as treatable. They are given in graphical form in the design triangle, Fig. 2.12.

A tradeoff must be found between mass (M), aerodynamic (A) and elastic (E) forces. The engine thrust as inner aerodynamic force has been assigned to the aerodynamic forces. Related to these forces are the familiar phenomena or problems which are listed in Table 2.1. Note that not included is the propulsion system. It is

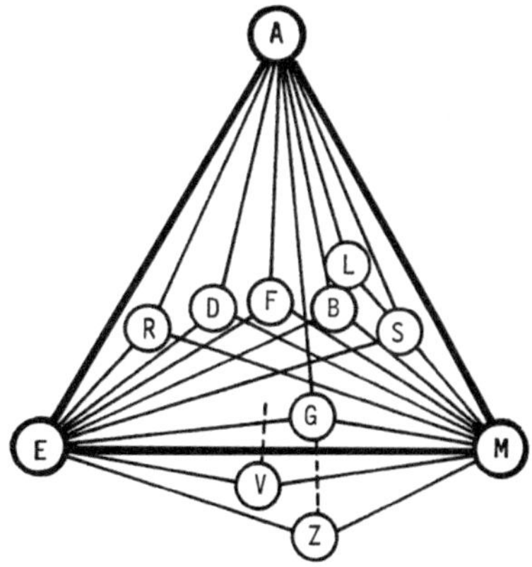

Fig. 2.12 Conventional aircraft design triangle [22]. See Table 2.1 for the definitions

Table 2.1 The loads and problems indicated in Figs. 2.12 and 2.13 [22]

Loads	Problem
A: aerodynamic forces	R: rudder reversal
E: elastic forces	D: divergence
M: mass forces	F: flutter
T: thermal loads	B: buffeting
	L: maneuver loads
	S: stability and control
	G: gusts
	V: mechanic vibrations
	Z: mechanical shocks (landing)

coupled weakly into the whole system in the sense of Cayley's design paradigm, see the two configurations at the left-hand side of Fig. 2.5.

In the left-hand column of Table 2.1 thermal loads, marked with T, do not apply to conventional aircraft. They come into play when considering hypersonic aircraft. They are included in the simplified tetrahedron shown in Fig. 2.13. Note, that the relationships shown in Fig. 2.12 are only a part of the—simplified—hypersonic tetrahedron.

The thermal loads, both as wall temperature T_w and as heat flux into the wall q_w, Sect. 3.7, are the biggest challenge in the design of hypersonic aircraft.[12] Note that the heat flux q_w into outer surfaces is not equal to the heat flux in the gas at the wall q_{gw}, because such surfaces are radiation cooled, Sect. 3.7.

[12] Thermal loads already were an issue with the SR-71 with a top flight Mach number $M_\infty = 3.36$.

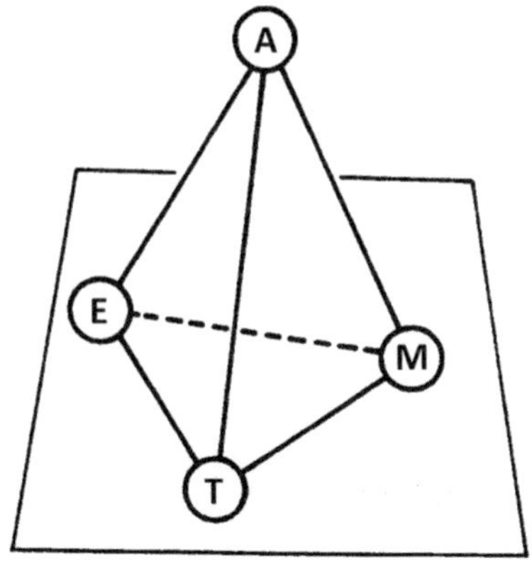

Fig. 2.13 Simplified hypersonic design tetrahedron [22]

Even higher than the thermal loads problem is with API-type 4 and in particular with type 5 flight vehicles the strong coupling of lift and propulsion at the lower vehicle side, see the two configurations at the right-hand side of Fig. 2.5. The coupling increases with increasing flight Mach number.

2.6 The Fuel and Its Two Major Roles

This section is devoted to a short comparative consideration of fuels which are candidates for hypersonic aircraft. When regarding fuel for hypersonic flight vehicles, first of all the generation of thrust comes into mind—role number one. Equally important is its second role, the cooling of the structure of the internal flow path—inlet, bleed and bypass channels, engine, nozzle—and also the cooling of the fuel tanks, the equipment and the payload.

Regarding the second role we observe that the external flow path of the vehicle is radiation cooled, also the outer parts of the inlet. At least partially cooled is the outer part of the single expansion ramp nozzle (SERN), the same holds for a bell nozzle.

The fuels of interest, Table 2.2, are liquid hydrogen (LH2), as well as hydrocarbons, i.e., liquid methane (LCH4) and jet propellant, there especially JP-7, which was developed in the USA for high-speed flight, and also JP-10, a fuel for airbreathing missiles, which is not in the table.

In view of the two roles of the fuel, the use of LH2 with its high heat of combustion and its high cooling capacity appears to be favored. The ramjet-propelled X-51A, however, with $M_{\infty_{max}} \approx 5$, was flown with JP-7. In contrast to that, the scramjet-propelled X-43A, with $M_{\infty_{max}} = 9.68$ had LH2 as fuel. Regarding the propulsion means and the respective flight speed domains see Fig. 2.1.

Table 2.2 shows that the combustion heat of LH2 is 2.8 times larger than that of JP-7, the heat sink capability is even 25 times higher. The density of LH2, however, is around 10% of that of JP-7. In view of the cooling capacity for a mission, LH2 has

Table 2.2 Characteristic data of relevant fuels [22]

Fuel	LH2	LCH4	JP-7
Heat of combustion [kJ/kg]	$12 \cdot 10^4$	$5 \cdot 10^4$	$4.3 \cdot 10^4$
Ratio against JP-7	2.8:1	1.16:1	1:1
Maximum heat sink capability [kJ/kg]	$1.5 \cdot 10^4$	$3.1 \cdot 10^3$	$0.6 \cdot 10^3$
Ratio against JP-7	25:1	5.17:1	1:1
Boiling temperature [K]	20	112	420
Ice point [K]	13.7	91	85
Density [kg/m^3]	70	435	770
Ratio against JP-7	0.091:1	0.565:1	1:1
Energy density [kJ/m^3]	$8.4 \cdot 10^6$	$21.7 \cdot 10^6$	$33.1 \cdot 10^6$
Ratio against JP-7	0.25:1	0.65:1	1:1
Maximum specific impulse at $M_\infty = 3$ [s]	6200	2500	2160
Ratio against JP-7	2.87:1	1.15:1	1:1

a roughly nine times higher ratio of 'heat sink capability' to 'heat of combustion' than JP-7. The cooling capacity is mainly used for the propulsion system. The energy density of LH2 on the other hand is only one fourth of that of JP-7.

Hence the design problem of hypersonic airbreathers with LH2 as fuel is the four times higher tank volume required to house the fuel needed to produce the necessary engine thrust. The higher tank volume leads to a larger volume of the flight vehicle—because of the needed slenderness (the wave drag as an issue!), the increase of the vehicle length comes in as a factor—and a larger surface area. The consequence is an increase of viscous and form drag.[13] LH2 as a fuel nonetheless appears to be the necessary choice for large hypersonic aircraft, despite the design implications to store it on board. Jet propellant on the other hand is the choice for airbreathing cruise missiles.

Results from an older study [24], which are still valid, are shown in Fig. 2.14. Given for SSTO systems is the ratio of needed fuel mass over total system mass against the achievable velocity. Single-stage rockets obviously cannot reach the orbital speed of 7.9086 km/s or $25.947 \cdot 10^3$ ft/s, because the needed fuel and oxidizer mass would exceed the system mass.

For SSTO airbreathers in this figure with JP-4 and even more with LH2 it seems that the orbital speed can be reached. However, with JP-4 fuel the cooling problem is unsolved, and with LH2 fuel the configuration problem to house the fuel becomes the major challenge.

We further demonstrate the design sensitivity of hypersonic flight vehicles with a rather extreme example, an ARV, i.e., an airbreathing winged SSTO device with

[13] The form drag is the viscous effects induced pressure drag, see, e.g., [23], and also the example in [4].

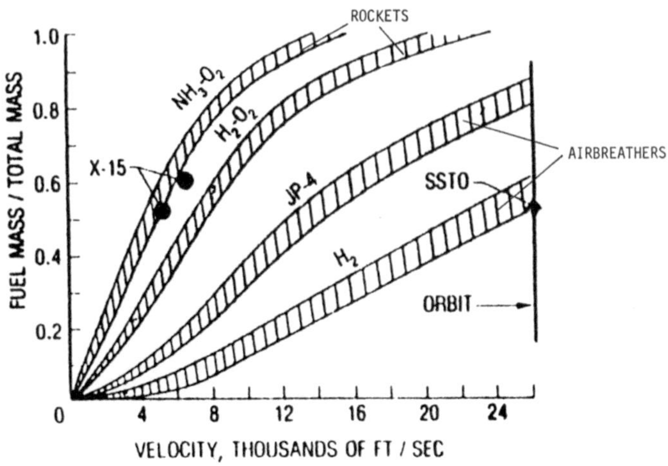

Fig. 2.14 Fuel-mass requirements of single-stage-to-orbit rockets and winged airbreathing systems [24]

LH2 as propellant [24]. Like the NASP it is to take off horizontally and to fly a plain accelerator-type mission.

Considered is the fuel/total mass ratio m_f/m_t, Eq. 2.7, as function of the orbital speed v_o to be reached at the orbital altitude h_o, the thrust-to-drag ratio T/D, the mass-specific energy of the fuel Q (heat of combustion in Table 2.2), and the total propulsion efficiency η_p [24][14]:

$$\frac{m_f}{m_t} = 1 - [1 - \frac{0.5v_0^2 + gh_0}{2\eta_p Q(T/D - 1)}]/[1 + \frac{0.5v_0^2 + gh_0}{\eta_p Q}(1 + \frac{1}{2(T/D - 1)})]. \quad (2.7)$$

The design sensitivity in terms of the fuel/mass ratio m_f/m_t is very high for small thrust to drag ratios T/D, Fig. 2.15: the fuel/mass ratio diverges for all parameters η_p (which is the total process efficiency of thrust generation). On the other hand this sensitivity vanishes for thrust to drag ratios larger than approximately four. Evidently it does not make sense to install a higher T/D, because the fuel/mass ratio is not further affected.

Figure 2.15 also shows that with a technology level in terms of η_p—at the times of the original study $\eta_p \approx 0.35$—a SSTO vehicle with $T/D \approx 4$ demands a LH2 fuel ratio of about 60% of gross take-off mass. This result, however, does not say anything about how to design this fuel ratio into a realistic configuration. It only says that about 35% of the maximum take-off mass is reserved for the vehicle, i.e., structure, propulsion system, equipment, ..., payload, which is not feasible.

[14] We note that Eq. 2.7 and Fig. 2.15 represent a mean value of v_0 from zero to orbital speed at h_o. But this simple formula can be used for many pairs of velocity and altitude, i.e., mission data of any acceleration vehicle. It gives an idea about the respective configuration, the 'thrust' to 'drag' ratio and the fuel plus the 'fuel' to 'mass' ratio.

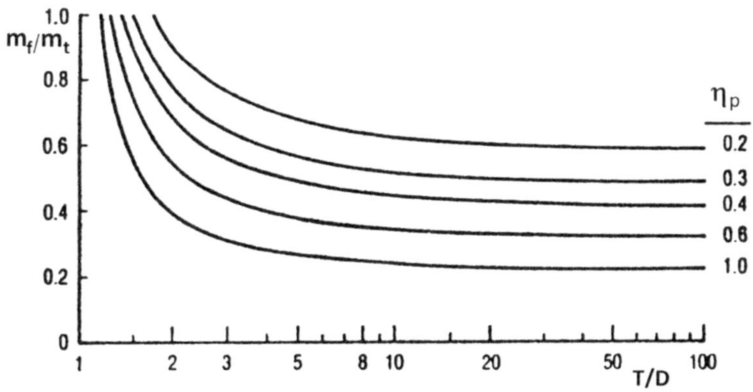

Fig. 2.15 SSTO: Fuel mass ratio m_f/m_t as affected by the total engine process efficiency η_p and the thrust to drag ratio T/D [24]

What is the way out? Two possibilities exist, both in the upper part of the ascent trajectory with a change to rocket propulsion. The first one is the TSTO system with an airbreathing lower stage and a rocket-propelled upper stage. The second one is the SSTO system with combined airbreathing and rocket propulsion. This is discussed in the following section.

2.7 SSTO System Versus TSTO System

With a TSTO system the above mass problem can be relaxed. This is demonstrated with Fig. 2.16 from [25]. The figure shows the fuel need of a winged rocket-propelled SSTO orbiter compared to that of an orbiter of a TSTO system. The figure is bound at the top by the orbital speed $v_{orbital} = 7.91$ km/s, which is to be reached, at the right by the conjectured construction limit $dm_L/d(m_f/m_L) \to \infty$. The payload mass m_{pay} is held constant, the specific impulse is $I_{sp} = 470$ s, and the thrust-to-launch-weight ratio is $T/m_L g = 1.25$. The graph v_{launch} gives the needed fuel-mass to launch-mass ratio m_f/m_L as function of the launch speed v. The graph v_{final} is the achievable final velocity of an orbiter, which is launched with $v_L = 0$.

The first result is that rocket propelled (LH2 plus LO2) winged SSTO vehicles with fuel-mass/launch-mass ratios $m_f/m_L > 0.87$ are not feasible, because their launch mass diverges. The rocket propelled single-stage orbiter does not reach the orbital speed. The maximum attainable speed is $v \approx 6$ km/s.

For the upper stage of a TSTO system the situation is relaxed, because of its non-zero launch speed $v_L > 0$. The minimum is $v_L \approx 1$ km/s, which leads to $m_f/m_L \approx 0.83$, which is just below the conjectured construction limit. If however $v_L \approx 2$ km/s is possible (the speed limit of the airbreathing first stage of the TSTO system SÄNGER), the launch-mass ratio of the upper stage is a feasible $m_f/m_L \approx 0.78$.

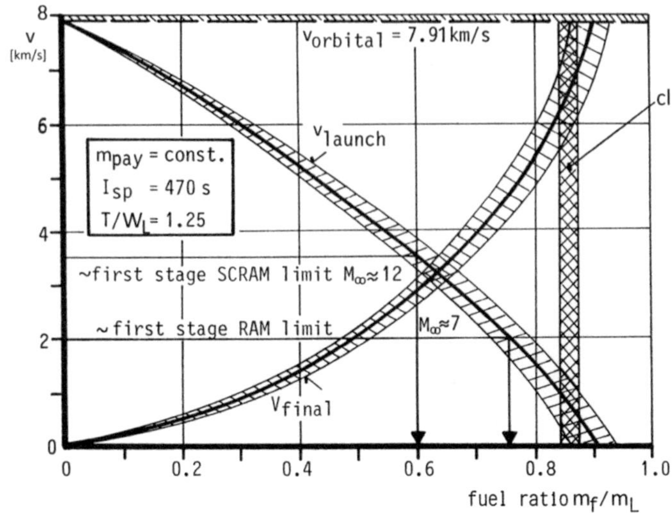

Fig. 2.16 Winged rocket propelled TSTO orbiter versus SSTO: necessary fuel-mass to launch-mass ratio m_f/m_L as function of the launch speed v [25]. The launch weight is $W_L = m_L g$, *cl* denotes the conjectured construction limit

This, however, implies an increase of the design problem of the lower stage, because in addition to the turbojet propulsion now ramjet propulsion is necessary in order to reach $M_\infty \approx 7$.

If moreover the airbreathing first stage of the TSTO system can reach $v_L \approx$ 3.5 km/s—which implies additional scramjet propulsion up to $M_\infty \approx 12$—the fuel-mass/launch-mass ratio of the orbiter drops to $m_f/m_L = 0.6$.

This result, which is in favor of the upper stage, is bought with big challenges regarding the lower stage. Two questions arise:

(1) How much is it possible to relax the design challenges of the upper stage at the expense of the—airbreathing—lower stage?

(2) Where is the optimum of the design of the overall system in view of risks, total mass, cost, etc.?

2.8 The Lower Stage of the TSTO Space Transportation System SÄNGER as API-Type 5 Reference Concept

In the following sections and in the next chapters we use as reference concept of API-type 5 flight vehicles the lower stage of the TSTO space transportation system SÄNGER. That itself was the reference concept of the German Hypersonics Technology Programme in the 1980s and 1990s [26]. A fully reusable TSTO system like

SÄNGER was seen as a possible successor of the partially reusable HERMES [27]. But in the background was also the desire to have an access to space from European territory.

The turbo-/ram propelled lower stage of SÄNGER is a large system of Airframe-Propulsion Integration Type 5, Sect. 1.4. Hence it is chosen as reference concept, because that poses the highest technological challenges. Moreover much information and data from that concept are available, even though it is more than 30 years old.

The SÄNGER configuration is given in Fig. 2.17, the major data are collected in Table 2.3. Note that in the literature the numbers fluctuate to a certain extent.

The desired access to space from Europe led to the mission with the take-off in middle/southern Europe as sketched in Fig. 2.18. The trajectory consists of a mix of

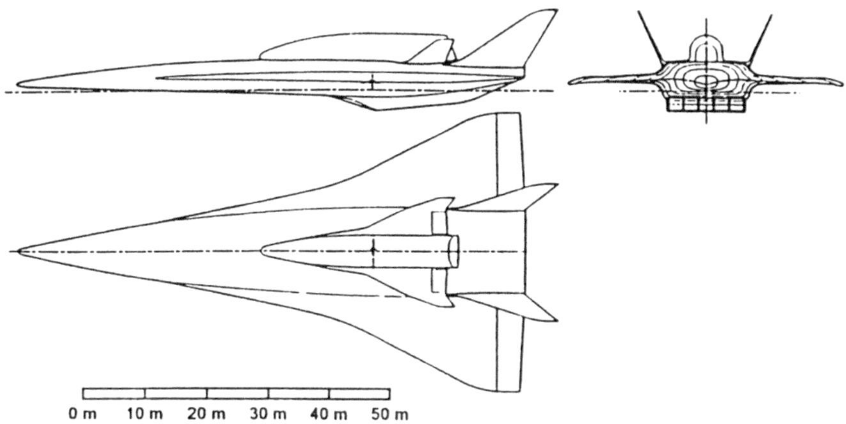

Fig. 2.17 Three-view drawing of the TSTO space transportation system SÄNGER [26, 28]

Table 2.3 Major data of the TSTO space transportation system SÄNGER configuration 4/1992 [26], see also [28]. The reference areas are approximate ones

	Lower stage	Upper stage
Fuselage length	82.5 m	32.45 m
Fuselage height	4.5 m	5.4 m
Fuselage width	14.4 m	5.2 m
Span width	45.1 m	17.7 m
Reference area	1,500 m^2	200 m^2
Engines	Turbo/ram (LH2/air)	Rocket (LH2/LOX)
Number of engines	5	1
Thrust	Max 500 kN each	1,500 kN
Propellant	LH2: 134 to	LH2/LOX: 83 to
Payload	115 to	Unmanned 7 to
Gross take-off mass	410 to (with upper stage)	115 to

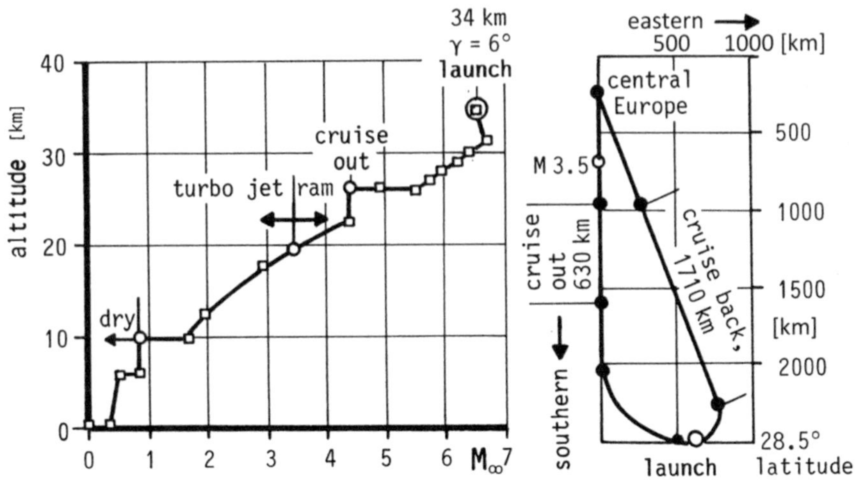

Fig. 2.18 The SÄNGER TSTO system ascent trajectory (left) and the mission foot print (right) [22]

acceleration and cruise elements. The cruise part at $M_\infty \approx 4.5$ in a sense is spread over the whole mission. The upper stage is transported first to the South and then to the East to the launch location at 28.5 latitude (right part of the figure). At 34 km altitude and $M_\infty \approx 6.5$ with a flight-path angle of $\gamma \approx 6$ the release of the upper stage into orbit is initiated, Sect. 3.12.

The ascent trajectory underlies some constraints. They are indicated at the left of the figure. The acceleration part up to 10 km altitude for noise contamination concerns is flown with dry turbojet engines, i.e., without afterburner. At about 10 km altitude the transonic region with its acceleration pinch-point, Sect. 2.1.1, is reached and afterburner flight is initiated up to about 20 km altitude and $M_\infty = 3.5$. How to cross the vulnerable ozone layer between about 19 km and 25 km, see Figs. 2.7, 2.8 and 2.9, was not established. Cruise out and ascent to the launch altitude then is made in ramjet mode. For a more detailed presentation and discussion of the SÄNGER trajectory see [13] and [26].

Of course, SÄNGER as a reference concept does not cover the whole Mach number range up to $M_\infty \approx 12$, which is the topic of this book. Indicated in Fig. 2.5 is that the higher the Mach number the higher is the coupling of the lift and the propulsion system of API-type 5 flight vehicles. For all integration types scram propulsion becomes necessary, Fig. 2.1, and thermal loads increase with the flight Mach number. Nevertheless, with SÄNGER the important issues of hypersonic airbreather design can be covered to a fair degree.

2.9 Inevitable Performance Trends

Under inevitable performance trends we understand basic flight-vehicle issues, i.e., flight performance and vehicle behavior. The reader gets a short introduction to these topics. For a deeper understanding the study of the corresponding literature is necessary. Regarding issues of stability and control we point to [29].

2.9.1 *Lift-To-Drag Ratio L/D*

The aerodynamic quality of a—trimmed—aircraft, the lift-to-drag ratio L/D, is a measure of the performance also of hypersonic aircraft. Inevitable is the trend of L/D to decrease with increasing flight Mach number M_∞.

We consider L/D starting with the simple symmetric drag polar with K being the factor of the induced drag

$$C_D = C_{D_0} + K C_L^2. \tag{2.8}$$

There is no restriction regarding slender hypersonic wings, because with them camber effects are small.

This leads to the well-known relation for the aerodynamic quality

$$\frac{L}{D}|_{opt} \approx \frac{1}{2\sqrt{C_{D0}\frac{K}{\pi \Lambda}}}, \tag{2.9}$$

with Λ being the wing's aspect ratio.

Consider now Fig. 2.19. The upper part shows at the left the zero-lift drag coefficient C_{D_0}—dominated by the slenderness of the wing (wave drag) and the surface (viscous drag)—and at the right a normalized K factor, both as function of the flight Mach number M_∞.

At the lower part of the figure trends of $(L/D)_{opt}$ are given. The upper hatched strip is representative for a number of designs for a given Mach number with the technology levels at their time.[15] The design of the CONCORDE is inserted with the square, and a re-design of it with the black triangle. The general result is that L/D decreases with increasing flight Mach number.

The lower broken line represents the design of a SÄNGER-type TSTO system with the design flight Mach number $M_\infty = 7$. Important is that $(L/D)_{opt}$ more or less is constant in the whole Mach-number interval up to M_{design}. The tendency of a decrease with increasing Mach number could not be verified. The consequence for that TSTO system is that after mastering the critical design point $M_\infty = 7$ most

[15] At the time of the drawing of this figure it was anticipated that by the year 2020 due to laminar-flow technology for transonic transport aircraft $L/D \gtrapprox 20$ will be reached.

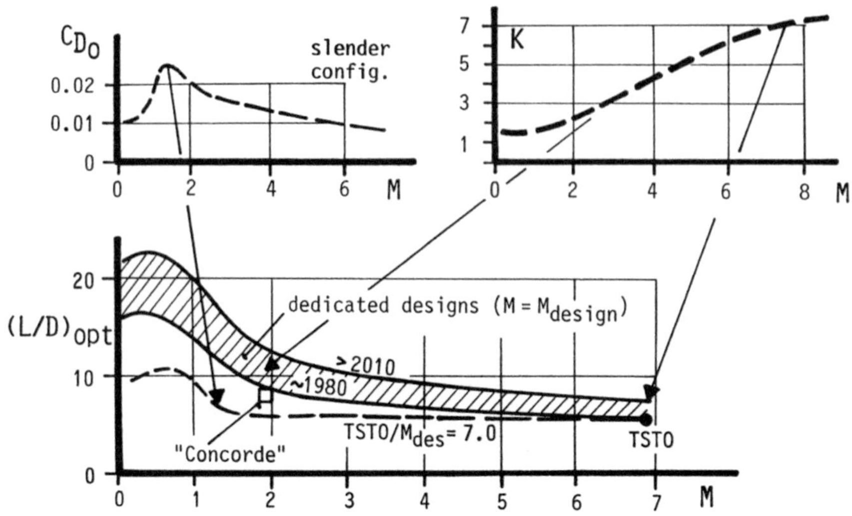

Fig. 2.19 The trend of the aerodynamic quality L/D with increasing flight Mach number M_∞ [25]

probably other critical design points will show up. These could be in the transonic regime—acceleration pinch-point 1—or at the Mach number where the switch from turbo to ramjet propulsion happens—acceleration pinch-point 2—at $M_\infty \approx 3.5$, Fig. 5.36.

With waverider configurations L/D also decreases with increasing flight Mach number. Figure 2.20 shows the results of two osculating-cone designs [10]—each inviscid and viscous—for $M_{\infty,design} = 8$ and 12. The triangles represent results from [30] and the circles from [31].

The inviscid picture shows high and nearly constant values of L/D over the Mach number range. Regarding viscous effects, L/D drops to values, which for $M_\infty = 7$ resemble those of the TSTO-system in Fig. 2.19. In the studies viscous effects were modeled with fully turbulent flat-plate boundary layer flow.[16]

Coming back to Fig. 2.5, we must realize that the apparent increase of vehicle slenderness with design flight Mach number has its limits. In [32], see also [4], it is noted that the slenderness of highly swept wings (half span over chord length) should not be smaller than $(b/2)/L \approx 0.3$, because low-speed properties become critical: too low $dC_L/d\alpha$, dutch-roll behavior and lateral/directional stability problems at high angle of attack. This holds for both classical and waverider shapes.

[16] In predesign work this is acceptable, but the problem of the determination of the location of laminar-turbulent transition remains, which in reality is of high importance, Sect. 3.11. Moreover on free-stream surfaces, i.e., surfaces without inclination to the free stream, the boundary layer can induce an oblique shock wave with the associated wave drag.

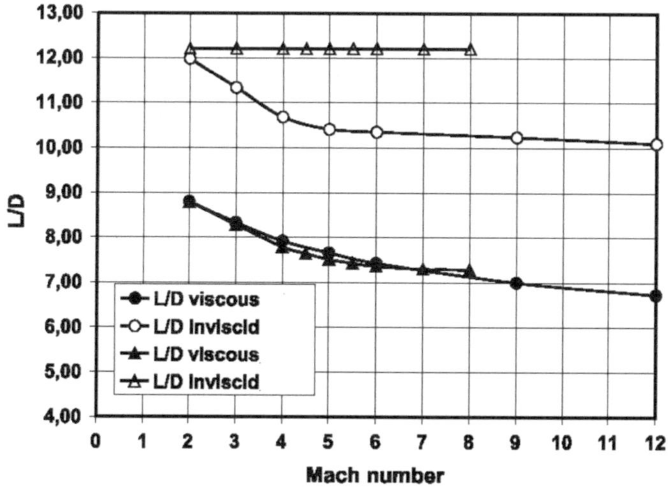

Fig. 2.20 The aerodynamic quality L/D as function of the flight Mach number of two waverider configurations without and with regard of viscous effects [13]

2.9.2 Fundamental Effects Regarding Stability and Control

Besides the aerodynamic quality L/D the characteristics of stability and control devices are decisive regarding the reliability and the success of the design. When talking about stability always the response of the flight vehicle—with fixed control effectors, here control surfaces—on an angle-of-attack or sideslip perturbance is meant.[17]

Trim means the return of the total pitching moment to zero by means of single or multiple trim surfaces. Magnitude and direction of the needed deflection of the trim surface(s) depend on the stability level of the system, either stable or unstable. They also are a function of the given zero-lift pitching moment (longitudinal trim). In contrast to trim surfaces, control-surfaces serve to change the vehicle's flight trajectory or flight attitude.

Before we treat issues of longitudinal stability we have a look at some definitions.

2.9.2.1 Aerodynamic Force Vector, Center of Pressure, Neutral Point

Sketched only are some elements of static stability of the longitudinal motion of an aircraft [13]. Longitudinal stability is given for $\partial C_m/\partial \alpha < 0$ at the angle of attack α_{trim}. This, however, is not sufficient to fly the vehicle stable. Necessary is that this condition holds for a sufficiently large range of the angle of attack.

[17] Control effectors can be aerodynamic surfaces, reaction-control devices, thrust-vectoring devices.

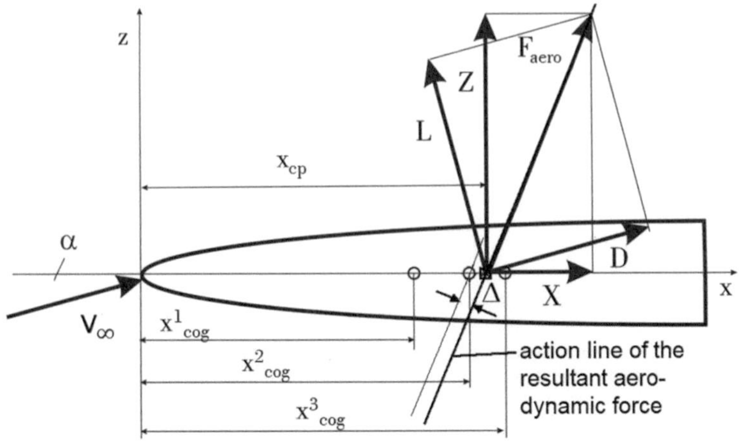

Fig. 2.21 Sketch of the of center of gravity (cog) locations relative to the center of pressure [13]

The resultant aerodynamic force vector F_{aero} in relation to three locations of the center of gravity (circles on the x-axis) is given in Fig. 2.21. For the rather flat and symmetric configuration the center of pressure (cop square) is located on the x-axis. What matters in reality is the location and the distance Δ of the action line of F_{aero} relative to the location of the center of gravity $cog(x, y, z)$, as well as the direction and the magnitude of the vector F_{aero}.

Consider Fig. 2.21. The location of the vector F_{aero} is fixed. If the cog-location is ahead of it with x^1_{cog} and x^2_{cog}, static stability is given. This can be considered as weather-cock stability. If the cog-location is behind the location of F_{aero}, e.g., at x^3_{cog}, the situation is unstable. Indifference is given, if F_{aero} lies on x_{cog}.[18]

For airfoils and wings with small thickness to length ratio, in general for slender configurations, the neutral-point concept is of interest. The neutral-point location x_n is the location where $dC_m/d\alpha = 0$.

The neutral-point concept is used now to consider longitudinal stability of slender hypersonic aircraft configurations.

2.9.2.2 Longitudinal Stability

The wing planform dominates the longitudinal stability of hypersonic aircraft. This is shown in Figs. 2.22 and 2.23 for the flight Mach-number interval $0 \leqq M_\infty \leqq 7$. Depending on the wing planform different trends of the basic neutral-point location x_N/l_i—l_i being the length of the wing—with $C_L = 0$ and $\alpha = 0$ are present.

A significant influence of the leading-edge sweep angle distribution over the length axis can be observed. Reference in all cases is the arbitrarily chosen indifferent design

[18] For the topic of dynamic stability see, e.g., [13, 29].

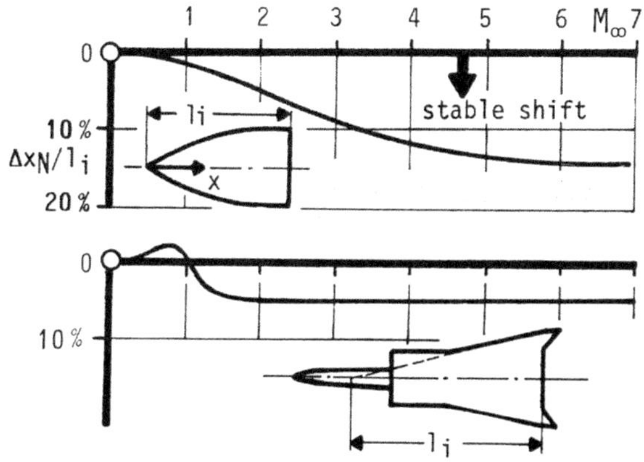

Fig. 2.22 Effect of the wing planform on the location of the neutral point as function of the flight Mach number M_∞ [33]. Wind-tunnel data, $\alpha = 0$, upper part: Gothic wing, $\Lambda = 0.9$, lower part: L1 airplane with two large ramjet engines

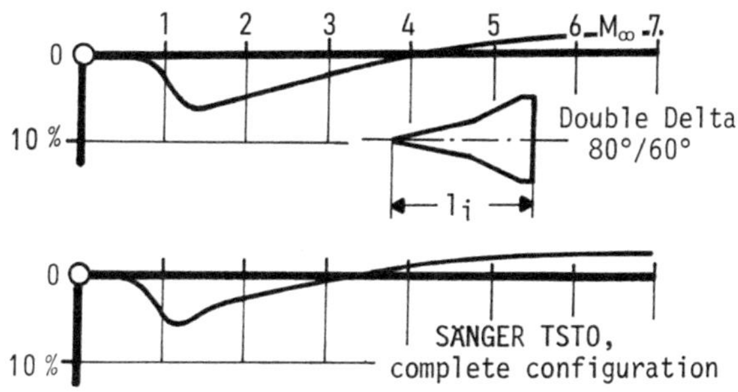

Fig. 2.23 Effect of the wing planform on the location of the neutral point as function of the flight Mach number M_∞. Wind-tunnel data, $\alpha =$ 0, upper part [33], lower part [34]

for $M_\infty = 0$ with $dC_m/dC_L = 0$. To a degree it is possible with a combination of the shown planforms and a redistribution of the wing surface parts to generate—within a certain range—desired trends of the location of the neutral point over the Mach number.

Noteworthy for the Gothic wing is the backward shift of the neutral point of up to 15%, upper part of Fig. 2.22. The L1 airplane on the other hand shows a virtually constant stability in the supersonic/hypersonic regime, lower part of the figure.

The double-delta wing, Fig. 2.5, and that of the lower stage of SÄNGER, however show above $M_\infty \approx 4$ the smallest neutral-point shift and hence permit a drag-favorable trim, Fig. 2.23. Yet it was required to adjust the zero-stability at $M_\infty = 0$.

Besides the basic stability at high Mach numbers and small angles of attack—with the design point in the hypersonic regime—it is mandatory to ensure the necessary stability properties at low-speed flight with high angles of attack, as given in the take-off and landing flight domains.

The pitch-up tendency with increasing angle of attack is related to the wing planform characteristics 'leading-edge sweep' and 'aspect ratio' as shown in Fig. 2.24. Wing planforms beyond the statistically found pitch-up boundaries may exhibit these tendencies.

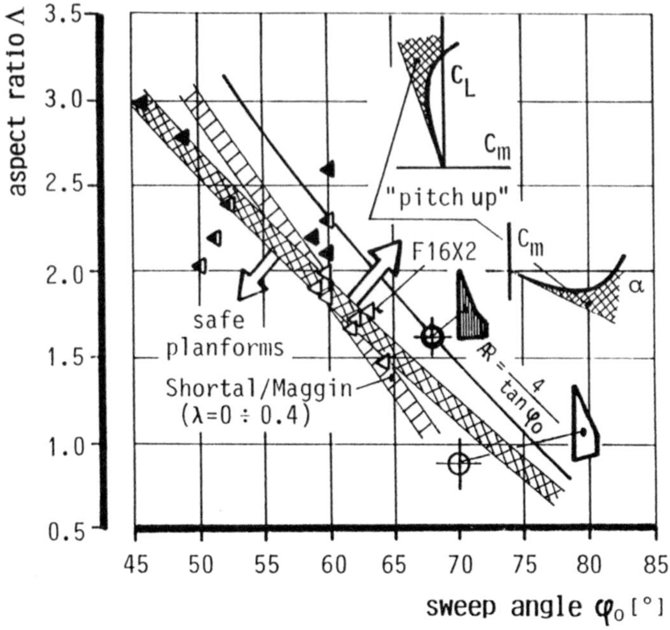

Fig. 2.24 Pitch-up boundaries of slender wings (subsonic domain) [4, 35]

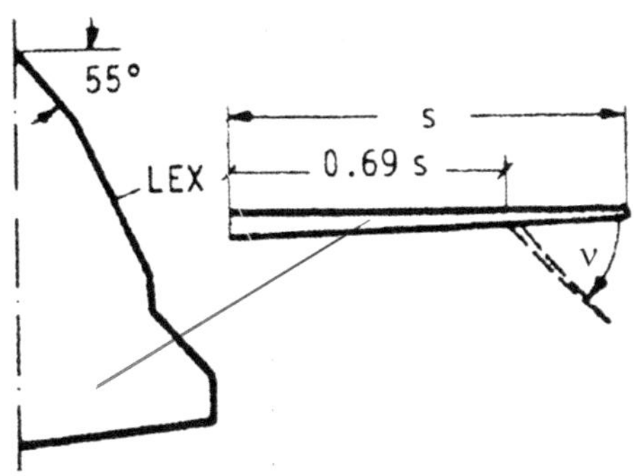

Fig. 2.25 Sketch of the wing and the anhedral alteration ν of the outer wing [4, 35]

A promising approach to avoid this unwanted phenomenon and to shift the pitch-up boundaries to higher aspect ratios and leading-edge sweeps is to analyze the topology of the appearing lee-side vortex system, see e.g., [4], already during the early design work. It is possible then to modify the respective geometrical wing properties in order to prevent the unwanted behavior.

In [25], see also [4], the lee-side vortex-flow topology of a delta wing was changed with a modification of the wing's planform and in particular with the introduction of an anhedral of the outer wing section, Fig. 2.25.

The original wing design had shown an intolerable pitch-up behavior. The initially present tendency to pitch up was equivalent to an upstream movement of the

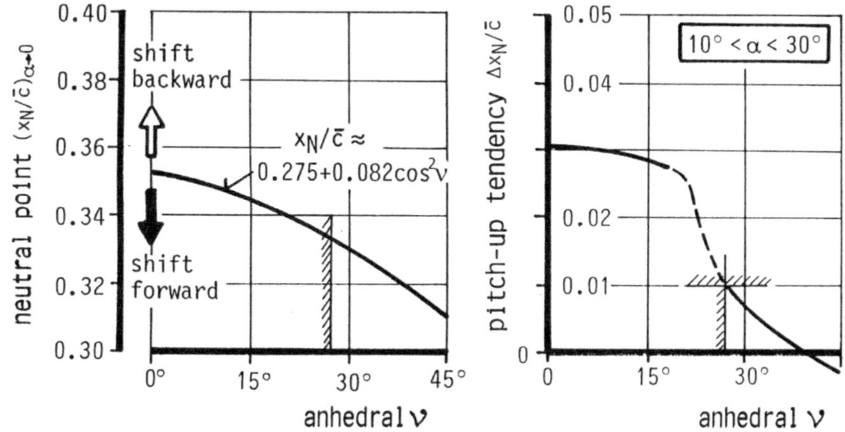

Fig. 2.26 Influence of the outer wing anhedral: left on the neutral-point location, right on the pitch-up tendency [4, 35]

neutral point—with increasing angle of attack—of about nine per cent of the mean aerodynamic wing chord $\bar{c}$. Geometrical modifications of the wing, but in particular the introduction of the anhedral of the rear outer wing portion, and the subsequent analysis of the flow topology of the lee-side vortex system led to a complete solution of the problem. In the following Fig. 2.26 the two major results are illustrated.

The figure shows at the left the influence of the angle of anhedral ν on the neutral-point location $x_N/\bar{c}$ at zero angle of attack. Hence regarding the stability, the shift of the neutral point with increasing anhedral angle follows exactly the $cos^2\nu$ law.

At the right of Fig. 2.26 evidence is given that the topological structure of the lee-side vortex system abruptly changes from an originally unstable structure into the—increasingly stable—structure due to the anhedral. That happens at the anhedral angle $\nu \approx 20$. This effect is triggered by the cross flow over the edge.

2.9.3 *Effectiveness of Stabilization, Trim and Control Surfaces*

In the whole flight domain, for all Mach numbers and flight attitudes flyability and controllability of the aircraft is mandatory, see, e.g., [29]. Stabilization, trim and control surfaces are the means to ensure and to control the moment household around the three vehicle axes, see, e.g., [13].

All control surfaces, especially trim surfaces must be highly efficient in order to avoid adverse drag increments in particular due to high flap deflections. Regarding trim surfaces the propulsion thrust-vector angle—due to the single expansion ramp nozzle (SERN), Sect. 5.7—strongly reigns in.

In addition to the drag issue the problem of thermal loads arises, notably at large flap deflections at high Mach numbers. Hence, high flap efficiency also at such Mach numbers is mandatory. Important geometrical parameters are the flap deflection angle, the hinge line orientation (the optimum case of orientation is orthogonal to the ramp's onset flow), the width and length of the flap, and the width of the hinge-line gap between the vehicle parent structure and the flap, Sect. 3.6.5.

The efficiency of a flap, in particular the needed deflection angle, depends on the surface of the flap, i.e., its width and length. Large flap surfaces, in particular the length of the flap, are a critical issue because of the ensuing large hinge moments. These demand high actuator performance, which goes along with large size and mass of the device.

Unavoidable is the loss of flap efficiency with increasing flight Mach number, Fig. 2.27. The data stem from wind-tunnel experiments with the SÄNGER configuration. The reference data are the following: the trailing-edge flap area is six per cent of the wing area, the lift gradient at $M_\infty = 1$ is $C_{L_{\delta flap}}/C_{L_\alpha} = 0.25$.

The figure shows that the flap-specific coefficients of the pitching moment $C_{m_{\delta_f}}$ and of the lift $C_{L_{\delta_f}}$ decrease faster with the flight Mach number than the lift coefficient of the wing C_{L_α}. Between the reference Mach number $M_\infty = 1$ and the final Mach number $M_\infty = 7$ about 90% of the flap effectiveness, with a very significant drop already between $M_\infty = 1$ and 2—at small angles of attack and small flap deflections—effectively disappears. The wing's lift coefficient, on the other hand, loses only 65%.

However, one has to keep in mind that the complete picture is given only after one has taken into account the dynamic pressure q_∞ at the considered flight Mach numbers.

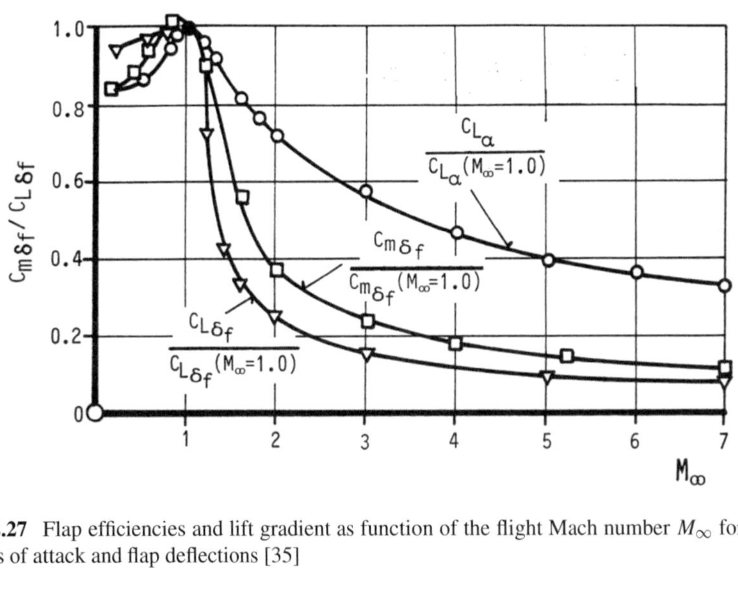

Fig. 2.27 Flap efficiencies and lift gradient as function of the flight Mach number M_∞ for small angles of attack and flap deflections [35]

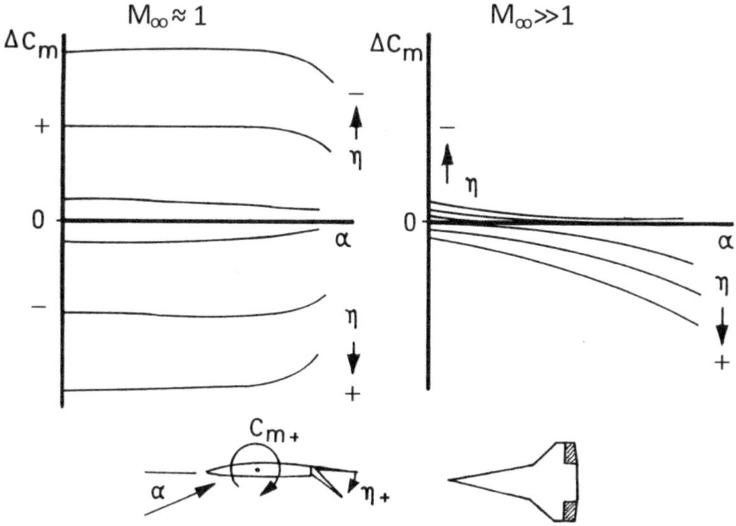

Fig. 2.28 Flap efficiency ΔC_m as function of angle of attack α and flap setting angle η. Left small, right high flight Mach numbers M_∞ [35]

Why does the flap efficiency reduce with increasing Mach number? The effect is emphasized in Fig. 2.28. At low Mach numbers, $M_\infty \approx 1$ in the left part of the figure, the efficiency practically is constant—in terms of ΔC_m—over a large angle of attack α range—as long as the flow is attached—as well as over a large range of the flap deflection angle η.

At high flight Mach numbers flap efficiency in both η directions is given only at small angles of attack, right part of Fig. 2.28. At higher α the flap effect is shifted completely to the lower, the pressure side of the flap. The reason for this effect is that the upper side increasingly is in the hypersonic shadow, where the vacuum limit of the pressure coefficient is approached: $c_{p_{vac}} = c_{p_{p=0}} = -p_\infty/q_\infty = -2/\gamma M_\infty^2$.

This effect is the reason, why winged re-entry vehicles, which enter the atmosphere at high angle of attack, fly in a longitudinal unstable mode: the trim flap is effective only at downward deflections. During the first re-entry of the Space Shuttle Orbiter, which happened at α = 40 —a value, which became standard for all later projects of winged re-entry vehicles—the location of the center of pressure was wrongly predicted.

The deflection of the trim flap was two times larger as predicted and the flap came near to its maximum possible deflection angle. This so-called hypersonic pitching moment anomaly, see, e.g., [13], with the accident barely averted, today appears to be forgotten.

The fluid dynamical phenomena of the ramp-type flow—shock/boundary-layer interactions—are complex. The original investigations by B. Edney, prompted by observations of structural damage at the pylon of a ram engine on the X-15, led to the identification of six interaction types. For a discussion of these cases see, e.g., [5, 13].

2.9.4 X-15: the Wedge-Shaped Vertical Stabilizer

The X-15 venture led to another find, too, the wedge-shaped vertical stabilizer. During the development of the aircraft a solution became necessary regarding the flap efficiency [36].

The problem was that the classical thin, all-moving tailplanes had not enough effect regarding stability and control. Hence, in order to ensure enough longitudinal and in particular directional stability at high flight Mach numbers—up to $M_\infty = 6.7$—and high flight altitudes—up to around 100 km—very large horizontal and vertical tailplanes appeared to be necessary. This also had to be seen in view of the shift of the vehicle's center of gravity during emptying of the fuel tank, and also the very low dynamic pressure found at the extreme flight conditions.

The tail volume—tail surface times distance to the center of gravity of the vehicle—could not be enlarged by moving the cruciform empennage back. The solution then was, for the vertical stabilizer, to give it a 10 wedge shape. The effect of a wedge shape is that the vacuum limit on either side is not reached for the given deflection angles in the Mach number domain. Figure 2.29 shows the general effect in terms of the ratio 'tail area' to 'wing area' for a classical thin airfoil in comparison to a split flap with angle δ as function of the flight Mach number M_∞.

At low Mach numbers the effect is small for both the vertical and the horizontal tail. For the vertical tail at $M_\infty \approx 7$ the ratio comes down from 0.3 for the thin airfoil to about 0.1 for the $\delta = 15$ split flap. The effect is similar for the horizontal plane.

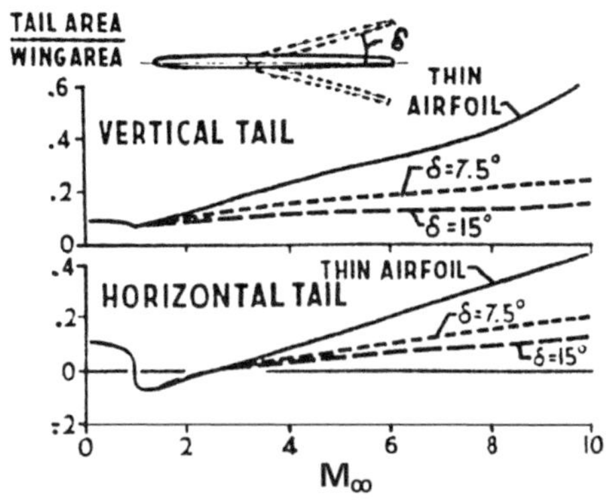

Fig. 2.29 Effect of a split flap on the ratio 'tail area' to 'wing area' for a thin airfoil as function of the flight Mach number M_∞ [36]. The upper part shows the effect for a vertical tail, the lower part for a horizontal one

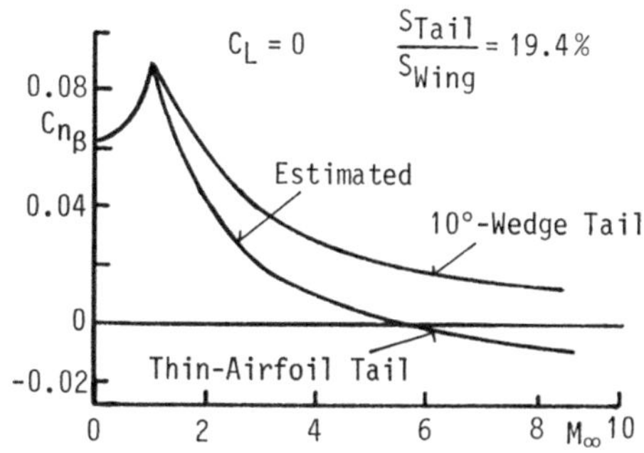

Fig. 2.30 Experimental verification of the split-flap effect [36]

In Fig. 2.30 the β-derivative of the yawing moment coefficient, C_{n_β}, as function of the flight Mach number for a thin airfoil tail and a 10 -degree wedge tail—instead of a split flap—at zero lift of the wing is given. The flap to wing surface ratio was 19.4%. At $M_\infty \approx 6$ C_{n_β} for the thin airfoil (estimated value) was zero, whereas that for the wedge tail (experiment) was about 0.02.

This result led to the wedge-shaped vertical tail plane of the X-15 with both an upper and a lower surface. Half of each surface was the all-moving rudder. Directional stability was secured in the whole Mach number domain. Flights were made with and without the lower moving tail plane. The wing and the horizontal tail plane had slender supersonic airfoil sections.

For the hypersonic airbreathers considered here it is questionable, whether wedge shapes are needed as for the X-15. For ARV-type vehicles it could be possible, also in view of the encountered thermal loads.

2.9.5 A Note on the Kutta-Joukowsky Condition

The wedge-shaped tail of the X-15 led to a thick blunt trailing edge. The Kutta-Joukowsky condition for lifting airfoils/wings assumes a sharp trailing edge. In reality the trailing edges of aircraft wings and stabilization, trim and control surfaces—other as assumed by Kutta and Joukowsky—are always blunt, of course not to the degree as with the X-15. For manufacturing and handling reasons thicknesses of the order of one to two centimeters can be found [4].

At the Common Research Model the thickness of the trailing edge is 1.31 cm at the Yehudi break. The matter of sharpness of trailing—and leading edges of swept wing configurations—in terms of the question "when can an edge be considered as being aerodynamically sharp" is discussed to a degree in [4].

2.9.6 A Look at Different Tail Configurations

In closing this section a comparison is made of approaches regarding tail configurations, Fig. 2.31. At the top sketched is the aft vertical tail volume principle. Δx is the distance from the tail surface to the center of gravity.

In the middle left the central vertical tail of the Space Shuttle Orbiter is shown. Due to the high angle of attack during re-entry it became effective only at $M_\infty \lessapprox 5$ and $\alpha \lessapprox 20$ to 25. Above these values a reaction control system was employed in order to facilitate directional stability, see, e.g., [13].

Below the Orbiter, HYTEX is mentioned, a study of an experimental airbreathing hypersonic flight vehicle in the frame of the German Hypersonics Technology Programme, see, e.g., [26]. In contrast to the Orbiter its flight angle of attack was always very small. At the right the above dealt with X-15 is sketched. Both the dorsal and the ventral wedge-shaped flaps were all-moving, the lower one was ejected for landing.

The lower part of the figure shows the twin stabilizer of the SÄNGER lower stage configuration, of the National Aerospace Plane, of RADIANCE, of the SR-71, of HERMES and HORUS, the upper stage of SÄNGER. Both surfaces of the SR-71 are tilted inward in order to lead the resulting force vectors through the x-axis of the

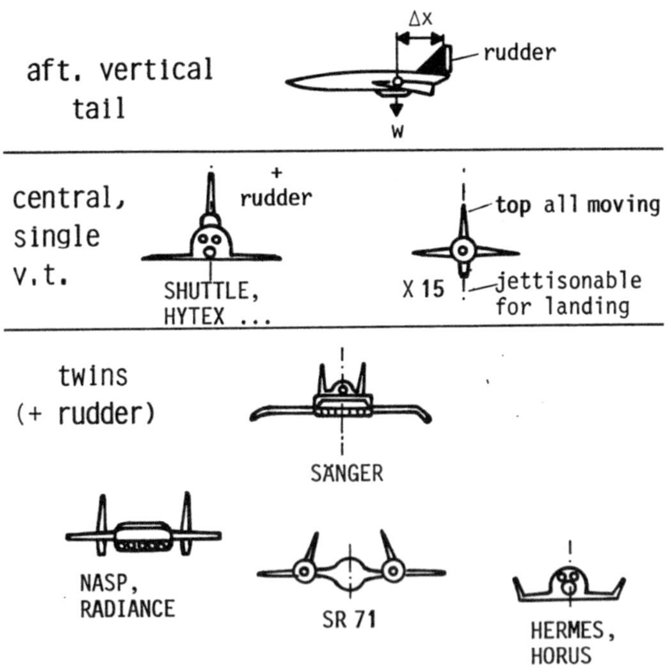

Fig. 2.31 Directional stability and control: examples [25]

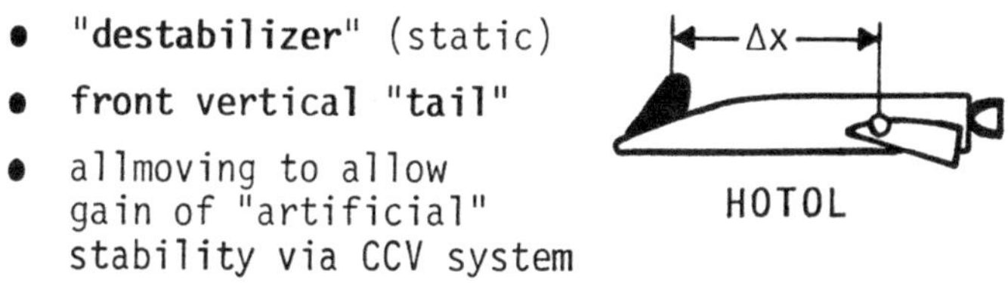

Fig. 2.32 HOTOL, a particular example [25]

center of gravity of the vehicle, thus avoiding induced rolling moments during one engine inoperative (OEI) situations.

A seemingly strange vertical tail configuration is that of HOTOL, Fig. 2.32. Because the vertical stabilizer could not be moved back to the needed degree, it was placed at a most forward position. In this way HOTOL was a control-configured vehicle (CCV) with artificial directional stability.

2.10 The Reference Concept: Design Sensitivities and Growth Factors

In Sect. 1.6 particular development problems were addressed. They are, besides the high complexity of hypersonic flight vehicles, the topics requirements creep, obsolescence and mass increase. The present section is devoted to a discussion of design sensitivities and mass increase of large hypersonic airbreathers in terms of the API-type 5 reference concept.

The figures of merit and also the other terms below are used for analysis of the design of flight vehicles and also as input for cost analysis, see, e.g., [37].

2.10.1 Figures of Merit and Other Terms

In order to judge the feasibility of a flight vehicle design, figures of merit are used in order to obtain information about the design. These figures concern the feasibility of the design in terms of performance, mass, robustness, risk and so on, and finally the cost of the flight system and its mission [14, 22, 25].

Figures of merit are

- Take-Off Gross Mass, TOGM [g] TOGM does not distinguish between vehicle hardware, propellant and payload. Hardware is extremely expensive to develop and maintain. TOGM is not a strong figure of merit, but it permits to receive statements about operation costs. Note that one metric tonne [to] is equal to 1×10^6 g = 1 Mg = 1 to.

- Take-Off Gross Weight, TOGW [N] TOGW is corresponding to TOGM multiplied by the gravitational constant g and also used.
- Operational Mass Empty, OME [g]
  OME is the mass of the airframe and of the engines, plus the permanently installed operating equipment with fixed locations, including ballast, hydraulic fluid, non-propulsive fuel, and engine oil.
- Operational Weight Empty, OWE [N] OWE is corresponding to OME and also used.
- Dead mass [g]
  Dead mass is a "friendly" mass, whose increase primarily does not directly lead to an increase of the flight-vehicle volume. This is in contrast to an increase of structural mass or fuel mass. However, secondarily volume and mass of the flight vehicle will rise, although usually with a relative small growth factor. Examples of dead mass are equipment and also internal structures.
- Growth factor and sensitivities These figures measure margins to the design limit, i.e., the divergence of the vehicle mass, and indicate design risks and performance limits. Growth factors and sensitivities are determined by variations of primary design parameters like mass, drag, thrust, fuel consumption. They reflect the configurational consequences due to uncertainties with respect to properties (e.g. mass, aerodynamics, dimensions) and performance (e.g. specific thrust) of components or systems.
- System complexity Overall system complexity indicates development and acquisition costs and risks. The number of major subsystems can be a measure of complexity, say of the amount of interfaces/interactions.

Other terms and abbreviations are:

- Dry thrust: Thrust of the turbojet without afterburner.
- SEP: Specific excess power [m/s], i.e., the power available to climb or to accelerate the flight vehicle.
- SFC: (thrust) specific fuel consumption [(kg/hr)/N]. Low SFC: high efficiency, high SFC: low efficiency of the propulsion system.
- SLS: Sea level standard, i.e., atmospheric data at sea level. For the standard atmosphere see Appendix A.
- SPC: Specific thrust, i.e., thrust per unit air mass-flow rate.
- SPS: Secondary power system.
- Thrust-to-weight ratio T/W [-] The thrust-to-weight ratio is a measure of the acceleration capability of the flight vehicle and also is the key parameter to drive SEP.
- TO: take-off.
- Wetted area [m^2] The wetted area is a measure of the viscous drag and an indicator for structures and thermal protection related expenses. The latter primarily regards flight systems with re-entry flight, like ascent and re-entry (ARV) systems.
- Wing loading W/A [N/m^2] The wing loading governs the size of the wing, takeoff and landing distances as well as maneuverability. A is either the area of the wing only or the downward projected area of the whole configuration.

These figures of merit and the other terms, e.g., payload fraction, fuel fraction, of the considered flight system eventually can be incorporated into an evaluation matrix, into cost models and so on. They further permit identification of potentials and risks.

All results in the following subsections were obtained with the CADEHYP scaling tool—Sect. 2.3—for a SÄNGER-like TSTO concept [22], see also [25].

2.10.2 Sensitivities and Growth Factors

There are many contributors to mass sensitivities, the most "friendly" one is the "dead mass", see above. The sensitivities due to an increase of dead mass of a TSTO flight vehicle are shown in Fig. 2.33 by means of characteristic mass fractions.

The mass fractions shown here are the take-off gross mass (TOGM), the operational mass empty (OME), the fuel mass (LH2), the mass of the structure and the mass of the propulsion system.

Note that the wing loading (W/A) and the thrust-to-weight ratio (T/W) are kept constant for this approach. The gradient of the straight lines—due to the small percentage of dead mass relative to the total TOGM—is the (constant) respective growth factor, showing that 1 ton of additional dead mass will increase TOGM/OME/fuel/structure/propulsion mass by 2.45/1.85/0.6/0.5 and 0.30 tons. About half of the increase of OME is due to the increase of the dead mass, the other half comes from the rise of the involved items.

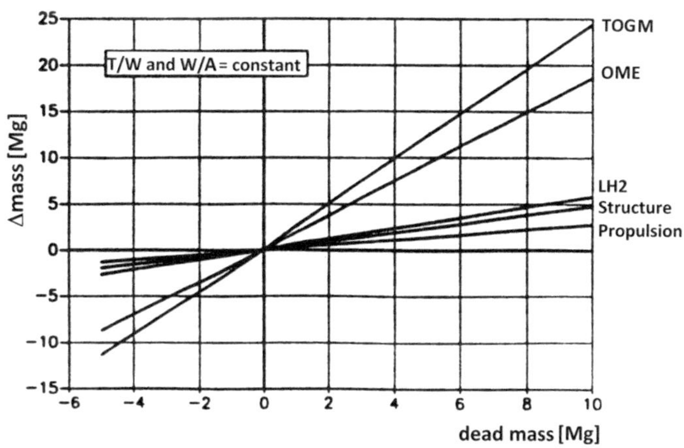

Fig. 2.33 Dead mass sensitivity [22]

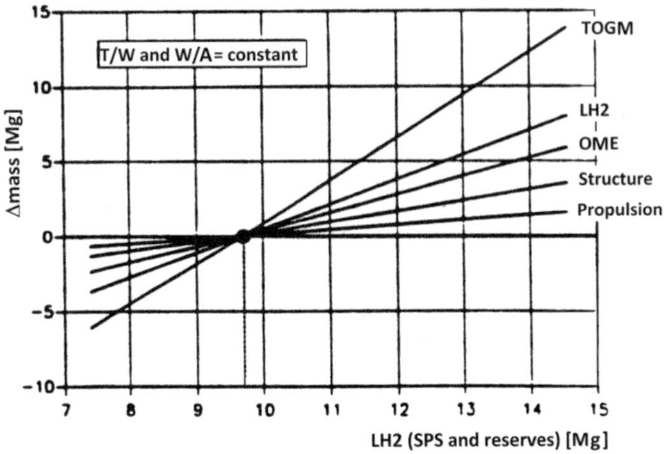

Fig. 2.34 Sensitivity due to non-propulsive LH2 [22]. SPS is the secondary power system

In contrast to the influence of dead mass is the effect of fuel mass (liquid hydrogen). An increase of LH2 mass—due to the corresponding increase of the tank volume—gives a highly vital mass. Additional mass especially is induced when LH2 is non-propulsive (secondary power/cooling/fuel reserves), Fig. 2.34.

The non-propulsive LH2 due to the additional volume is producing additional flight vehicle volume and surface and hence increases the drag. The higher drag has to be compensated by an increase of thrust, fuel (LH2) and this in turn will demand more volume, surface ... .

Comparing with Fig. 2.33—note that the scales on the ordinate are different—one sees that the mass sensitivity due to non-propulsive LH2 is even higher than that for empty mass OME and now is the highest sensitivity of all partial masses, resulting in almost 60% of the additional total mass.

These growth factors are shown in Fig. 2.35. The dominant effect of non-burned LH2 is apparent—growth factor 1.65 for itself, meaning 1 ton of non-propulsive LH2 additionally induces 0.65 tons of LH2 for the complete configuration—and a total TOGM growth factor of 2.75 is found.

This effect is resulting from the "food chain" mass → volume → surface → drag → additional fuel consumption. The "drag-production" of this non-propulsive use of LH2 is depending on the size of the carrier, say on its ability to allow additional volume without a—relative—high generation of wetted surface. Remember the "square-cube law" for the mass of the wing.[19]

The take-off gross mass TOGM can be used as a measure of size when comparing flight vehicles. Here we use the global growth factor. We follow back the

[19] The square-cube law, first formulated by Galileo in 1638, states that if a shape grows in size, its volume grows faster than its surface area.

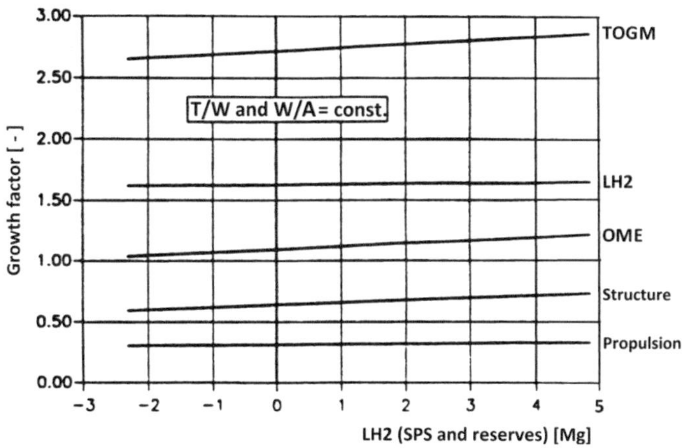

Fig. 2.35 Growth factors due to non-propulsive LH2 [22]

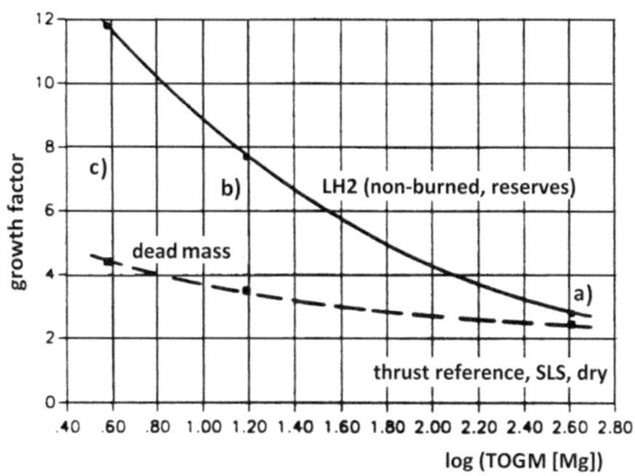

Fig. 2.36 Effect of the vehicle size on the growth factors of dead mass and non-burned LH2 and reserves [22]. **a** stands for a TSTO space-transport system with 400 tons take-off mass, **b** for a medium sized experimental vehicle with 15 tons, and **c** for a small one with 4 tons

effect of TOGM from (a) the large TSTO system (400 tons) to (c) a small demonstrator with about 4 tons take-off mass, in-between is a 15 tons demonstrator (b). Figure 2.36 represents the relevant effects of dead mass and non-burned LH2. The 'helpful' effect of the size of the air vehicles is evident, especially when regarding liquid hydrogen. Smaller vehicles underly the punishment for their less "compact" design possibility.[20]

Hence, when we scale hypersonic flight vehicles powered by LH2, an "inverted square-cube law" is found. Remember that the structural mass for instance of a wing varies with the third power of the length scale, but the aerodynamic load only with

[20] An additional contributor to the high sensitivity due to LH2 for small vehicles is due to the fact that, the smaller the vehicle, the relatively thicker is the insulation, thus resulting in a less "dense" and consequently heavier design.

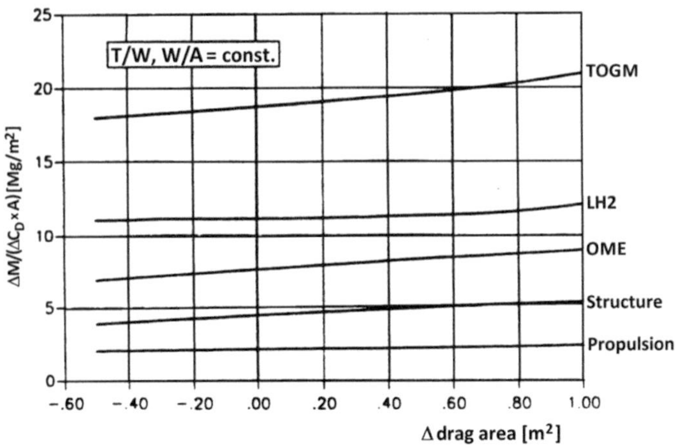

Fig. 2.37 Mass sensitivity due to additional drag [22]

the square. Therefore, the structural mass will increase proportional to $TOGM^{1.5}$ (empirical 1.43). From the indirect effect of the drag increase we now turn to the direct one.

Figure 2.37 presents the mass increases/sensitivities due to an increase of drag. For our example the "drag area" $C_D \times A$ is based on $A = 1{,}500\,m^2$. The drag increase will demand—to balance it—a more powerful, heavier engine and more fuel. Therefore, structural mass will increase too, in consequence, so as to carry the additional mass (fuel, tank, propulsion system). OME, volume, surface and TOGM rise and the "drag food chain" is found again.

For the scaling process the thrust/weight-ratio and the wing loading again were kept constant. Obviously the lion's share of the increase of TOGM is found for the fuel (LH2), the smallest for the propulsion system.

Note that in our case $1\,m^2$ additional drag area results in less than 10% of the subsonic zero-lift drag. But this increment generates—effective over the total trajectory—about 5% of TOGM, that is just over 20 Mg for a 400 Mg flight vehicle. Of course this increase of drag is not equally distributed over the whole flight trajectory. It is acting strongly near the acceleration pinch-points—points with minimum excessive thrust—around $M_\infty \approx 1$ and $M_\infty \approx 7$, thus heavily influencing the complete design in terms of TOGM.[21]

Another high sensitivity of the TSTO design is found regarding its consumption of fuel, Fig. 2.38.

From Fig. 2.37 we learned that a drag increase gives rise to a total increase of mass, 55% of which is caused by higher fuel mass. For the case of an increase of specific fuel consumption (SFC), as shown in Fig. 2.38, this percentage now is close

[21] The acceleration pinch points, see Fig. 5.36, indicate minima of excessive thrust.

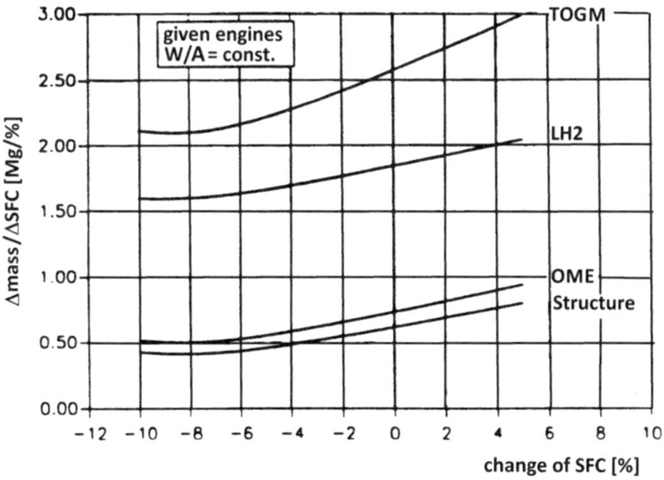

Fig. 2.38 Mass sensitivities due to change in specific fuel consumption SFC [22]

to 70% (fixed thrust and wing loading). A 5% increase of SFC is resulting in a 15 per cent increase of TOGM, more than two thirds of which is additional LH2 mass.

This increase is rapidly driving the design to divergence, the constant level of thrust cannot counteract anymore the increase of drag, the wave drag and the friction drag in the acceleration pinch-points.

Following the curves of Fig. 2.38 for reduced SFC, at the left a minimum of Δmass/ΔSFC is found for a reduction of 8 to 10%. This is the result of the increased thrust-to-weight ratio of the now 20% lighter configuration.

The flat minimum of fuel mass (LH2) is even more shifted to the left. This is due to the then excess thrust of the design, which allows a longer flight range with reduced—but not efficient—power setting.

More details concerning the effect (and the efficiency) of two types of engines—turbojet with afterburner for $M_\infty \lessapprox 3.5$—, then transition to ramjet up to $M_\infty \approx 6.8$ are presented in Fig. 2.39.

We concentrate on the growth factor of TOGM as function of the per cent change of SFC. Evidently the basic design burns twice the mass of fuel in the ramjet mode than in the turbojet mode. This ratio is increasing with decreasing total SFC.

It is clear from Fig. 2.39 that there should be an early maturing of ramjet technology in order not to get stuck with unexpected and insurmountable problems during the development of the complete design. This is also and still more true for the scramjet technology.

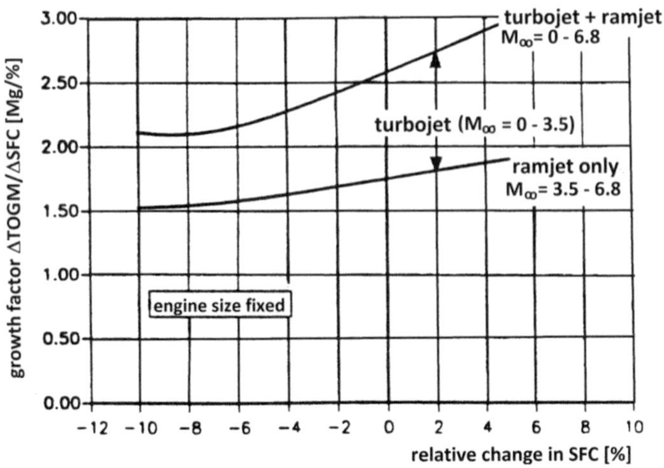

Fig. 2.39 Growth factor of TOGM affected by turbojet and ramjet efficiency in terms of the specific fuel consumption SFC [22]

2.10.3 Influence of the Launch Mach Number

Up to now the results are representing the effect of selected technologies on the development of mass with regard to the complete TSTO system (lower and upper stage). Additional mass is only a problem of and for the lower stage. The upper stage is not affected because the mission of the lower stage is assumed to provide always an identical launch condition for it. If this would be changed, the influence of the now to be modified upper stage on the design of the lower stage and hence on the complete configuration has to be investigated.

The influence of the design drivers LH2 (fuel for the lower stage) and the orbiter (the upper stage) launch mass as function of the launch Mach number is given in Fig. 2.40. The 100% reference is based on data for launch at $M_\infty = 6.6$ at an altitude of 34 km and a given flight-path angle γ_i at injection into orbit. Engines and wing loading are fixed. The variation of fuel mass of the first stage (LH2) shows a minimum for $M_{launch} = 5.8$, whereas the orbiter mass—as expected—decreases with increasing launch Mach number.

Figure 2.41 shows the relative mass of the complete system—based on the launch at $M_\infty = 6.6$—as a function of the launch Mach number.

Any change of the (optimized) launch point—$M_\infty = 6.6$—is increasing TOGM. The sensitivity of the rocket-driven upper stage to a reduction of its launch speed is by far exceeding the fuel gain—less LH2—of the lower stage. It becomes more difficult to integrate the growing mass and dimensions of the upper stage into the decreasing size of the lower stage. On the other hand an increase of the launch Mach number

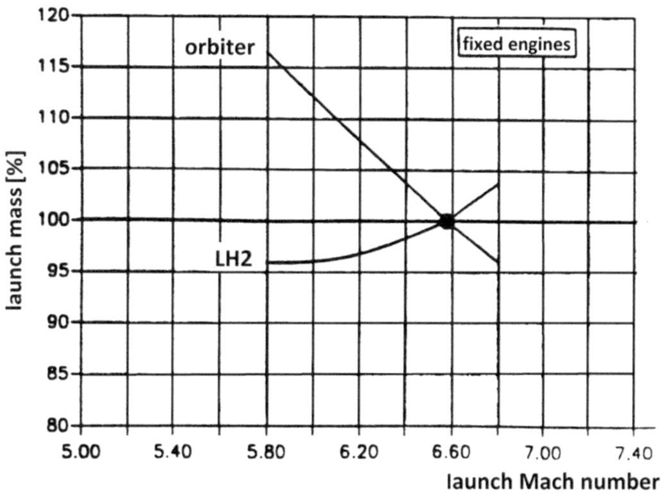

Fig. 2.40 Effects of the orbiter (upper stage) launch Mach number on the mass of the orbiter and the LH2 consumption of the lower stage [22]

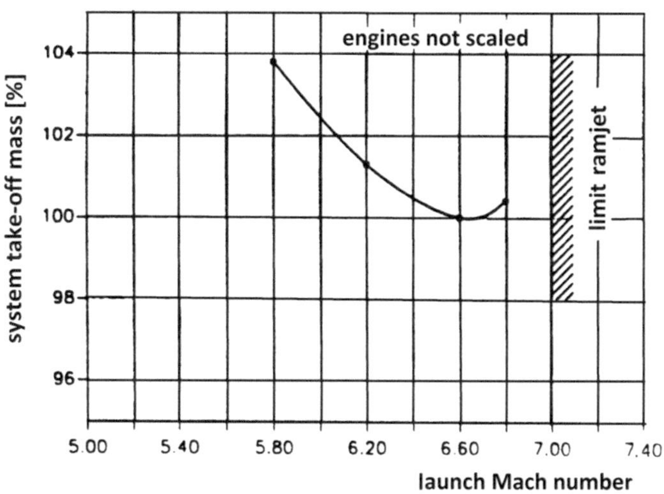

Fig. 2.41 System take-off mass depending on the launch Mach number [22]

rapidly drives the lower stage to design divergence due to the Mach number limit of the ramjet, as indicated in Fig. 2.41. Additionally we then would have a thermal problem for the lower stage—cooling system and abandoning of the hot structure concept.

2.10.4 *Influence of Mission Type*

The original mission and design philosophy of the reference concept SÄNGER of the German Hypersonics Technology Programme, Sect. 2.8, was based on three technical premises: (1) complete reusability, (2) operation like an air transport system (technical security, horizontal take-off and landing, logistics .... .), and (3) the mission starts from central Europe: political independence and security.

From premise (3) distance and duration of the mission are specified, hence the mass of fuel and the size of the lower stage. The expression "transport system" is correct insofar, as the distance from Europe to the launch of the upper stage (for instance 28.5 of latitude) stands for an inevitable transport mission in view of its size and costs.

In FESTIP, see Sect. 1.1, a SÄNGER mission from Kourou was investigated. Expected was a reduction of the costs of the mission, but the amount of total operational expenses was unknown. We give a short comparison of the original mission and the Kourou mission.

The respective footprints of the original and the Kourou mission are compared in Fig. 2.42, resulting in trajectory lengths of about 5,500 km and 3,500 km. Note that the original SÄNGER launch Mach number is $M_\infty = 6.8$.

The question now is, how the mission length ratio of 1 to 0.64 affects the dimensions of the respective designs. The Kourou mission represents a clean acceleration process.

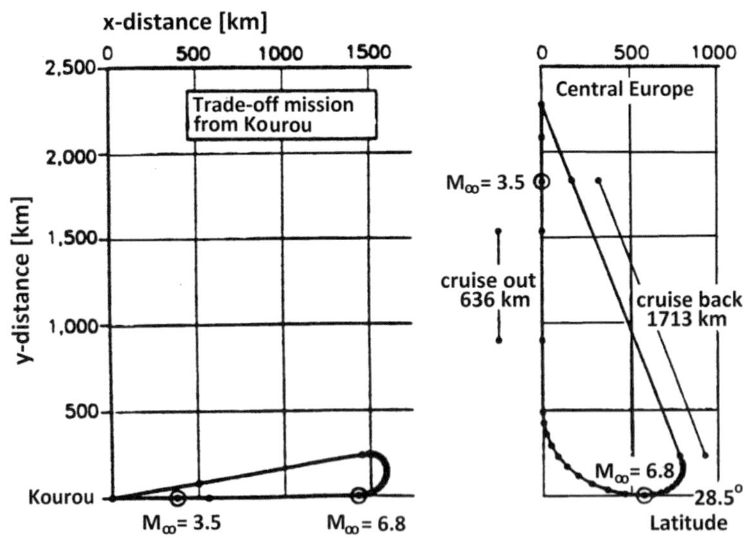

Fig. 2.42 Comparisons of the original SÄNGER mission at the right and the Kourou mission at the left in terms of their footprints [22]

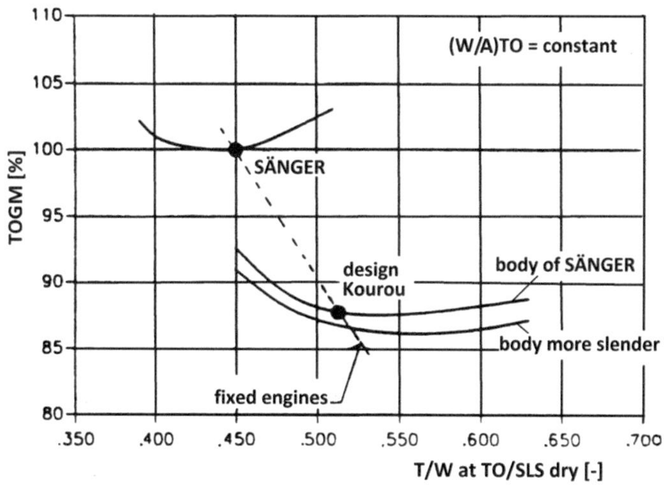

Fig. 2.43 Effect of the mission types [22]: TOGM in per cent of the original SÄNGER mission as function of the thrust-to-weight ratio T/W

For the investigation the following characteristics were fixed: (a) take-off wing loading for both systems is the same and (b) the payload of the lower stage, i.e., the orbiter—the upper stage—with the mass of 115 tons, is the same, with identical conditions of separation.

Characteristic (a) was due to safety aspects: minimum speed to guarantee a safe possible abortion of take-off. Characteristic (b) of course was due to the general transport mission, i.e., the insertion of the given upper stage into orbit. The effects on the mass of the systems (TOGM) are shown in Fig. 2.43.

The comparison of the missions was made with the two characteristics as preconditions. In addition it was assumed that the thrust to weight ratio could be varied for the turbojets (dry).

Fortunately it was found that the two missions can be realized with the same engines. The lower system mass for the shorter Kourou mission allows to make use of the—consequently—given higher thrust-to-weight ratio due to its pure accelerator mission and the shorter trajectory.

In any case the resulting thrust-to-weight ratio is on or close to the minimum mass condition. Any variation for SÄNGER is needless here as its design is already optimized. The two lower curves of Fig. 2.43 are representing designs for the Kourou mission, the upper, slightly less efficient design is based on the body of SÄNGER, the lower curve is its adaptation for this mission by a more slender body.

Uncertainties for instance in total drag prediction lead to design margins, which make more engine thrust necessary. As a consequence bigger engines and a larger fuel tank volume are needed, hence a larger mass of engine and tank, a larger airframe volume and a larger wetted vehicle surface. This all leads to a larger total drag, and finally a larger take-off gross mass ensues.

2.10.5 The Design Diagram

Combining the results found so far we can state that the SÄNGER concept is feasible in view of the given mission. Hence the design diagram can be established. It will allow to find the—more or less—total optimum of the take-off mass TOGM, Fig. 2.44. Variables are take-off wing loading and the thrust to weight ratio of the dry engine case with turbojets at SLS. The (relative) TOGM is the parameter, 100% denoting its absolute minimum. Note that the low nominal values of wing loading are based on the complete, downward projected area of the configuration.

The shell-type closed curves of Fig. 2.44 are showing the well-known possibility to exchange the thrust-to-weight ratio against the wing loading. The dotted line at the bottom is the border of divergence. The design diagram can be divided into four quadrants between the points (a) to (d). Their axes—inclined and curved—are given by the maxima and minima of the wing loading (x-axis) and the thrust-to-weight ratio (y-axis).

The quadrants now run clockwise from one end to the other via a line of constant take-off mass. We arbitrarily choose the most exterior "shell" TOGM = 105.1%. In quadrant 1—from point (a) to point (b)—T/W is increasing from 0.44 to about 0.54 and the wing loading is increasing from the minimum given (2.2 kN/m^2) to 3.0 kN/m^2. This increase is transformed into a possible increase of thrust and hence engine mass. The same is true for a reduction of fuel consumption due to the higher acceleration performance.

We now turn from quadrant 1 to quadrant 2, i.e., from point (b) to point (c). In this domain we have $0.54 > T/W > 0.47$ and $3\ \text{kN/m}^2 < \text{TOGW/A} < 3.4\ \text{kN/m}^2$. Here T/W is decreasing despite of the increasing wing loading. Obviously there is

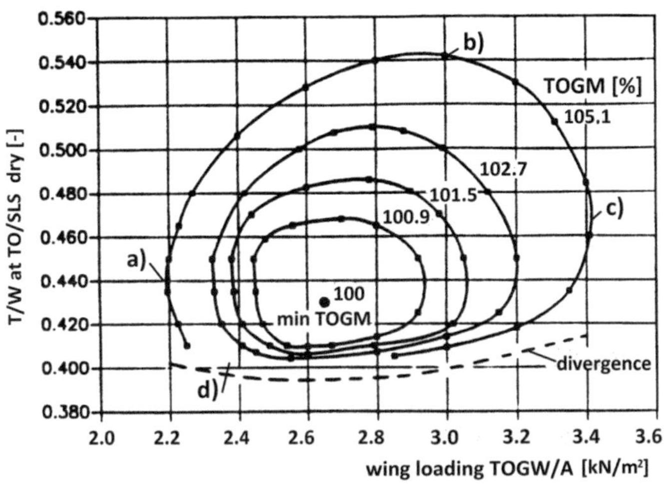

Fig. 2.44 The design diagram: determination of the minimum take-off mass [22]

another partial mass that increases even when the engine mass is decreasing. This only can be the fuel mass, because the payload is kept constant.

The high wing loading leads to a higher induced, i.e., lift-dependent drag, because the wing area A is reduced. That drag is proportional to $1/A$, and has to be balanced by more fuel. The lowest wing loading 2,2 kN/m^2 is resulting in the highest zero-lift drag (quadrant 1), the highest one (quadrant 2), with TOGW/A = 3.4 kN/m^2 is producing maximum induced drag.

With decreasing T/W and simultaneously decreasing wing-loading we now pass through the third quadrant from point (c) to point (d). At the latter point we have the minimum of the thrust-to-weight ratio (= 0.40) and TOGW/A = 2.35 kN/m^2 (not shown in the diagram). The mass of the wing is increasing now, but a reduction of the engine mass and the fuel mass are compensating the increase of the mass of the wing.

In quadrant (4), from point (d) to point (a), we have 2.35 kN/m^2 $>$ TOGW/A $>$ 2.2 kN/m^2 and 0.40 $<$ T/W $<$ 0.44. Wing loading and induced drag go down to their minimum and the wing area as well as the wing mass are increasing to maximum, the same is true for the zero-lift drag of the configuration.

Watching the balance of total mass a possibility is to introduce a 10% stronger engine (0.44 $>$ T/W $>$ 0.40). By that the line of design is completed and we come back to the starting point of our considerations. We note that the praxis of the engineering process is highly complicated. In reality all the considered design variables are interfering one with the other, there are secondary, tertiary, even higher order mechanisms and effects which only can be investigated and solved by using sophisticated numerical methods for design and analysis.

In Fig. 2.44 the borderline, noted as "divergence", needs a short concluding remark. Close to it are positioned the highly sensitive and risky designs with strong growth factors and increases of mass. Reliable predesign and design in this regime still is very challenging and will remain to be so.

2.10.6 *The Banana Effect*

A particular situation is present at the forebodies of hypersonic flight vehicles of API-type 4 and 5, Sect. 1.5. The lower side of the forebody is a pre-compression ramp for the inlet—it of course also provides lift—, the upper side in the limit is a free-stream surface. Consequently the Reynolds number is higher at the lower side, the boundary layer there thinner and the radiation-adiabatic wall temperature higher than at the upper side.

This situation is treated in Sect. 3.8. In Fig. 3.31 it is shown that for laminar flow the temperature at the lower side of the SÄNGER forebody is about 100 K higher than at the upper side. For turbulent flow the difference is about 200 K. The fact that these results were found for the flight altitude H = 33 km, whereas the following results are for H = 31 km, does not play a deciding role.

The results for $H = 31\,\text{km}$ are the last example of the high "entanglement" of effects and branches of knowledge in hypersonic flight vehicle design. Considered is a TSTO space transportation system like SÄNGER flying with $M_\infty = 6.8$ at the altitude of $H = 31\,\text{km}$.

Assume now that a hot primary structure is given with no internal insulation between the lower and the upper side of the forebody. For SÄNGER the material of the structure was assumed to be titanium almost throughout, Sect. 6.3. The temperature difference results in different strains and thus bending up the forebody to a banana-like shape. The nose is displaced upwards as shown in Fig. 2.45.

The primary effects are given by the first set of two columns of Fig. 2.45, differentiating between the laminar and the turbulent state of the boundary layer. Due to this "bananization" of the forebody its aerodynamic and temperature load distributions are varied (second set of two columns). The nose will bend up further until the shown displacements are reached.

In the next step the trim changes necessary because of the bananization are considered. They are giving—besides a positive increment of lift—a change of the (zero-lift) pitching moment. The increased pre-compression effect of the lower side of the forebody increases the thrust to some extent. Therefore we have to re-trim the configuration (angle of attack, trim-flap deflection, power setting) ending up with a slightly lower displacement of the nose (third couple of columns).

This is the final state, the shape of a flight vehicle—here the forebody—following the hypersonic design rule for structures, allowing to build up strain without producing intolerable stress. That is vital not only for hot structures.

The usefulness of this "pure" rule may be topped by good compromises as for instance the employment of a "cold structure". That will become possible by use of (partial) insulation at the lower side of the forebody, thus reducing the difference of

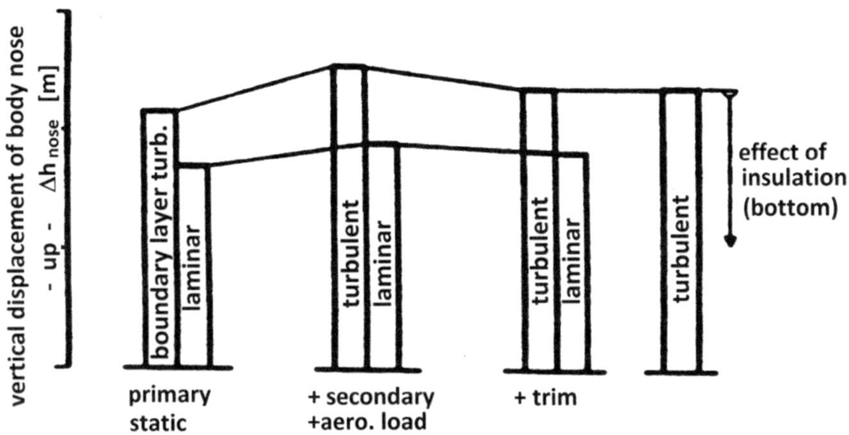

Fig. 2.45 The "Banana" effect [22]: impact of the temperature difference $T_{lower} > T_{upper}$ on the geometry of a TSTO-like forebody. $M_\infty = 6.8$, $H = 31\,\text{km}$, $\alpha = 6$, $L = 55\,\text{m}$

the temperature relative to that at the upper side. This effect is symbolized by the arrow on the right column of Fig. 2.45.

Insulation would be a possibility—transverse radiation must be taken into account, Sect. 3.10.5—but also surface emissivity tailoring, Sect. 3.8. All this must be seen with regard to an actual design of the forebody of such a lower stage of a TSTO-system. The conceptual design of the lower stage of SÄNGER envisaged a hot primary structure of the forebody of about 36 m length. The following forebody section of about 20 m length was to house the LH2 tanks. For that a cold primary structure was foreseen, Sect. 6.3.2.

Perhaps the old aphorism is verified in this way: "a good design will be found by finding the best compromise".

2.11 Problems

Problem 2.1 (*Mach Number and Dynamic Pressure*) A hypersonic flight vehicle is cruising at an altitude of 30 km with a velocity of 2.5 km/s. Given that the atmospheric temperature at this altitude is 226.5 K, calculate the Mach number M_∞ and the dynamic pressure q_∞. Assume the ratio of specific heats $\gamma = 1.4$ and the gas constant $R = 287.06$ J/kgK.

Problem 2.2 (*Scramjet Propulsion*) At what Mach number range does the scramjet propulsion system operate effectively, and what are the primary advantages of using hydrogen as fuel for scramjets?

Problem 2.3 (*Cayley's Paradigm in Hypersonic Design*) Explain how Cayley's design paradigm loses its validity in the design of hypersonic airbreathing vehicles of API-type 4 and 5, and describe the major challenges of airframe-propulsion integration (API) with these vehicles.

Problem 2.4 (*Growth Factors*) Consider a hypersonic airbreather vehicle experiencing a 5% increase in non-propulsive liquid hydrogen (LH2). Use the concept of growth factors to explain how this affects the overall take-off mass.

Problem 2.5 (*Mach-Number Independence Principle*) What is the Mach number independence principle, and at what Mach number range does it typically become relevant for hypersonic flight?

Problem 2.6 (*Induced drag*) Is the induced drag rising with reduced wing area, but the same span width? Is that of importance for typical hypersonic flight vehicle shapes?

Problem 2.7 (*Mach numbers*) Consider flight at 2 km/s. What are the flight Mach numbers at the altitudes 20 km, 40 km and 60 km? Why are they changed that way? What are the two major effects of changes of the Mach number? Are they of importance in this case?

Problem 2.8 (*Dynamic pressure*) What are the dynamic pressures of the three cases of Problem 2.7? Are they acceptable?

References

1. Schlichting, H., Truckenbrodt, E.: Aerodynamik des Flugzeuges, vols. 1 and 2, Springer, Berlin/Göttingen/Heidelberg (1959). Also: Aerodynamics of the Aeroplane, 2nd edn (revised). McGraw Hill Higher Education, New York (1979)
2. Anderson, J.D., Jr.: Fundamentals of Aerodynamics, 5th edn. McGraw Hill, New York (2011)
3. Rizzi, A., Oppelstrup, J.: Aerodynamic Design of Aircraft—A Computational Approach with Software. Cambridge University Press (2021)
4. Hirschel, E.H., Rizzi, A., Breitsamter, C., Staudacher, W.: Separated and Vortical Flow in Aircraft Wing Aerodynamics. Springer, Heidelberg (2021)
5. Hirschel, E.H.: Basics of Aerothermodynamics: Second, Revised. Springer, Berlin (2015); Progress in Astronautics and Aeronautics, AIAA, Reston, Va, vol. 204. Springer, Heidelberg (2004)
6. Reubush, D.E., Nguyen, L.T., Rausch, V.L.: Review of X-43A return to flight activities and current status. NASA-2003-RTF (2003)
7. Hornung, M.: Entwurf einer luftatmenden Oberstufe und Gesamtoptimierung eines transatmosphärischen Raumtransportsystems (Design of an Airbreathing Upper Stage and Global Optimization of a Transatmospheric Space Transportation System). Doctoral thesis, Universität der Bundeswehr, Neubiberg/München, Germany (2003)
8. Anderson, J.D., Jr.: The Airplane–A History of Its Technology. AIAA, Reston, VA (2002)
9. Hirschel, E.H.: Towards the virtual product in aircraft design? In: Periaux, J., Champion, M., Gagnepain, J.-J., Pironneau, O., Stoufflet, B., Thomas, P. (eds.) Fluid Dynamics and Aeronautics New Challenges. CIMNE Handbooks on Theory and Engineering Applications of Computational Methods, pp. 453–464. Barcelona, Spain (2003). See also www.ehhirschel.com (2024)
10. Hirschel, E.H., Weiland, C.: Design of hypersonic flight vehicles: some lessons from the past and future challenges. CEAS Space J. **1**(1), 3–22 (2011)
11. Hirschel, E.H.: Historical perspective on programs, vehicles and technology issues. In: Proceedings of RTO/AVT/VKI Lecture Series AVT-116 "Critical Technologies for Hypersonic Vehicle Development", Rhode-Saint-Génése, Belgium, May 10 to 14, 2004. RTO-EN-AVT-116, pp. 1-1–1-22 (2005)
12. Edwards, C.L.W., Small, W.J., Weidner, J.P.: Studies of scramjet/airframe integration techniques for hypersonic aircraft. AIAA-Paper 75-58 (1975)
13. Hirschel, E.H., Weiland, C.: Selected Aerothermodynamic Design Problems of Hypersonic Flight Vehicles. Progress in Astronautics and Aeronautics, AIAA, Reston, Va, vol. 229. Springer, Berlin, Heidelberg (2009)
14. Lentz, S., Hornung, M., Staudacher, W.: Conceptual design of winged reusable two-stage-to-orbit transport systems. In: Jacob, D., Sachs, G., Wagner, S. (eds.) Basic Research and Technologies for Two-Stage-to-Orbit Vehicles, Final Report of the Collaborative Research Centers 253, 255 and 259 of the Deutsche Forschungsgemeinschaft. Wiley-VCH, Weinheim (2005)
15. Hirschel, E.H., Hornung, H.G., Mertens, J., Oertel, H., Schmidt, W.: Aerothermodynamik von Überschallflugzeugen. MBB/LKE122/HYPAC/1/A, Ottobrunn, Germany (1987)
16. Koelle, D.E., Kuczera, H.: SÄNGER—an advanced launcher system for Europe. IAF-Paper 88-207 (1988)
17. Williams, R.M.: National aerospace plane: technology for America's future. Aerosp. Am. **24**(11), 18–22 (1986)
18. Weiland, C.: Mechanics of Flow Similarities. Springer Nature, Switzerland (2020)
19. Vincenti, W.G., Kruger, C.H.: Introduction to Physical Gas Dynamics. John Wiley & Sons, New York, London, Sydney (1965). Reprint edition, Krieger Publishing Company, Melbourne, FL (1975)
20. McRuer, D.: Design and modelling issues for integrated airframe/propulsion control of hypersonic flight vehicles. In: Proceedings of the American Control Conference, Boston, Massachusetts, pp. 729–735 (1991)

21. Weiland, C.: Aerodynamic Data of Space Vehicles. Springer, Heidelberg, New York (2014)
22. Staudacher, W., Wimbauer, J.: Design sensitivities of airbreathing hypersonic vehicles. AIAA-Paper 93-5099 (1993)
23. Hirschel, E.H., Cousteix, J., Kordulla, W.: Three-Dimensional Attached Viscous Flow. Springer, Berlin, Heidelberg (2014)
24. Jones, R.A., Donaldson, C.D.: From earth to orbit in a single stage. Aerosp. Am. **25**(8), 32–34 (1987)
25. Staudacher, W.: Entwurfsproblematik luftatmender Raumtransportsysteme. Space Course Stuttgart (1995)
26. Kuczera, H., Sacher, P.W.: Reusable Space Transportation Systems. Springer, Berlin, Heidelberg (2011)
27. Abeelen, L. van den: Spaceplane HERMES—Europe's Dream of Independent Manned Spaceflight. Springer (2017)
28. Hirschel, E.H.: The technology development and verification concept of the German Hypersonics Technology Programme. AGARD-R-813, pp. 12-1–12-15 (1996)
29. Chudoba, B.: Stability and Control of Conventional and Unconventional Aerospace Vehicle Configurations. Springer Aerospace Technology (2019)
30. Eggers, T., Sobieczky, H., Center, K.B.: Design of advanced waveriders with high aerodynamic efficiency. AIAA-Paper 93-5141 (1993)
31. Eggers, T., Strohmeyer, D., Nickel, H., Radespiel, R.: Aerodynamic off-design behavior of integrated waveriders from take-off up to hypersonic flight. ESA SP-367 (1995)
32. Küchemann, D.: The Aerodynamic Design of Aircraft. Pergamon Press, Oxford; Also AIAA Education Series, Va, p. 2012. AIAA, Reston (1978)
33. Ceresuela, R.: Problems de Stabilite et de Controle d'Avions Hypersoniques. ICAS paper 70-17 (1970)
34. Kraus, W.: Analyse der SÄNGER Windkanalmessungen, MBB-FE127-TN-HYP-87 (1989)
35. Staudacher, W.: Die Beeinflussung von Vorderkantenwirbelsystemen schlanker Tragflügel (The Manipulation of Leading-Edge Vortex Systems of Slender Wings). Doctoral thesis, University Stuttgart, Germany (1992)
36. Becker, J.V.: The X-15 project. Astronaut. Aeronaut. **2**, 52–61 (1964)
37. Koelle, D.E.: Handbook of Cost Engineering and Design of Space Transportation Systems with TransCost 8.2. TransCostSystems, Ottobrunn, Germany (2013)

Chapter 3
Aerothermodynamic Features of the External Flow Path

This chapter is devoted to an overview of the aerothermodynamic features present at airbreathing hypersonic flight vehicles. The chapter primarily concerns the external flow path. First the tasks of the technical discipline aerothermodynamics are sketched. It follows an overview of the flow phenomena, which can be found at the flight vehicles. Similarity rules are considered, some of the flow phenomena are discussed in more detail.

The importance of the thermal state of a vehicle surface together with the thermal surface effects are demonstrated and the all-deciding role of laminar-turbulent transition and turbulence is sketched. The chapter closes with a look at the aerothermodynamic aspects of upper-stage separation of TSTO space transportation systems.

The objective of the chapter is to show which aerothermodynamic features are present at the external flow path of hypersonic flight vehicles. That is needed in order to master the associated flight-vehicle design problems. No detailed presentations are given, basic facts are highlighted, literature is listed.

3.1 Preliminaries

First we look at the involved reference systems, Fig. 3.1. At the left of the figure the flight situation is shown, at the right the wind-tunnel system, which also is the system of theory and computation.

Both systems are compatible, because the flow problem is Galilean invariant. In reality the aircraft flies through the atmosphere, case (a), in the ground-facilities and in the mathematical models the air-stream is flowing past it, case (b).

E. H. Hirschel et al., *Elements of Hypersonic Airbreather Design and Development*,
https://doi.org/10.1007/978-3-031-94219-8_3

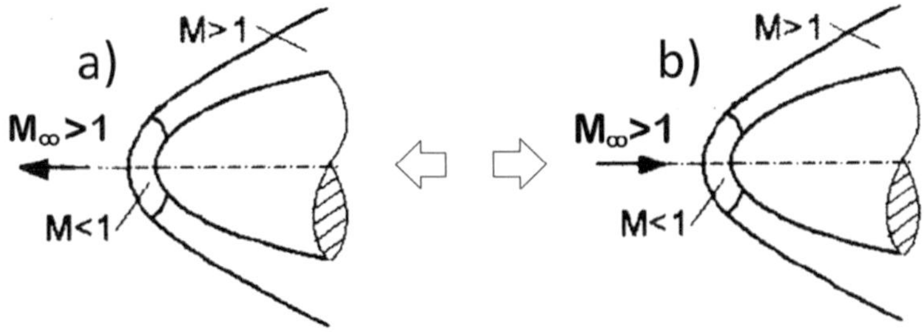

Fig. 3.1 Two reference systems: **a** flight situation, the vehicle moves through the atmosphere at rest (gusts to be kept in mind), **b** wind-tunnel and theoretical/computational situation, the vehicle is at rest, the air or the wind-tunnel gas moves past it (fluctuations to be kept in mind)

For the flight vehicles under consideration steady motion in general is more or less an exception. However, usually quasi-steady motion of the vehicle can be assumed and, hence, steady macroscopic flow phenomena, see, e.g., [1]. Truly unsteady flow phenomena can be present, but are not treated here. They comprise—apart from transitional and turbulent flow phenomena—shock/boundary-layer interactions, stall, flutter, panel flutter, vortex/structure interaction and others, all partly very important in view of structural fatigue.

Two observations are important, because they regard the aerothermodynamic model building of the flow fields under consideration and in particular also laminar-turbulent transition and turbulence.

The first observation is that with high flight Mach numbers the flight Reynolds numbers become lower. This is due to the dynamic pressure constraints of both the airframe and the propulsion system. The higher the flight speed and the lower these constraints, the higher is the flight altitude and, hence, the lower the Reynolds number, see Figs. 2.7 and 2.8.

The second observation is that locally the flow-field Mach number M is not necessarily the flight Mach number M_∞. This effect is very strong when considering the windward side of winged re-entry vehicles, which fly at high angle of attack during re-entry. The flight of the Space Shuttle Orbiter initially was with $M_\infty \approx 24$ at an angle of attack $\alpha \approx 40°$. Over the windward side of the vehicle only $M \approx 2$ was present, see, e.g., [2].

With the flight vehicles considered here low Mach numbers are present for instance in the nose region, where, however, the flow is laminar. Along the slender airframe with turbulent flow, high Mach numbers are present, which locally, for instance at parts of the wing, can distinctly exceed the flight Mach number.

This also holds for the unit Reynolds number $Re^u = \rho u/\mu$ and any other relevant similarity parameter.

3.2 The Tasks of Aerothermodynamics

The aerothermodynamic design process of hypersonic airbreathers is embedded in the general vehicle design process. Aerothermodynamics has, in concert with the other disciplines, the following tasks, see also [1].

3.2.1 Aerodynamic Shape Definition

Aerodynamic shape definition of hypersonic flight vehicles in particular has to take into account the thermal state of the surface, Sect. 3.7. That strongly influences via the thermal-surface effects, Sect. 3.10, the aerodynamic performance, in particular the drag of CAV and ARV vehicles. Governed by it too is the inlet performance, the performance of trim and control surfaces, and the thermal loads on the structure. The particular tasks of aerodynamic shape definition are:

1. Supplying the aerodynamic data set—some examples can be found in [3]—which must enable performance, flyability and controllability on the whole flight trajectory.
2. Layout of trim and control surfaces, also in view of actuator performance.
3. Aerothermodynamic airframe-propulsion integration.
4. For ARV-type vehicles the integration of reaction control systems.
5. Aerothermodynamic upper stage integration and separation for TSTO space transportation systems.

3.2.2 Determination of External Mechanical and Thermal Loads

The determination of the aerothermodynamic loads for the layout of the structure and materials concept includes the sizing of the structure, and the thermal protection in view of either a cold or a hot primary structure, including possible active cooling of parts of the airframe:

1. Determination of mechanical loads (surface pressure, skin friction), both as static and dynamic loads, including in particular also acoustic loads.
2. Determination of thermal loads on both external and internal surfaces/structures, also in view of the thermal management of the flight vehicle.
3. Determination of the aerothermoelastic properties of the airframe.

3.2.3 Definition of Surface Properties

The definition of the surface properties regards the external and the propulsion flow path.

1. The most important surface property is the radiation emissivity in view of the external surface-radiation cooling. Radiation cooling is necessary to cope with the tremendous thermal loads, which are encountered during hypersonic flight. High surface emissivity is produced with a coating, which gives the flight vehicle a nearly black appearance.
   Re-entry vehicles like the X-37B, the Dream Chaser and also the former Space Shuttle Orbiter etc., which fly and flew their re-entry—braking—mission at high angle of attack, have this coating at the windward side, at the nose and the leading edges only. The upper side of the vehicles is white and in orbit turned toward the Sun for the purpose of thermal management. The airbreathers considered here fly at very low angles of attack and therefore look black all around.
   The surface-radiation emissivity governs the thermal loads on structure and materials, but also the thermal-surface effects regarding viscous-flow and high-temperature real-gas phenomena.
2. Other very important surface properties are surface imperfections like roughness, waviness, steps, gaps etc.[1] They play a deciding role regarding laminar-turbulent transition and turbulent boundary-layer flow. They must be sub-critical in order to avoid unwanted increments of viscous drag, and of the thermal state of the surface. Premature triggering of laminar-turbulent transition by all means must be avoided. In turbulent flow surface irregularities must be sub-critical, otherwise skin friction and heat transfer are enlarged, Sect. 3.11. On the other hand it is necessary to know how much irregularities can be tolerated in order to reduce manufacturing effort.
3. Another surface property, if strong high-temperature real-gas effects are present, is the surface catalycity. It can lead to adverse increments of the thermal loads on the structure. It concerns the recombination of the oxygen atoms in the flight regime of the CAV vehicles considered here, Sect. 2.4. But there it should not be so important. However, it can become important for ARV vehicles. Usually the surface catalytic behavior, together with the surface-radiation emissivity and also the possible anti-oxidation protection are properties of the surface coating of the airframe or the TPS material.
4. Of utmost importance is that each of the coatings must withstand erosion and contamination because of the dirt and insects in the lower atmosphere and, in general, of rain and hail phenomena. The same holds for erosion due to the skin-friction, in particular if the boundary-layer flow is turbulent. This all, of course does not necessarily hold for missile applications.

[1] A particular issue are expansion joints. They are present where different flight vehicle surface materials come together, see Sect. 6.3.2.

3.3 Aerothermodynamic Phenomena Present at Hypersonic Airbreathers

At hypersonic airbreathers a large number of aerothermodynamic phenomena is present, see, e.g., [1–4]. The designer of such vehicles must cope with them. Here we look at them, taking the API-type 5 lower stage of the SÄNGER TSTO system [5], Fig. 3.2, as our choice of a reference vehicle, Sect. 2.8. The presented phenomena and design issues assigned to the vehicle components are just for orientation, therefore not complete. They can be found also at other locations, and—looking for instance at $M_\infty = 12$ airbreathers—may have another degree of importance, or other phenomena may show up.

With these limitations in mind and in view of the fact that the anticipated phenomena and problems depend on the actual configuration and mission, we find the following listing, [7], see also [8].

1. General aerothermodynamic issues:

   - shock waves,
   - shock-wave interactions,
   - wave drag,
   - radiation cooling,
   - non-convex effects,
   - thermal loads,
   - hot-spot and cold-spot phenomena,

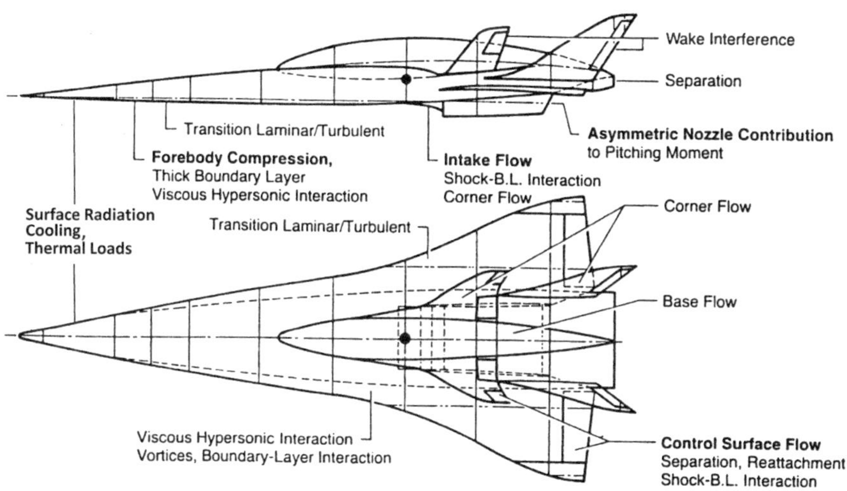

Fig. 3.2 Aerothermodynamic phenomena present at hypersonic airbreathers, here at the API-type 5 lower stage of the TSTO-system SÄNGER with $M_{\infty_{max}} = 6.8$ at $H = 34$ km [6]

- contradictory demands at vehicle nose and leading edges: small radii for low wave drag, large radii for lower thermal loads and radiation cooling,
- high-temperature real-gas effects,
- boundary layer flow,
- corner flow,
- gap flow,
- shock/boundary-layer interaction,
- viscous drag (skin friction drag, form drag)
- laminar-turbulent transition,
- turbulent separation,
- base drag,
- coupling of aerothermoelastic shape deformations into flow paths,
- ... .

2. Nose of the forebody:

   - bow shock,
   - wave drag,
   - entropy layer,
   - finite laminar boundary-layer thickness,
   - thermal loads,
   - ... .

3. Lower side of the forebody:

   - lift-, drag-, pitching-moment increments,
   - forebody pre-compression (inlet capture area),
   - flow pattern (between the attachment lines two-dimensional onset flow of the inlets),
   - entropy-layer swallowing,
   - hot-spot situation along the attachment lines,
   - aerothermoelastic coupling into the propulsion system
   - ... .

4. Inlet:

   - external inlet ramp flow,
   - boundary-layer diverter,
   - ramp shocks, shock-on-lip situation, inlet-cowl lip shock,
   - shock/boundary-layer interaction, boundary-layer bleed,
   - internal compression, shock-wave train, corner flow
   - ... .

5. Afterbody, nozzle:

   - afterbody/nozzle (SERN) flow,
   - secondary (boundary-layer bleed) nozzle flow,

- thermochemical effects,
- total thrust vector angle,
- contribution to pitching-moment household,
- afterbody: strong interaction/separation,
- base drag,
- … .

6. Upper side of the configuration:

- possible interaction of bow shock with vertical tail or winglets,
- corner flow fuselage/vertical tail and wing or winglets/vertical tail,
- afterbody flow,
- base drag,
- with TSTO-system: flow interference forebody/upper stage,
- corner flow fuselage/upper stage,
- corner flow upper stage/vertical tail,
- wake interference at vertical tails of upper and lower stage,
- … .

7. Wing:

- interaction of fuselage bow shock with leading-edge bow shock of outer-wing delta part,
- interaction flow phenomena at trim and control surfaces, gap flow,
- lee-side vortex flow and interaction with vertical tail flow,
- … .

Subsequently some of the above aerothermodynamic phenomena are discussed in some detail. Their choice does not imply a rating.

3.4 Hypersonic Vehicle-Shape Similarity Rules

Similarity rules give general guidance and, moreover, help to assess design data of in our case hypersonic airbreathing aircraft. We have a short look at three of them, the Oswatitsch Mach-number independence principle, the Tsien hypersonic similarity parameter and the Weiland mapping rule. All basically assume thermally and calorically perfect gas.

Experimental observations show for instance for spheres that the drag coefficient becomes independent of the Mach number for $M_\infty \gtrapprox 5$. For a 80°-cone-cylinder configuration the independence begins at $M_\infty \gtrapprox 9$. When looking at the relations for the flow across normal and oblique shock waves, one observes some finite flow properties for $M_\infty \to \infty$, see, e.g., [1].

For blunt bodies, or the blunt nose of a slender body, the theory of Oswatitsch, [9], see also [1], shows that for high Mach numbers not only the drag coefficient, but all force and moment coefficients become independent of the Mach number. That also holds for the pressure coefficient, the shapes of the bow shock, the sonic surface and the characteristics.

The theory assumes perfect gas. In [2] the discussion of the pitching-moment anomaly observed at the first re-entry mission of the Space Shuttle Orbiter led to the conjecture that a benign inviscid surface Mach-number interval $1 < M_{wall} \lessapprox 2.2$ exists, where high-temperature real-gas effects only weakly violate the Mach-number independence principle.

For slender, sharp-nosed bodies H.S. Tsien introduced the hypersonic similarity parameter [10]:

$$K = M_\infty \sin\beta \gg 1. \tag{3.1}$$

Here β is the maximum deflection angle the flow undergoes on the considered configuration. In terms of the ratio 'body thickness' to 'body length' τ the parameter reads

$$K = M_\infty \tau \gg 1. \tag{3.2}$$

Without going into details we note that for the surface pressure of a body with the thickness ratio τ the surface-pressure coefficient reads $c_p \propto \tau^2$ and the wave-drag coefficient $C_{D_w} \propto \tau^3$ [10, 11].

These relations can be expected to hold also for the slender, but slightly blunt-nosed configurations of the here considered CAV- and ARV-type vehicles.

A recent very significant development is the Weiland mapping rule [11], with a more detailed presentation of the results in [12, 13]. Originally devised for blunt-shaped re-entry vehicles, Fig. 3.3, it probably holds to a degree also for the slender airbreather configurations.

We show in Fig. 3.4 as function of the angle of attack α the lift coefficients C_L of the Space Shuttle Orbiter at $M_\infty = 8$ and 10 and HERMES at the same Mach numbers, and also those of the X-33 at $M_\infty = 8$ and 10 and of the X-38 at $M_\infty = 5.47$ and 10.

The curves show well that the Mach-number independence principle holds. Further we observe that the curves in the interval $0 \leqq \alpha \lessapprox 30$ have a weak quadratic non-linearity, then have a point-of-inflection at $\alpha \approx 30$, and finally a sinusoidal pattern in the interval $30 \lessapprox \alpha \lessapprox 50$.

The mapping rule now makes use of a homogenization of the reference areas S_{ref} of the re-entry vehicles [11–13]. Without going into the details we show the result in Fig. 3.5. The surprising result is that now all curves collapse into one.

Not exactly the same applies when considering the drag coefficient. The original data in Fig. 3.6 again show the to be expected increase of the coefficient with increasing angle of attack, as well that the Mach-number independence principle

Fig. 3.3 The considered re-entry vehicles: the lifting-body vehicles X-38 and X-33, and the winged vehicles HERMES and Space Shuttle Orbiter [11]

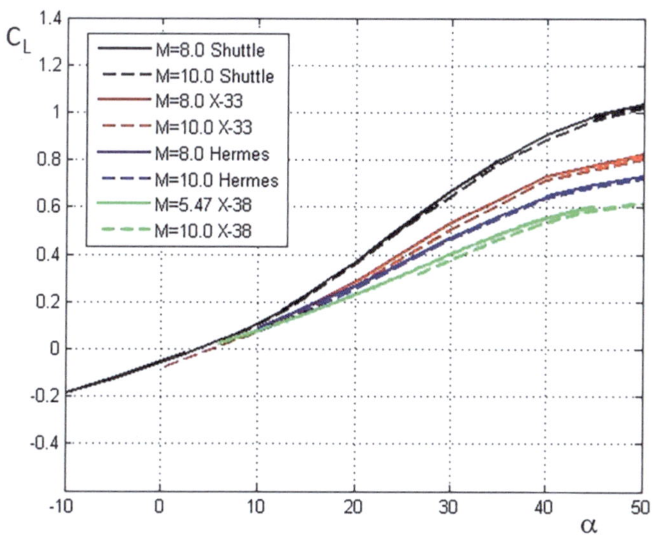

Fig. 3.4 Hypersonic lift coefficients C_L of re-entry vehicle configurations as function of the angle of attack α [11]

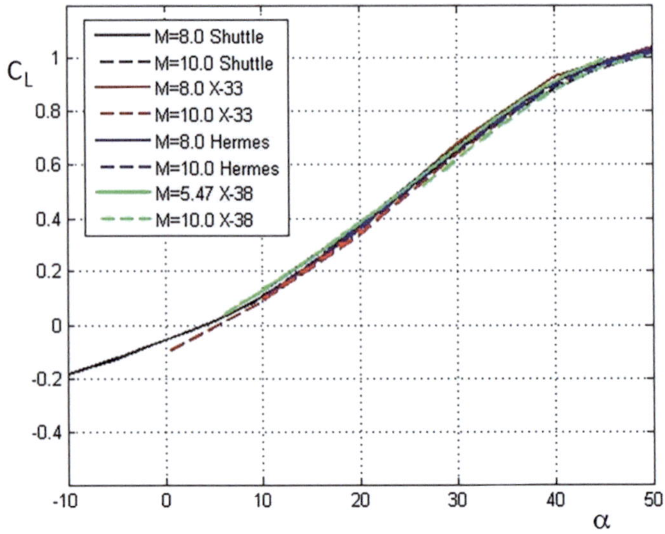

Fig. 3.5 Hypersonic lift coefficients C_L of re-entry vehicle configurations as function of the angle of attack α after application of the mapping rule [11]

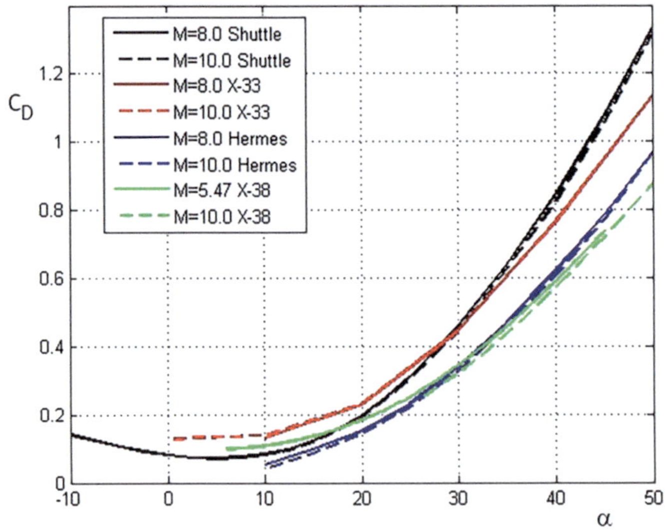

Fig. 3.6 Hypersonic drag coefficients C_D of re-entry vehicle configurations as function of the angle of attack α [11]

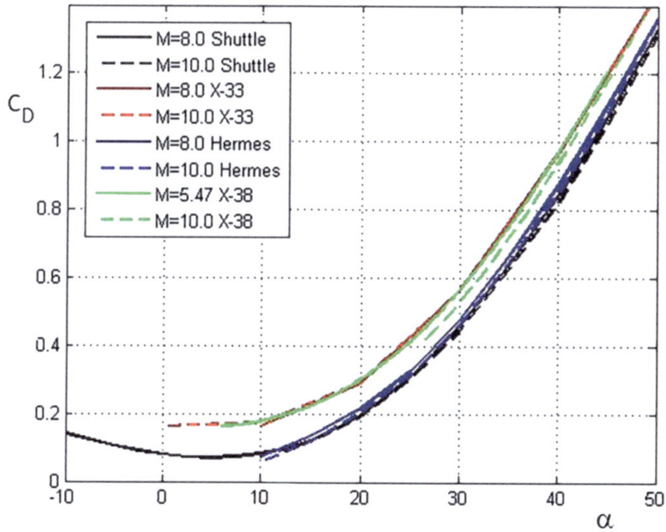

Fig. 3.7 Hypersonic drag coefficients C_D of re-entry vehicle configurations as function of the angle of attack α after application of the mapping rule [11]

holds. Present now is a grouping of the curves. Space Shuttle Orbiter and the X-33 have the highest drag coefficient with a crossing of the curves at $\alpha = 30$, HERMES and X-38 lie close together.

After the mapping the curves of the two lifting bodies X-33 and X-38 lie together but above the curves of the winged Space Shuttle Orbiter and HERMES, Fig. 3.7.

Regarding the pitching-moment coefficient, not shown here, the result basically is the same. The curves are grouped in a similar way as those of the lift and drag coefficients. The problem was to find the reference points for the pitching moment of the different configurations [11].

An interesting result was obtained while including data of the SÄNGER TSTO API-type 5 space-transportation system. The lower stage is a CAV-type airbreather with a slender low-drag configuration. In the angle-of-attack range $-3 \leqq \alpha \leqq 20$ aerodynamic data are available. The lift coefficient now is included with the data of Fig. 3.5 and fits well into picture with also a weak quadratic non-linearity, Fig. 3.8.

The drag coefficient in Fig. 3.9 shows the expected difference to the re-entry configurations. Those are blunt on purpose in order to fly their braking mission. The lower stage of SÄNGER, on the other hand is a typical slender airbreather with small radii of the fuselage nose and the leading edges.

We note that the mapping rule was also applied to some supersonic data of the considered flight vehicles [12, 13]. For the Mach numbers 2, 3, and 4 a good mapping was obtained, showing the large range of application.

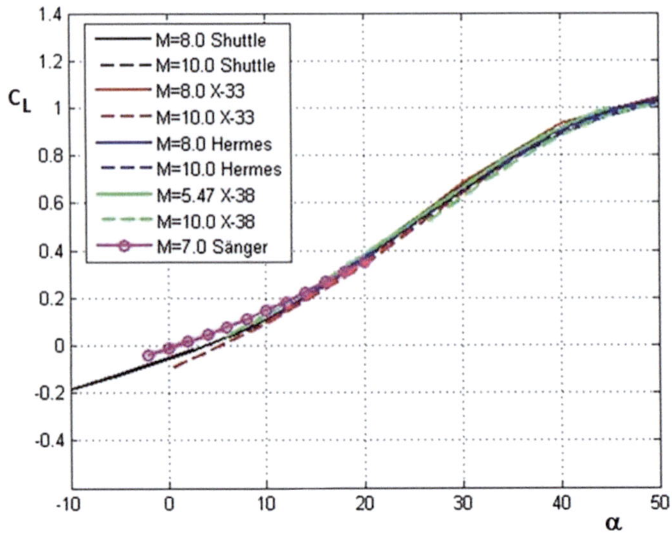

Fig. 3.8 Hypersonic lift coefficients C_L of re-entry vehicle configurations and of the SÄNGER TSTO space-transportation system at $M_\infty = 7$ as function of the angle of attack α after application of the mapping rule [11]

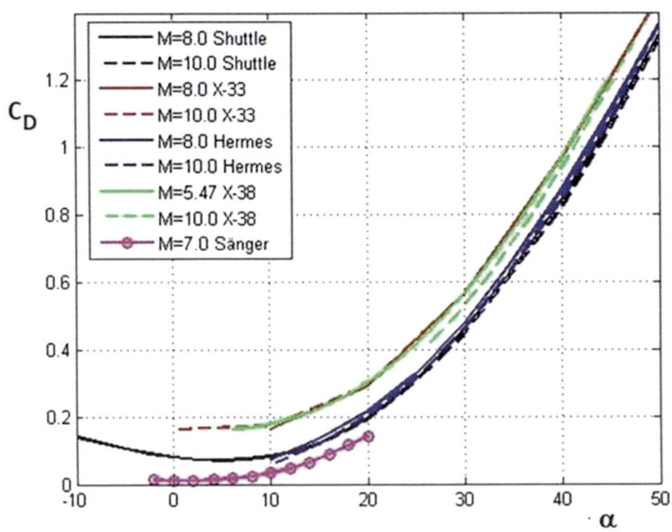

Fig. 3.9 Hypersonic drag coefficients C_D of re-entry vehicle configurations and of the SÄNGER TSTO space-transportation system at $M_\infty = 7$ as function of the angle of attack α after application of the mapping rule [11]

The value of the mapping rule is not that one is able to save effort to generate the aerodynamic data set. The real value is that one is able to check generated data. The basis of the rule, however, should be expanded by adding data from a broader range of configurations.

Regarding the slender hypersonic airbreather configurations, the lone SÄNGER data are not enough to assess the value of the mapping rule for such type of CAV vehicles. Here more aerodynamic data sets are needed to build a reliable foundation of the principle.

3.5 About the Bluntness of Vehicle Noses and Leading Edges

The typical hypersonic airbreathers under consideration, right part of Fig. 2.5, have slender shapes and either a blunt spherical nose, or in the case of waveriders oval nose sections, or even a—only slightly—blunt and unswept leading edge, like the X-43A and the X-51A, see the schematic in Fig. 3.10. Wing leading edges, side edges of waveriders, and leading edges of stabilizers are also blunt, but highly swept.

The bluntness in all cases should be as small as possible in view of the wave drag at supersonic and hypersonic flight Mach numbers. On the other hand a certain bluntness is always necessary in order to cope with the thermal loads.

Regarding in particular the blunt nose or the unswept leading edge of the configuration, two cooling modes are possible: (a) active cooling via an inner cooling circuit or effusive/transpiration cooling through the surface, (b) passive cooling via surface radiation cooling, Sect. 3.8.

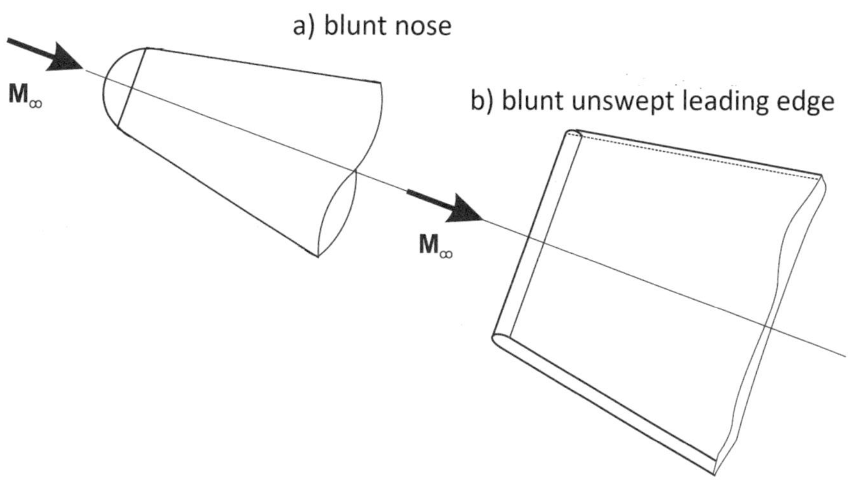

Fig. 3.10 Schematic of blunt nose and blunt unswept leading edge

The active cooling modes need inner volume and hence a certain nose or leading-edge radius. In any case the systems complexity of the devices is high. Passive surface radiation cooling needs a sufficiently thick boundary layer of the nose or leading edge.[2] Then it is a question what material is needed.

Take for instance the X-43A with the for about 12 s reached maximum Mach number $M_\infty = 9.68$ at the altitude $H = 33.36$ km. The stagnation temperature at this flight condition is about $T_0 \approx 3{,}800$ K, Fig. 2.8. The nose section of the vehicle was made from a tungsten alloy as ballast to achieve the needed longitudinal stability properties of the vehicle, see [15] and Sect. 6.4. The melting temperature of tungsten is about 3,700 K.

The unswept leading edge of the vehicle was made from carbon-carbon composite with a very small nose radius of less than one millimeter. Carbon-carbon composite has properties which allow to reach temperatures up to about 3,300 K. This number indicates that for the short overall flight time—separation from the booster to $M_\infty \approx 10$ about 75 s plus scramjet mode 12 s—the material could stand the resulting heat load. Surface radiation cooling with the small leading-edge radius certainly did not contribute much.

A simple consideration shows how the dimension of a spherical nose affects the wave drag and also the radiation-adiabatic wall temperature at the stagnation point. Two cases are studied: (a) $M_\infty = 6.8$ at $H = 32$ km, (b) $M_\infty = 12$ at $H = 36$ km. The wave drag reads

$$D_W = C_{D_w} q_\infty A_{ref}, \tag{3.3}$$

with C_{D_w} being the wave-drag coefficient, $q_\infty = 0.5\,\rho_\infty v_\infty^2$ the dynamic pressure and $A_{ref} = \pi r^2$ the reference area with the radius r of the spherical nose.

Since the influence of the bluntness—in terms of the nose radius—on the wave drag is of interest, instead of Eq. (3.3) the reduced wave drag

$$\frac{D_W}{C_{D_w}} = q_\infty A_{ref} \tag{3.4}$$

is used. In this way the coefficient of the wave drag must not be specified.

For comparison besides the spherical nose also a blunt unswept (rectangular) leading edge with the width $b = 0.5$ m is considered. The result is given in Fig. 3.11.[3] For the actual computation see Problem 3.2. The figure shows, as to be expected, that the reduced wave drag drops with decreasing radius—note that on the abscissa the radius decreases from the left to the right. It appears that in view of the reduction of the reduced wave drag the spherical nose has an advantage over the blunt unswept leading edge, if the radius r is downsized.

---

[2] Note that the boundary layer at a blunt nose or unswept leading edge always has a finite thickness [14]. That thickness increases with an increase of the nose radius.

[3] We gratefully acknowledge that Figs. 3.10, 3.11 and 3.12 were made by C. Weiland.

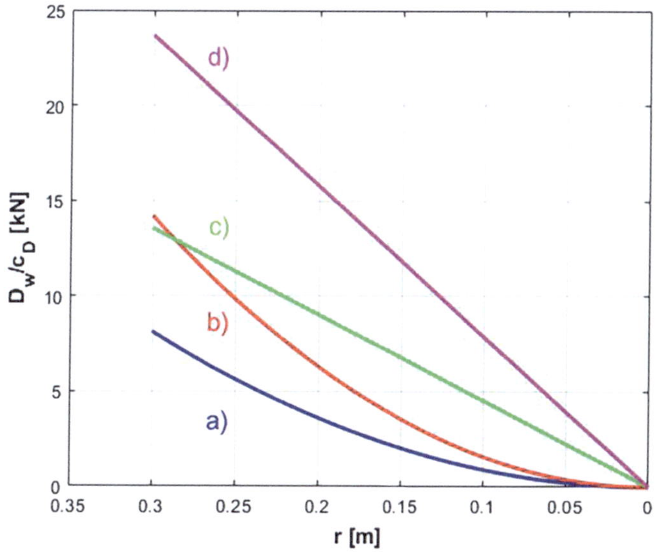

Fig. 3.11 The reduced wave drag D_w/C_{D_w} as function of the radius r: spherical nose: **a** $M_\infty = 6.8$, **b** $M_\infty = 12$, blunt unswept leading edge: **c** $M_\infty = 6.8$, **d** $M_\infty = 12$.

The influence of the bluntness of the nose on its thermal state is studied in terms of the radiation-adiabatic wall temperature, Sect. 3.7. A simple relation for the determination of the heat flux in the gas at the wall is

$$q_{gw} = \frac{5.1564 \cdot 10^{-5}}{\sqrt{r_N}} \rho_\infty^{0.5} v_\infty^{3.15}, \tag{3.5}$$

which was found from correlation of data from blunt-nosed hypersonic vehicles [16, 17], see also [2]. The input dimensions are m for the radius r_N, kg/m^3 for the density ρ_∞ and m/s for the flight speed v_∞. The resulting q_{gw} has the dimension W/m^2.

By equating Eq. (3.5) with Eq. (3.15) from Sect. 3.7 for the surface radiative heat flux, one finds for the wall temperature, which in this case is the radiation-adiabatic wall temperature

$$T_w = T_{ra} = \left[\frac{q_{gw}}{\sigma\,\epsilon}\right]^{0.25} = \left[\frac{5.1564 \cdot 10^{-5} \rho_\infty^{0.5} v_\infty^{3.15}}{\sqrt{r_N}\sigma\,\epsilon}\right]^{0.25}. \tag{3.6}$$

This result holds for spherical noses. For unswept cylindrical leading edges no data are available. In the correlations of van Driest, Lees, Fay and Ridell, and Cohen, see [2], the velocity gradient at the stagnation point $du_e/dx|_{st}$ is the deciding entity. Hence the ratio of the gradients of the two shapes can be used to determine the resulting wall temperature. Taking provisionally the relations from potential theory with

$$\frac{du_e}{d(x/R)}|_{stag\ point} = \frac{3}{2}u_\infty \tag{3.7}$$

for the sphere, and

$$\frac{du_e}{d(x/R)}|_{stag\, line} = 2u_\infty \tag{3.8}$$

for the cylinder, the resulting wall temperature ratio approximately is

$$\frac{T_{w,cylinder}}{T_{w,sphere}} = 1.33. \tag{3.9}$$

Figure 3.12 gives the radiation-adiabatic wall temperature T_{ra} at the spherical nose and at the blunt unswept leading edge, for the latter obtained with the above temperature ratio. Again on the abscissa the radius decreases from the left to the right. Additionally it should be mentioned, that the situation at the side edges of the body with the blunt unswept leading edge is not covered.

In each case the temperature rises strongly once the small radii are reached. When looking at Fig. 2.5 in Chap. 2, it appears that the temperature levels in Fig. 3.12 are

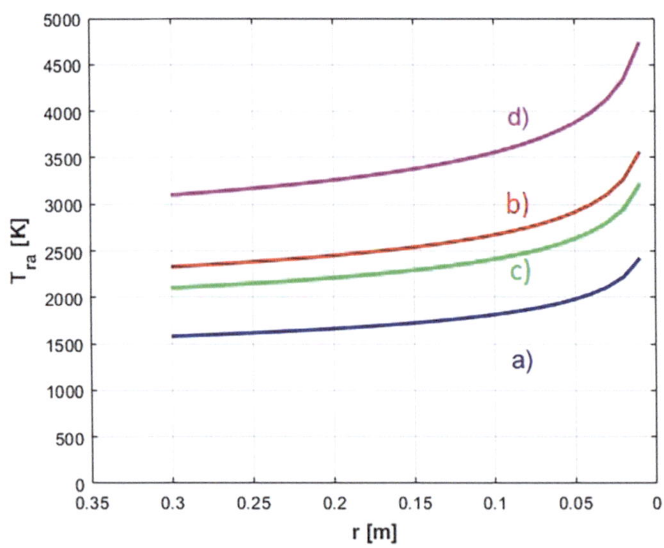

Fig. 3.12 The radiation-adiabatic temperature T_{ra} as function of the radius r: spherical nose: **a** $M_\infty = 6.8$, **b** $M_\infty = 12$, blunt unswept leading edge: **c** $M_\infty = 6.8$, **d** $M_\infty = 12$

too high, in particular those for the blunt unswept leading edge. This indicates that Eq. (3.9) gives an unrealistically high ratio.

In any case, Figs. 3.10 and 3.12 demonstrate the design challenge in terms of wave drag and thermal loads at a spherical vehicle nose in contrast to that at a blunt unswept leading edge. It appears that the spherical nose is the better solution. However, one has to take into account what such a nose means in view of the forebody shape: lengthy slender forebody with aerothermoelastic challenges and the need to obtain a parallel inlet onset flow.

Radiation cooling of either a spherical nose or a blunt unswept leading edge at very high flight Mach numbers anyway needs special structure and material approaches, Chap. 6. Active cooling is a technically and in view of complexity very demanding means. The situation at the oval nose sections of waveriders would need an extra consideration.

In closing this section a design example is presented. The basic design of the wing of the SÄNGER lower stage assumed a nose radius, which matched the material properties of the swept leading edge—the carbon-ceramic composite CSiC. The determining flight condition was that at the launch of the upper stage with M_∞ = 6.8 at H = 31 km altitude with the lower stage's angle of attack α = 6. The surface radiation cooling emissivity was assumed to be ϵ = 0.8.

Figure 3.13 shows the basic design at the right of the figure with the relative nose radius (100%).

The temperature is 1,400 K, which is matched by the CSiC structure. During the analysis of the design it was found that a large part of the zero-lift drag was due to the blunt nose and the blunt leading edge of the wing.

A sensitivity analysis was performed in view of the influence of the nose radius on the flight performance, the mass of the flight vehicle, and the wall temperature at the wing's attachment line ('stagnation point' in the figure). The subsequent choice of carbon-carbon composite material with a maximum allowable temperature of 2,000 K permitted to reduce the nose radius to 10% of the original radius.

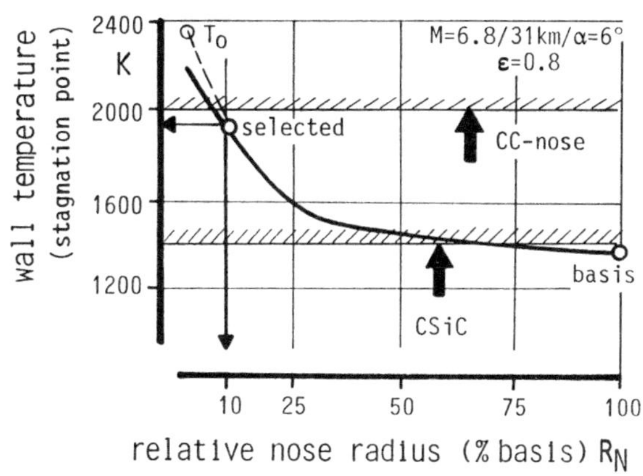

Fig. 3.13 Design example: links between nose geometry, thermal load and choice of material [7]

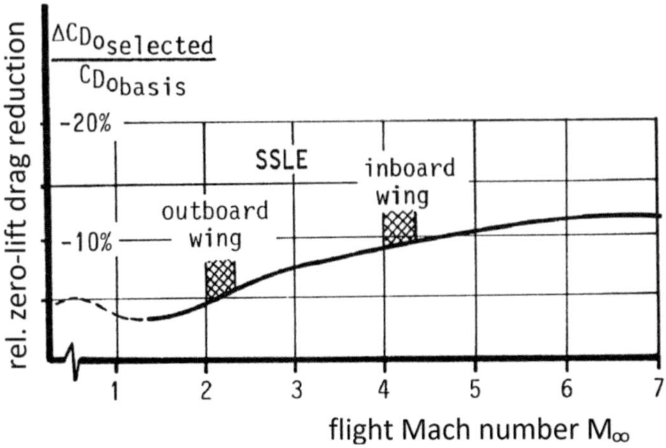

Fig. 3.14 The final picture: influence on the zero-lift drag of the complete configuration [7]

The influence of this reduction on the zero-lift drag of the configuration is shown in Fig. 3.14.

The hatched Mach-number areas around $M_\infty = 2$ and 4 indicate the Mach numbers at which the two parts of the double-delta wing—Fig. 3.2, lower part—just become supersonic leading edges (SSLE).

The full line below the hatched areas show the relative zero-lift drag reduction in per cent. That is rather low at low flight Mach numbers, but increases with increasing Mach number. The mean yield of the material change was a reduction of the zero-lift drag by about eight per cent, which led to a reduction of the take-off mass of about 16 tonnes.

Figure 3.15 finally shows how the bluntness of vehicle noses and—here— leading edges must be seen in view of drag, wall temperature and material choice of the flight vehicle.

All is highly interconnected and needs the truly multidisciplinary design and optimization approach discussed in Sect. 2.2.

Fig. 3.15 The links between material, geometry, temperature, aerodynamic drag and their connection to the performance and mass of the flight vehicle [7]

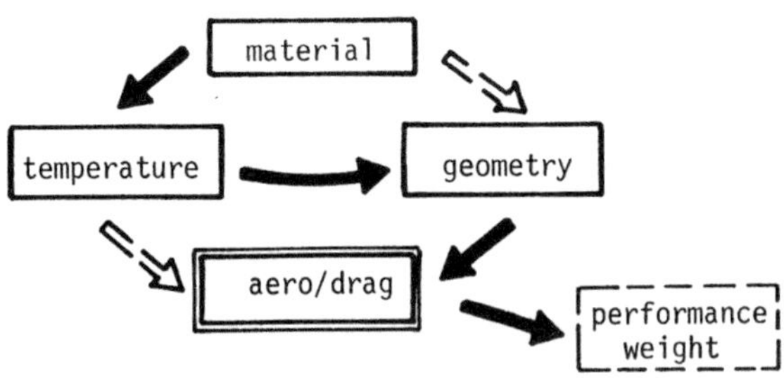

3.6 Hypersonic Flow Issues

This section gives a short introduction to some issues of inviscid and viscous hypersonic flow that are important for hypersonic flight vehicles. Phenomena are discussed and trends are identified. For details see the cited literature. Not treated in depth are high-temperature real-gas effects. The reader is referred to Sect. 2.4. Their importance is high and they must be regarded in any kind of design and simulation.

First a few phenomena of shock waves in supersonic and hypersonic flow are considered. For detailed presentations see, e.g., [1], but also [4, 18] and others.

3.6.1 Shock Waves

For a hypersonic airbreather, specifically a TSTO-system, the bow-shock wave, embedded shock waves and the subsonic flow pocket at the blunt nose are sketched in Fig. 3.16. For the shock system at the outer flow path of the inlet see Figs. 5.13 and 5.14 in Sect. 5.2.

Of interest now are the stand-off distance of the bow shock at the blunt vehicle nose and the asymptotic behavior of the bow shock. The bow shock at the nose with the subsonic pocket is shown schematically and enlarged in Fig. 3.17.

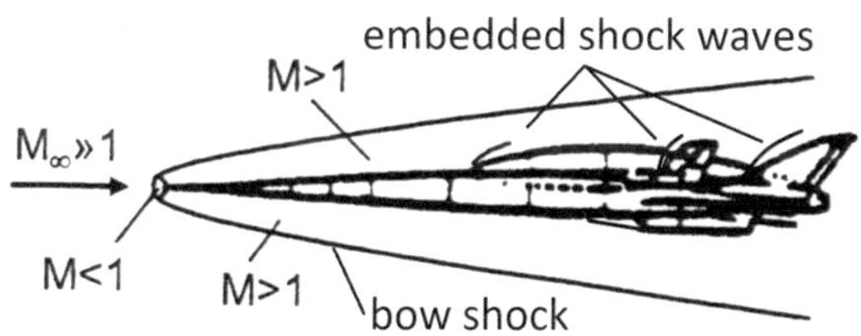

Fig. 3.16 Hypersonic TSTO airbreather: schematic of the bow shock, embedded shock waves and the subsonic nose pocket ($M < 1$)

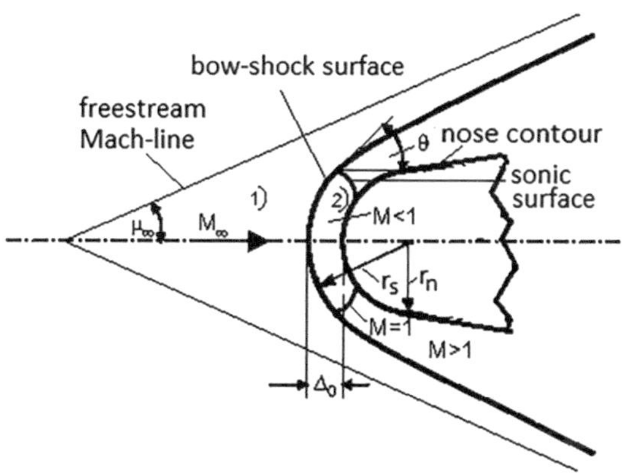

Fig. 3.17 Schematic of blunt vehicle nose with bow shock

The shock wave is a convex surface, wrapped around the nose. At and in the vicinity of the very nose point the flow is subsonic up to the sonic surface. The shock stand-off distance Δ_0—depending on the shape of the nose—governs downstream locations where interactions of the shock wave with configuration parts can occur.

For the spherical nose an approximate analytical relation is available for sufficiently high Mach numbers ($M_\infty \gtrapprox 5$) [19], see also [1]. Assumed is that the bow-shock surface with radius r_s lies close to the body surface with radius r_n. Also assumed is a constant density behind the shock wave in the shock layer (region 2). The stand-off distance Δ_0 then is approximately

$$\Delta_0 = r_s - r_n \approx \frac{\epsilon \, r_s}{1 + \sqrt{\frac{8\epsilon}{3}}}, \tag{3.10}$$

with ϵ being the inverse of the density rise across the normal part of the bow shock [1][4]

$$\epsilon = \frac{\rho_\infty}{\rho_s} = \frac{\rho_1}{\rho_2} = \frac{(\gamma - 1)M_\infty^2 + 2}{(\gamma + 1)M_\infty^2}. \tag{3.11}$$

By means of the γ_{eff} approach, see, e.g., [2], the asymptotic behavior of the flow variables across the normal shock portion can be studied, Fig. 3.18.[5] High-temperature real-gas effects are regarded by assuming different values of the ratio of specific heats γ in the shock layer, region (2) in the figure. Of interest here are the curves for the free-stream value γ_1 = 1.4 and the shock-layer values γ_2 = 1.4, 1.3, 1.2, 1.1, representing increasing degrees of vibrational excitation and dissociation of the air molecules. For the Mach-number range up to M_∞ = 12 values down to γ = 1.2 can be reached.

We observe, with all reservations regarding the γ_{eff} approach, that a strong dependence on the real-gas assumption exists for all flow property ratios across the normal and near normal part of the bow shock wave, except for the pressure ratio. The density ratio increases with decreasing γ_2, correspondingly decreases the temperature ratio and with that the Mach number in the shock layer. In the here interesting flight Mach-number domain up to $M_\infty \approx 12$, Mach-number independence is barely reached only for ρ_2/ρ_1 and M_2.

The situation is different for the stagnation pressure coefficient

$$c_{p_{stag}} = \frac{p_{stag} - p_\infty}{q_\infty}, \tag{3.12}$$

with $q_\infty = 0.5\, \rho_\infty\, v_\infty^2$ being the dynamic pressure.

[4] This ϵ is not to be confused with the surface emissivity coefficient ϵ, Sect. 3.7.

[5] The γ_{eff} approach assumes perfect gas ahead of the shock wave and a value of γ—i.e., γ_{eff}—behind the shock wave in the shock layer. It permits to quantify, parameterize, and illustrate high temperature real-gas effects in a simple, but approximate way.

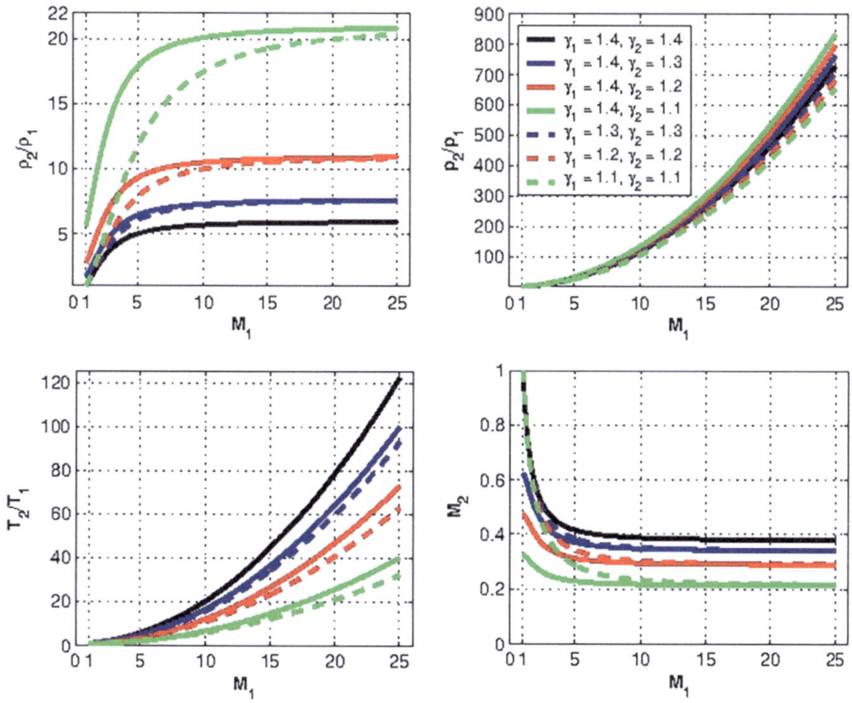

Fig. 3.18 γ_{eff} approach for the normal shock wave: dependence on the free-stream Mach number M_1 ($= M_\infty$) of the density ratio (upper left), the pressure ratio (upper right), the temperature ratio (lower left) and the Mach number in the shock layer [2]

The well-known value for incompressible flow is $c_{p_{stag}} = 1$. For compressible flow and in particular for compressible flow over a normal shock wave ahead of the stagnation point, the following relation holds, assuming isentropic compression behind the shock wave, and p_{t_2} being the total pressure behind the shock wave [1, 2, 18]

$$c_{p_{stag}} = \frac{p_{t_2} - p_1}{q_1} = \frac{2}{\gamma M_1^2} \left\{ \left[\frac{(\gamma+1)^2 M_1^2}{4\gamma M_1^2 - 2(\gamma-1)} \right]^{\frac{\gamma}{\gamma-1}} \left[\frac{2\gamma M_1^2 - (\gamma-1)}{\gamma+1} \right] - 1 \right\}. \tag{3.13}$$

For $\gamma = 1.4$ this relation at large Mach numbers reaches $c_{p_{stag}} = 1.838$. Applying the γ_{eff} approach for different values in the shock layer, the curves in Fig. 3.19 result. The higher the degrees of vibration excitation and dissociation, the higher is the stagnation-point pressure coefficient. The upper limit is $c_{p_{stag}} = 2$, which is reached in the Newton limit with $\gamma = 1$. All curves become more or less constant for $M_1 \gtrapprox 5$.

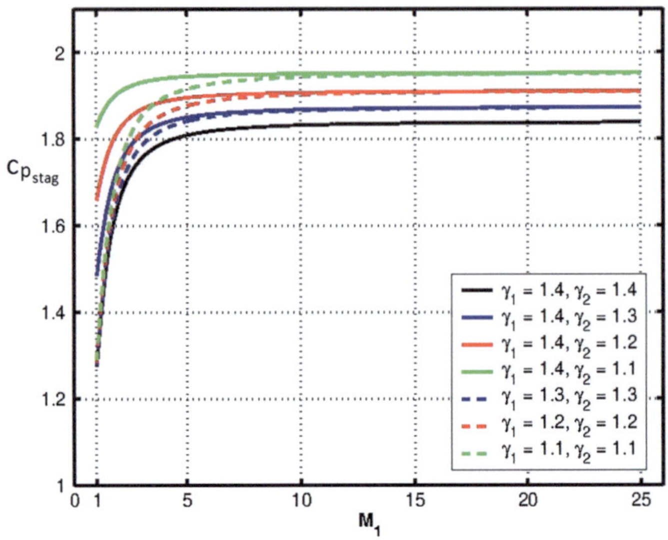

Fig. 3.19 The stagnation pressure coefficient $c_{p_{stag}}$ at the blunt nose as function of the free-stream Mach number and several values of γ_{eff} [2]

Figure 3.19 also points to a special problem of hypersonic flight: flight velocity measurements with conventional pitot-static probes become inaccurate, even impossible for Mach numbers $M_1 \gtrapprox 3$, which is due to the small and finally vanishing gradient $dc_{p_{stag}}/dM_1$.

Coming back to the bow-shock stand-off distance, we note that it and the free-stream Mach-line angle govern downstream phenomena at the flight vehicle. The Mach-line angle μ_∞ as function of the Mach number M_∞ is given in Fig. 3.20.

The bow shock in Fig. 3.17 dissipates far downstream and its inclination asymptotically reaches the free-stream Mach-line angle. The figure shows that this angle decreases very fast with the Mach number. At $M_\infty = 5$ we find $\mu_\infty \approx 10$, at $M_\infty =$ 12 we are down to $\mu_\infty \approx 5$. Actually the Mach line in Fig. 3.17 is the trace of a cone, the Mach cone with opening angle μ_∞. If the nose bluntness and the shock standoff distance increase, the "effective origin" of the Mach cone moves upstream.

Regarding the shock stand-off distance we use Eq. (3.10) to obtain an—approximate—picture. Figure 3.21 gives the ratio 'stand-off distance Δ' to 'sphere diameter $2r_n$' as function of the Mach number M_∞. Assumed is air with $\gamma = 1.4$.

At the right of the figure the asymptotic value of $\Delta/2r_n = 0.0555^{\cdots}$ for $M_\infty \to \infty$ with $\epsilon = 0.166^{\cdots}$—Eq. (3.11)—is well approached. For the small Mach numbers the values are somewhat below experimentally found values, see for instance [20] and also the discussion in [1].

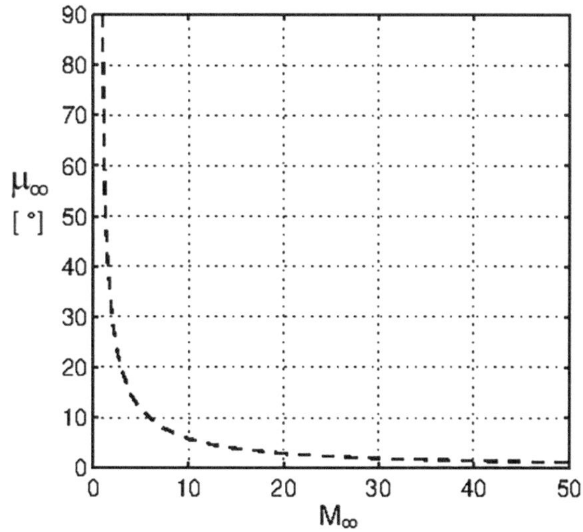

Fig. 3.20 The Mach-line angle μ_∞ as function of the free-stream Mach number M_∞

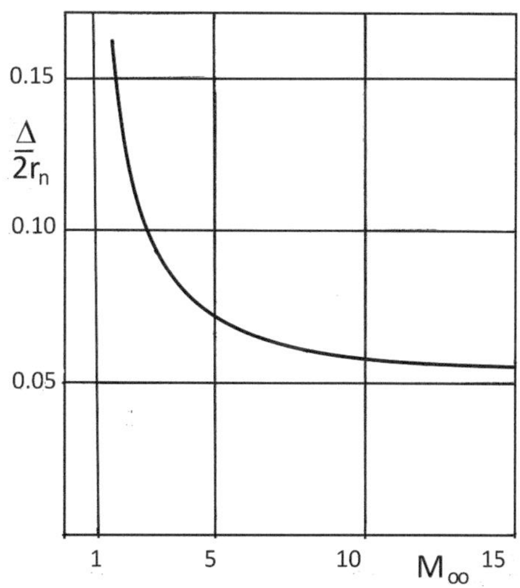

Fig. 3.21 Bow-shock stand-off distance at a sphere as function of the Mach number, air, $\gamma = 1.4$

Anyway, the result is that the bow-shock distance decreases with increasing Mach numbers. The γ_{eff} approach then shows that high-temperature real-gas effects lead to even smaller values of $\Delta/2r_n$. For high Mach numbers we find from Fig. 3.18 for the free-stream $\gamma_1 = 1.4$ and the shock-layer $\gamma_2 = 1.2$ the value $\rho_2/\rho_1 \approx 11$. This amounts to $\epsilon \approx 0.091$ and for $M_\infty \to \infty$ to the asymptotic value $\Delta/2r_n \approx 0.0324$. This result indicates that with increasing real-gas effects the shock stand-off distance decreases.

Both the Mach cone and the stand-off distance have their relevance, which is indicated in Fig. 3.22. The wing planform shown has a primary and a secondary

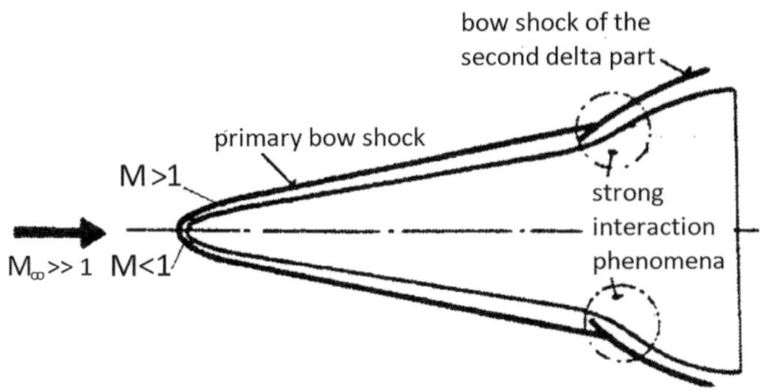

Fig. 3.22 Slenderness of the wing and strong interaction of the vehicle bow shock with the bow shock of the second delta wing-part, schematic situation at a CAV: strong interaction phenomena leading to hot-spot situations

delta part. In order to avoid drag increments, the wing's primary leading edge should lie within the Mach cone, i.e., it should have a subsonic leading edge. The second delta part is to yield a sufficiently large aspect ratio. That is needed to achieve the needed low-speed aerodynamic properties, see, e.g., [21]. The second part has a supersonic leading edge.

The stand-off distance governs the location of the bow shock with regard to downstream interaction regions, in the figure on the second delta part that is denoted with 'strong interaction phenomena'. There a Edney type VI interaction occurs with extreme pressure and thermal loads, see, e.g., [1]. The actual location of this region is governed by the flight Mach number, the vehicle attitude in terms of angle of attack and yaw, and the aerothermoelastic deformation of the airframe. All this also holds for the vertical stabilizers of the flight vehicle.

Even if the stand-off distance of the bow shock at the nose is very small compared to the dimensions of the vehicle, interaction locations must be known exactly in order to apply proper thermal protection and to avoid unnecessary mass addition.

With the waverider the situation in principle is different. This is best seen with the caret wing as epitome of the waverider, see, e.g., [2]. The sharp leading edges literally sit—ride—on the by them created shock surface. There is no spill-over to the upper side of the wing. The upper side is a free-stream surface. There, of course, also a boundary layer develops, which induces an oblique shock wave. The situation is different when looking at off-design situations and at possible modifications at the rear of the vehicle in order to ensure the necessary low-speed performance. That holds for any hypersonic flight vehicle configuration.

3.6.2 The Entropy Layer

At supersonic and hypersonic Mach numbers blunt vehicle noses induce a particular phenomenon, the entropy layer.[6] The total-pressure loss over a shock surface locally is governed by the inclination of the shock surface against the local flow direction, see, e.g., [1]. In our case the bow shock surface is curved and the streamlines passing through it take a total-pressure loss, which is highest across the locally normal shock surface. Away from that location the total-pressure loss diminishes. The—inviscid—flow past the nose then resembles a slip-flow boundary layer $v(n)$, which is called entropy layer, Fig. 3.23a.

The viscous boundary layer, having a finite thickness at the stagnation point [14], grows along the body surface in downstream direction and swallows the entropy layer. The entropy-layer swallowing is a phenomenon with importance for hypersonic vehicles. It influences the boundary-layer growth, the characteristic boundary-layer thicknesses, the heat flux in the gas at the wall q_{gw}, the wall shear stress τ_w and also the laminar-turbulent transition, see, e.g., [1].

Figure 3.23b shows the asymmetric case—the blunt cone at angle of attack—which also seems not to have found much attention so far. At the upper side in the sketch the situation is like that in Fig. 3.23a. The entropy layer at the lower side, however, has a wake-like profile. Whether this second form of the entropy layer—for some details see [2]—has an influence, in particular regarding the laminar-turbulent transition process, is not known.

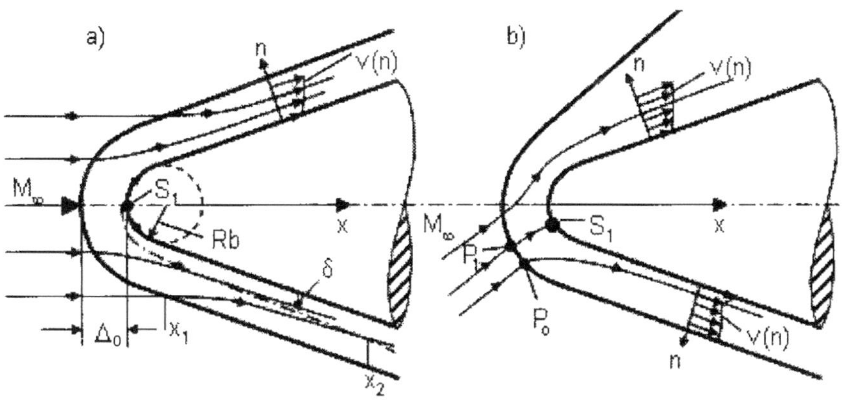

Fig. 3.23 The entropy layer at a blunt nose [1]. **a** the symmetric case, **b** the asymmetric case

[6] This phenomenon in principle appears also at swept blunt leading edges, but apparently sofar has not attracted much attention.

3.6.3 Creation of Lift

Creation of lift in hypersonic flight needs a short consideration. The basic situation is shown in Fig. 3.24. At subsonic flow most of the lift of a wing stems from the upper side of the wing. With increasing flight Mach number the creation of lift successively shifts to the lower side of the wing, or in general to the lower side of the flight vehicle.

This effect is due to the fact that the flow expansion on the upper side has its limits. That is given when the pressure drops to zero. The vacuum pressure coefficient $c_{p_{vac}}$ reads, [1]

$$c_{p_{vac}} = c_{p_{p=0}} = -2/\gamma M_\infty^2. \tag{3.14}$$

Figure 3.25 shows with curve (a) how the absolute value of $0.8\, c_{p_{vac}}$ drops with increasing Mach number. The factor 0.8 is a safety factor. For $M_\infty = 0.5$—not in the figure—we obtain with $0.8\, c_{p_{vac}} = -4.57$ the typical low-speed value and for $M_\infty = 10$ with $0.8\, c_{p_{vac}} = -0.011$ the typical hypersonic value.

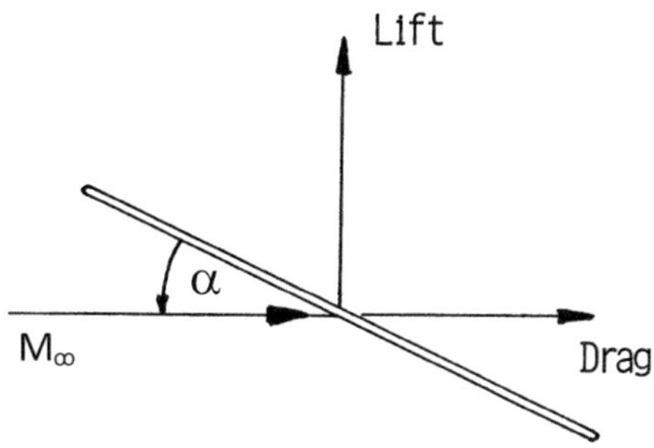

Fig. 3.24 The basic situation: lift and drag at the flat plate

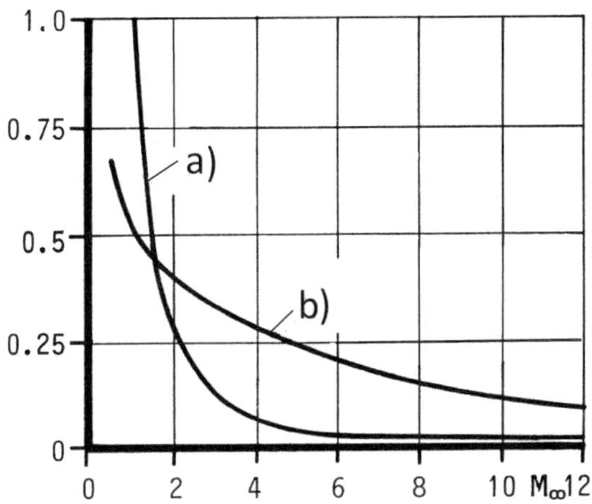

Fig. 3.25 The trend: curve **a**—$0.8\, c_{p_{vac}}$ and curve **b** $C_{L_{upper}}/C_{L_{totel}}$ of a flat plate at angle of attack $\alpha = 10$ as function of the Mach number M_∞ [7]

Curve (b) indicates with the flat plate at $\alpha = 10$ as example, how the lift contribution of the upper side drops with increasing Mach number. At $M_\infty = 12$ it is down to about 10%. A hypersonic airbreather flies at much lower angle of attack. In combination with the drag development, the waverider concept here finds its justification: the contribution of the upper side to lift and drag is negligible, therefore the upper side is designed as a free-stream surface.

3.6.4 Base Drag

A particular aerodynamic phenomenon is the transonic drag divergence of airfoils and wings. Above the lower critical Mach number a sudden strong rise of the drag occurs. For the involved and determining fluid-mechanical phenomena see, e.g., [14, 21].

With the base pressure, respectively the base drag, the phenomena are somewhat different, but the problem is the same. In [22] it is shown that the measured base pressures of two-dimensional and three-dimensional configurations as function of the flight Mach number behave quite different. The two-dimensional configurations are blunt wing trailing edges as well as waverider shapes. The three-dimensional ones are fuselage shapes.

In the subsonic/transonic Mach number range the drag coefficients of the two-dimensional shapes are about three times higher than those of the three-dimensional shapes. At high Mach numbers, say above $M_\infty \gtrapprox 3$, the coefficients converge. In view of the size of the respective base surface of the flight vehicle, the base drag can dominate the total drag in the transonic regime. From the beginning of design work this must be kept in mind, because in the transonic regime—acceleration pinch-point 1, Fig. 5.36—the difference between engine thrust and vehicle drag anyway is small, like in the hypersonic regime.

The transonic base drag of the upper stage of a TSTO space transportation system during the ascent flight of the whole system is a particular problem. To minimize it, an afterbody fairing becomes necessary. Figure 3.26 shows this for the upper stage HORUS of the SÄNGER system. It is similar to the tail cone of the Space Shuttle Orbiter and also of BURAN used for ferry flights on large subsonic aircraft.

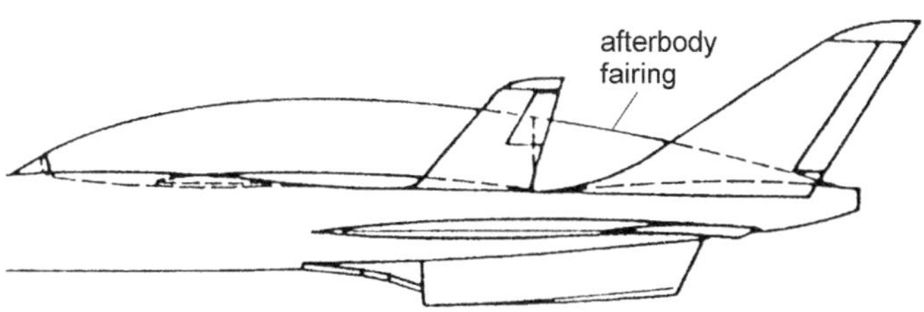

Fig. 3.26 Afterbody fairing of the HORUS Orbiter [23]

Above we have seen that lee-side or base surfaces of flight vehicles at hypersonic Mach numbers are drag uncritical once the vacuum pressure coefficient is reached. An example is the thick, dorsal and ventral wedge-fin stabilizer of the X-15, Sect. 2.9.4. The X-15 was launched from a B-52 at about 14 km altitude and $M_\infty \approx 0.75$. Because the vehicle must cross the transonic regime, hence a contradiction to the above exists. However, because of the high thrust of the rocket engine the transonic drag rise was not a problem in that case.

3.6.5 Flap/Ramp Flow

Flap flow and ramp flow have common features. Flap flow on the one hand is that occurring at aerodynamic trim, control and lift-enhancement surfaces, Sect. 2.9.3. Ramp flow is that at the external compression ramps of inlets of the propulsion system, Sect. 5.2.

A major issue of flap/ramp flow is the situation in the vicinity of the flap's hinge line, respectively around the ramp's foot. Due to for instance Edney type VI interaction large pressure and the thermal loads can occur. An extensive treatment of the problem can be found in [1, 2].

Here we discuss only the flow at a generic flap configuration, Fig. 3.27, for inlet ramps see Sect. 5.2. The corner line is the hinge line, its location and that of the hinge axis are not the same. The strong-interaction domain extends upstream and downstream over distances, which as a rule are small compared to the flap length L_{flap}. In the literature usually the clean flap configuration is treated without the hinge-line gap.

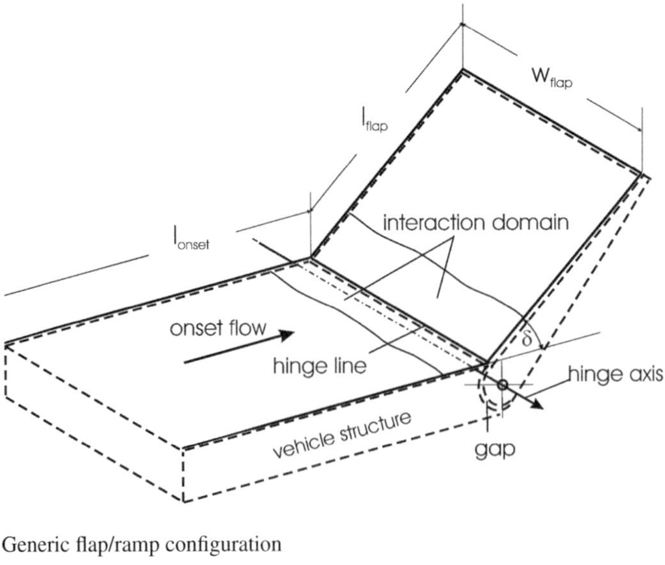

Fig. 3.27 Generic flap/ramp configuration

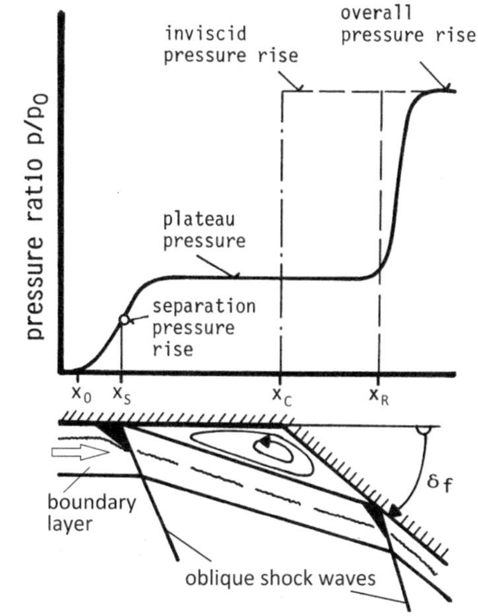

Fig. 3.28 Basic types of flap flow: pressure distribution with laminar flow [7]

The gap ensures the movability of the flap. Flow through this gap—similar to the bleed-flow at inlet ramps—can degrade flap effectiveness because it takes flow momentum away, and in particular is the source of large thermal loads on the structure surrounding the gap, since surface radiation cooling is not possible there, see, e.g., [2]. This is an extreme case of non-convex effects of surface radiation cooling, Sect. 3.9.

Two basic types of flap flow—two-dimensional and laminar, Fig. 3.28, as well as two-dimensional and turbulent, Fig. 3.29—are examined in view of the pressure distribution and the flap effectiveness, however, without consideration of the thermal-loads problem. For the latter see, e.g., [2].

Inviscid flow assumed, an oblique shock wave would form at the very corner (hinge) line x_C. The pressure jumps up instantly to a value depending on the Mach number of the onset flow and the flap angle δ_f.

Viscous flow in both cases—laminar and turbulent—leads to a separation bubble, which forms upstream of the corner at x_S, with laminar flow farther upstream than for turbulent flow. Reattachment happens on the flap at x_R, for laminar flow farther downstream than for turbulent flow. Instead of one oblique shock wave in inviscid flow, two oblique lambda shock waves result at x_S and x_R, which eventually converge depending on the Mach number of the onset flow.

The pressure rise over the hinge region is quite different for the two cases. In the laminar case the resulting pressure plateau to a degree compensates the pressure jump of the inviscid case. Regarding the lift component due to the flap deflection, virtually no loss is present compared to the inviscid case. The moment effectiveness however

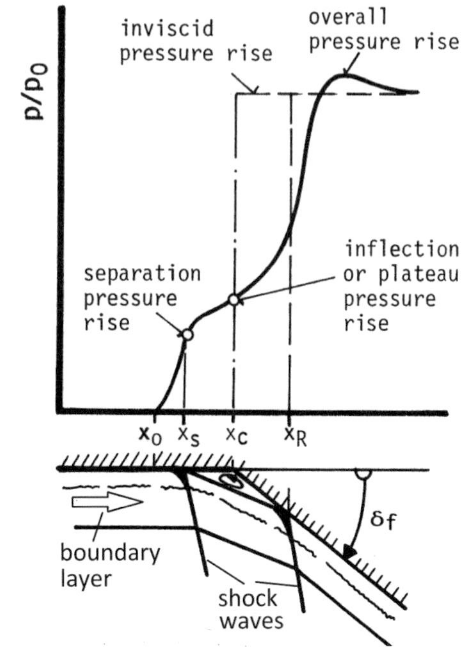

Fig. 3.29 Basic types of flap flow: pressure distribution with turbulent flow [7]

is reduced, because part of the flap-induced lift now acts ahead of the hinge line. In the turbulent case the pressure rises with a point-of-inflection and an overshoot to the inviscid pressure level on the flap.

The pressure level on the flap, possibly affected by the boundary-layer there, in general is not impaired by the situation around the hinge line. In any case a large l_{onset}, i.e., a large downstream location of the hinge line is beneficial for the vehicle's pitching moment. Regarding the needed actuator performance, a large flap width w_{flap} is to be preferred rather than a large flap length l_{flap}. However, in any case a trade-off is necessary, because with increasing flap length in general smaller deflection angles are needed with the benefit of smaller thermal loads. Further it holds that a large aspect ratio of the flap reduces flow three-dimensionality and associated losses in effectiveness.

All this must be seen in view of the general flight mechanical layout of the vehicle. Reducing the static stability margin, would reduce the trim moment. With the small lever arms of hypersonic vehicles, in contrast to classical aircraft with the empennage far aft, this means a rather large trim force and hence a considerable trim drag. Reducing the stability margin, however, must be seen in view of atmospheric and other perturbations.

3.7 The Thermal State of the Vehicle Surface

The outer flow path of hypersonic flight vehicles generally is radiation cooled, see, e.g., [1]. In ground-facility simulation this cannot be duplicated, hence there another thermal situation is present. This section gives a short introduction to the thermal state of a surface, which has a strong influence on the flow past it. That is sketched in the next sections.

Five basic cases of the thermal state of the surface can be distinguished, Fig. 3.30. It is assumed that no shock-layer radiation is present—which anyway occurs only at much higher Mach numbers, see, e.g., [1]—, that no non-convex effects are present, they are treated in Sect. 3.9, and that in the wall no heat transfer happens in x-direction, i.e., tangential to the surface.

The five cases are shortly characterized:

- Case (a) This is the general case of a radiation-cooled surface. The heat flux toward the wall is q_{gw}, radiated away from the surface is the flux q_{rad}. The heat flux into the wall is q_w. The general balance can be written as $q_w = q_{gw} - q_{rad}$, with the wall temperature T_w being the result.
- Case (b) If the wall is heated up such that q_w vanishes, the radiation-adiabatic wall is reached with $q_{gw} = q_{rad}$ with the corresponding radiation-adiabatic temperature $T_w = T_{ra}$.
- Case (c) Without radiation cooling the heat fluxes towards and into the wall are equal: $q_{gw} = q_w$. This is the situation found in a wind tunnel with the model temperature T_w.
- Case (d) Once the wall is heated up such that q_w vanishes, without radiation cooling the adiabatic wall is reached with $q_{gw} = 0$ and $q_w = 0$. T_w then is the recovery or adiabatic temperature T_r.
- Case (e) The last case is that of thermal reversal, Sect. 3.10.4. After a preceding high-velocity flight—hypersonic and even high supersonic flight—, i.e., at lower flight velocity, the vehicle surface is hotter as it would correspond to the actual

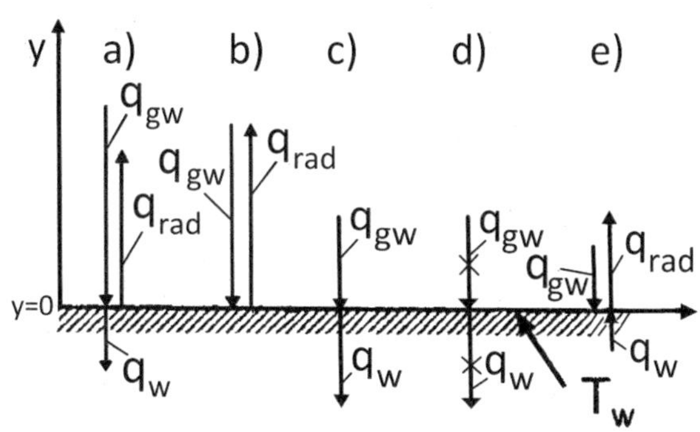

Fig. 3.30 The thermal state of the body surface in terms of the involved heat fluxes. Basic cases regarding the surface: **a** radiation cooled, **b** radiation-adiabatic, **c** without radiation cooling, **d** adiabatic, **e** thermal reversal. The surface temperature T_w in each case is a result of the heat fluxes

flight condition. That is due to the thermal inertia of the either hot structure or the thermal protection system. The heat fluxes are $q_{gw} + q_w = q_{rad}$ and T_w is the thermal reversal temperature. Transverse radiative heat transport, Sect. 3.10.5, can be considered as a special case of thermal reversal.

The radiation heat flux can be written as

$$q_{rad} = \epsilon \sigma T^4|_w, \tag{3.15}$$

where ϵ is the emissivity coefficient with $0 \leqq \epsilon \leqq 1$. Its magnitude depends on the surface coating, see Sect. 3.2.3. Usually the emissivity coefficient is assumed to be $\epsilon \approx 0.85$. The constant $\sigma = 5.6704 \cdot 10^{-8}$ W/m^2 K^4 is the Stefan-Boltzmann constant.

Generally for hypersonic flight the radiation-adiabatic wall temperature—Case (b)—can be considered as conservative estimate of the real situation. This was seen to be the case for re-entry flight of the Space Shuttle Orbiter, see e.g., [24], and also [1]. Of course the situation on the whole flight trajectory must be considered.

For the airbreathers considered here that is the flight segment from take-off up to the maximum flight Mach number, as well as the segment down to approach and landing. On the latter segment thermal reversal happens with potentially important thermal surface effects.

The matter of conservative estimates further is supported by a study of T. Eggers for the stagnation-point situation of several configurations [25]. There also the situation on the whole flight trajectory was considered.

If the radiation-adiabatic case is a good approximation of a given reality, we can write

$$k_{gw} \frac{\partial T}{\partial y}|_{gw} = \epsilon \sigma T^4|_w, \tag{3.16}$$

where k_{gw} is the thermal conductivity of the gas at the wall.

After introducing a finite difference it follows

$$T_{ra}^4 \approx \frac{k_{gw}}{\epsilon\sigma} \frac{T_r}{\Delta} (1 - \frac{T_{ra}}{T_r}), \tag{3.17}$$

with Δ being a characteristic boundary-layer thickness, which for laminar flow is the thickness of the thermal boundary layer δ_T and for turbulent flow that of the viscous sub-layer δ_{vs} [14].

From this with some assumptions, which are not repeated here, limiting properties are obtained [1]:

$$
\begin{aligned}
&\varepsilon \to 0 &&: \ T_{ra} \to T_r, \\
&Re^u_{ref} \to 0 &&: \ T_{ra} \to 0, \\
&Re^u_{ref} \to \infty &&: \ T_{ra} \to T_r, \\
&L \to 0 &&: \ T_{ra} \to T_r, \\
&\frac{x}{L} \to 0 &&: \ T_{ra} \to T_r, \\
&\frac{x}{L} \to \infty &&: \ T_{ra} \to 0, \\
&T_r \to 0 &&: \ T_{ra} \to 0.
\end{aligned}
\tag{3.18}
$$

Further the qualitative results follow, that the radiation-adiabatic wall temperature T_{ra} on a flat surface approximately

- is inversely proportional to the surface emissivity: $\varepsilon^{-0.25}$,
- is inversely proportional to the characteristic boundary-layer thickness $\Delta^{-0.25}$,
- is proportional to the recovery temperature $T_r^{0.25}$, and with that in perfect-gas flow proportional to $M_\infty^{0.5}$,
- falls with laminar flow with increasing running length: $(x/L)^{-0.125}$,
- falls with turbulent flow less with increasing running length: $(x/L)^{-0.05}$,
- for laminar flow is proportional to $(Re^u_{ref})^{0.125}$ (note that the recovery temperature does not depend on the Reynolds number),
- for turbulent flow is proportional to $(Re^u_{ref})^{0.2}$, and hence is appreciably larger for turbulent flow than for laminar flow, because the turbulent scaling thickness, i.e., the viscous sub-layer δ_{vs} is smaller than the thermal boundary layer thickness δ_T, if the wall would be laminar at the same location.

3.8 Surface Radiation Cooling

An example of the effectiveness of surface radiation cooling is given in Fig. 3.31. It shows wall temperature distributions along the lower and the upper symmetry line of the forebody of the lower stage of the TSTO-system SÄNGER. The reference length L is 55 m, the flight data are $M_\infty = 6.8$ and altitude $H = 33$ km [26, 27]. For a detailed discussion see [1].

The perfect-gas stagnation-point temperature is about 2,300 K, and the recovery temperature, laminar and turbulent, about 2,150 K. The wall temperature at the stagnation-point with real-gas assumption is about 2,100 K, see also Fig. 2.7, the recovery temperature, again laminar and turbulent, is about 1,900 K. With radiation cooling, assuming $\epsilon = 0.85$, the stagnation-point temperature drops to 1,600 K, which corresponds to a nose radius of about 0.2 m, Fig. 3.12.

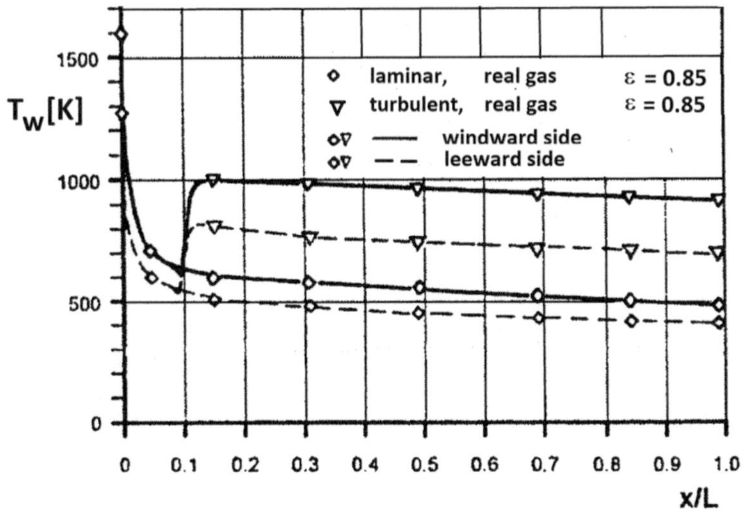

Fig. 3.31 Wall temperatures along the lower and the upper symmetry line of the SÄNGER lower-stage forebody

Along the symmetry lines the radiation-adiabatic temperature initially drops steeply. At the upper symmetry line all temperatures are lower compared to those at the lower symmetry line. The reason is that the upper surface of the forebody more or less is a free-stream surface, whereas the lower surface is a pre-compression surface for the inlet flow, Sect. 5.5. If this temperature difference would lead to structural problems, a reduction of the emissivity level at the upper side of the forebody would be a way out. Such a tailoring of the surface radiation emissivity via the surface coating of both metallic and ceramic material appears to be an option [28].[7]

Laminar-turbulent transition was arbitrarily assumed to occur at $x/L = 0.1$. The temperature immediately jumps up by about 400 K. This is a large increase with large influence on a number of vehicle design issues, e.g., aerodynamic performance, structure and materials concept, propulsion integration. A problem is the determination of the location and the extent of the transition zone, Sect. 3.11.

Surface radiation cooling is the key to prolonged hypersonic flight. This holds not only for hypersonic airbreathers, but also for winged and non-winged re-entry vehicles from Earth orbit. However, surface radiation cooling becomes ineffective at low flight altitudes. This and other important results can be obtained from Fig. 3.32, although it only covers the low Mach-number flight span of a hypersonic airbreather.

These results can be summarized as follows, for details and further related publications see [1, 2].

[7] For the structure and materials concept of the lower stage of the TSTO-system SÄNGER see Sect. 6.3.

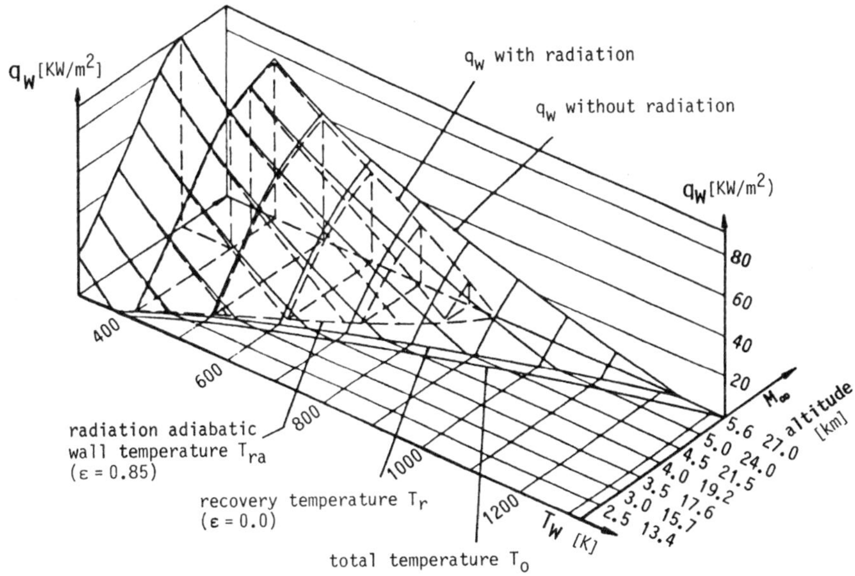

Fig. 3.32 Heat fluxes and temperatures at the lower symmetry line of a hypersonic flight vehicle at different flight Mach numbers M_∞ and altitudes [29]. The data point lies on the lower centerline 5 m downstream of the vehicle nose. $\varepsilon = 0.85$, turbulent flow. q_w is the heat flux into the wall

- Considering the M_∞-T_w surface at $q_w = 0$ we see that the recovery temperature is smaller everywhere than the total temperature. Radiation cooling reduces the temperature at $M_\infty = 5.6$ by approximately 350 K compared to the recovery temperature.
- At smaller flight Mach numbers, and hence flight altitudes, radiation cooling loses its effectiveness! This important result is due to the high unit Reynolds numbers there, which reduce the boundary layer thickness.
- Look now at the q_w-T_w surface at $M_\infty = 5.6$. To sustain a wall temperature of, for instance, $T_w = 1{,}000$ K, a heat flux into the wall of $q_w \approx 14$ kW/m^2 would be necessary without radiation cooling. With radiation cooling the radiation-adiabatic wall temperature would anyway only be $T_{ra} \approx 870$ K.
- Assume now an actual heat flux into the wall of $q_w \approx 10$ kW/m^2. This is approximately 27% of the radiation heat flux q_{rad} in the case of the radiation-adiabatic surface. This—in per cent not so small actual heat flux—reduces the wall temperature compared to the radiation adiabatic temperature only by $\Delta T \approx 70$ K.
- It was mentioned above that the radiation-adiabatic wall temperature under certain conditions is a conservative estimation of the wall temperature, and that the actual temperature will lie close to it, $T_w \approx T_{ra}$. Coupled flow-structure analyses appear to show that this in general is true.
- After prolonged hypersonic flight at approach and landing thermal reversal is present, Sect. 3.10.4. That means that T_w of the vehicle surface is higher than the

recovery temperature T_{ra} that corresponds to the actual flight speed and altitude. The aerodynamic data found in a low-speed wind tunnel would be incorrect to a certain degree. In particular with slender configurations this fact—wall-temperature effects—needs attention.

- A similar situation is present at flight vehicles with internal heat transfer mechanisms, namely transverse radiation cooling or heating through the structure of hot airframe elements or control surfaces, Sect. 3.10.5. Here also $T_w > T_{ra}$ can result.

3.9 Non-convex Effects

So far radiation cooling on fully convex shapes was considered. Real aircraft configurations have surface parts which face each other. This situation is found at wing roots, fins, at inlets, hinge gaps and so on. In such cases the surface parts receive radiated heat from each other and the radiation-cooling effect is severely reduced.

This non-convex effect is discussed in some detail in [1]. There also the approach reported in [30, 31] is presented. To illustrate the non-convex effect in terms of the fictitious emissivity coefficient ϵ_f, the situation at a generic fin configuration at M_∞ = 7 and H = 27 km is shown in Fig. 3.33.

The fictitious emissivity coefficient, which is the actual local emissivity coefficient, at the roots of the fins is strongly reduced down to $\epsilon_f \approx 0.1$. The—not shown—radiation-adiabatic wall temperature there is rising from about 800 K without the non-convex effect to about 1,500 K. The effect hence is of large importance

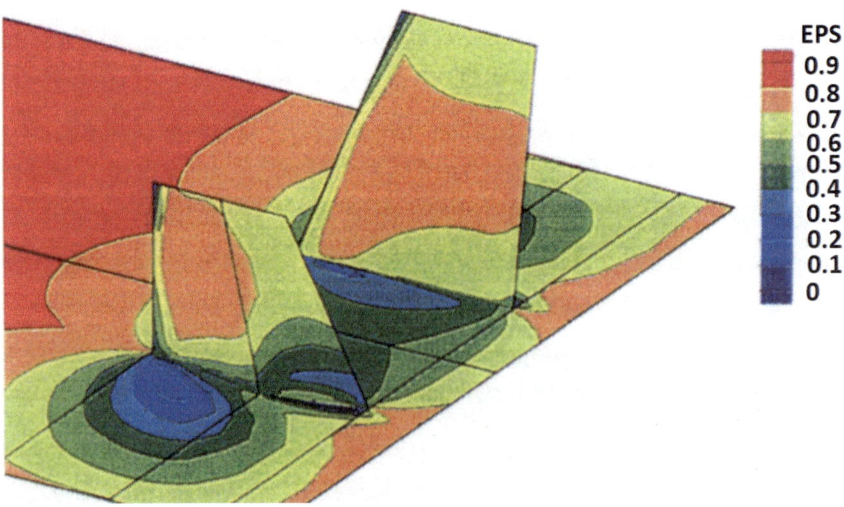

Fig. 3.33 The distribution of the fictitious emissivity coefficient ϵ_f at a generic stabilizer configuration [31]

in view of the structure and materials concept at such configuration parts. The actual wall temperature of course also affects the wall shear-stress as well as separation phenomena, i.e., the related thermal surface effects.

The extreme of non-convex effects is found at hinge-line gaps, Sect. 3.6.5. In the gap radiation cooling is not possible and therefore extreme thermal loads result, see, e.g., [2]. This requires high attention and multidisciplinary simulation in order to quantify the thermal loads as well as the actual flap effectiveness.

3.10 Thermal Surface Effects

Boundary-layer textbooks usually treat the effects of the wall temperature only in the context of the Mach-number influence on the wall shear stress. In view of the hypersonic airbreathers considered here a close look at thermal surface effects is needed.

Table 3.1 lists major thermal surface effects although without indicating explicitly the dependence of the various viscous flow and thermo-chemical phenomena on the wall temperature T_w and its derivative $\partial T/\partial y|_{gw}$ in the gas at the wall and its derivative $\partial T/\partial y|_w$ in the wall. Subsequently a few of the effects are shortly discussed, in general the reader is referred to [1, 2].

Table 3.1 Wall and near-wall viscous-flow/thermo-chemical phenomena, and structure and materials issues influenced by the thermal state of the surface [1, 27]. y is the direction normal to the surface, ( ) indicates indirect influence

Item	T_w	$\partial T/\partial y\|_{gw}$	$\partial T/\partial y\|_w$
Boundary-layer thicknesses (δ, δ_1, ...)	X		
Laminar Skin friction	X		
Heat flux in the gas at the wall q_{gw}	X	X	
Hot spot/cold spot phenomena	X		
Surface-radiation heat flux q_{rad}	X	X	(X)
Laminar-turbulent transition	X	X	
Turbulent skin friction	X		
Uncontrolled/controlled flow separation	X		
Shock/boundary-layer interaction	X		
Hypersonic viscous interaction	X		
Catalytic surface recombination	X	(X)	
Transport properties at/near the surface	X	X	
Wall heat flux q_w	X	(X)	X
Material strength and endurance	X		
Thickness of TPS or internal insulation (time integral of q_w)	X		X

3.10.1 Wall Shear Stress

Figure 3.34—referring to Fig. 3.31—shows one of the most important effects. The wall shear-stress coefficient along the lower side of the forebody of the SÄNGER lower stage reveals that a hotter surface leads to a lower skin-friction level. The effect is strong for turbulent flow, but small for laminar flow.[8]

Behind the effect is that the characteristic boundary-layer thicknesses become smaller with decreasing wall temperature [1, 14]. The dependence of the viscosity on the temperature nearly is canceled out and hence the skin friction increases with decreasing wall temperature. The effect is markedly stronger with turbulent flow than with laminar flow, see also [32, 33].

Note also that the values of the turbulent flow are much higher than those of the laminar flow. The result in general underlines the need to know exactly the location of laminar-turbulent transition. In this case the switch from the laminar to the turbulent state of the boundary-layer flow was modeled to occur instantly. In reality a longer transition zone—the intermittency region, Sect. 3.11.1—is present. It is wrapped around the forebody and depending on the actual flight situation may further complicate the situation and also the design implications.

Figure 3.35 shows the effect in more detail. The skin-friction coefficient decreases with increasing wall temperature ratio T_w/T_∞ and Mach number M_∞. The flow is fully turbulent at $Re = 10^7$. Included in the figure is the curve for the adiabatic wall, i.e., for zero surface radiation cooling and zero heat transfer into the wall, Case (d) in Fig. 3.30.

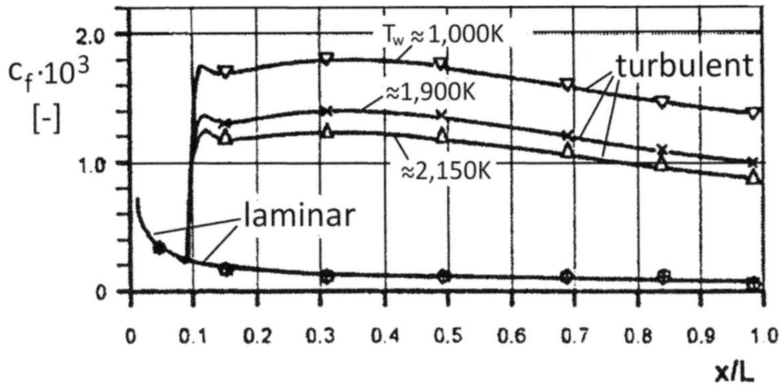

Fig. 3.34 Skin-friction coefficient distribution along the lower side of the forebody of the SÄNGER lower stage for the different wall-temperature levels without and with surface radiation cooling, see Fig. 3.31

[8] The data for laminar flow from [26] may need a reconsideration.

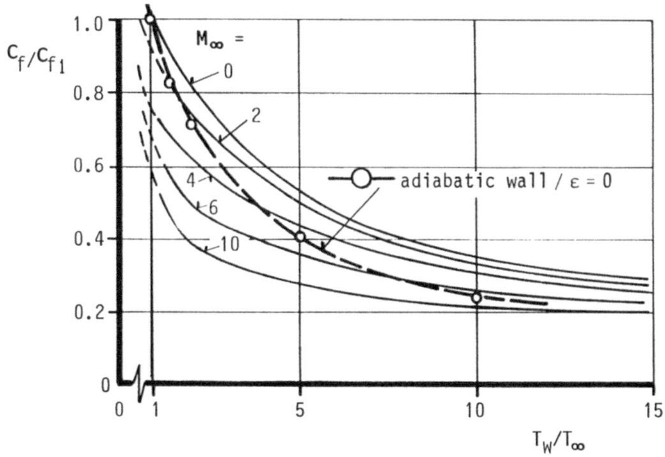

Fig. 3.35 Influence of wall temperature and the Mach number on the skin-friction coefficient of fully turbulent flow at $Re = 10^7$ [7]

3.10.2 Hypersonic Flight Vehicle Design Implications

The wall shear-stress behavior points to a very important design problem of hypersonic airbreathers: the hotter the vehicle surface, the lower is the skin-friction drag of the turbulent flow path![9] This means that the surface emissivity coefficient must be tailored to the structure and materials concept and vice versa, see the remark in Sect. 3.8.

The flight-vehicle design problem hence is to find an optimum between low turbulent skin-friction drag—the form drag must be considered too—and a suitable structure/surface-material concept to support the corresponding wall temperature. In the background still is the all-dominant uncertainty of the location of the onset of laminar-turbulent transition and its downstream extent.

The full picture is sketched in Fig. 3.36: the influence of the four essential parameters on the equivalent mean skin-friction coefficient at the lower centerline of the forebody of a SÄNGER-like configuration ahead of the inlet ramps:

- wall temperature (abscissa),
- state of the gas (perfect or real gas),
- surface radiation emissivity ϵ,
- state of the boundary layer (laminar or turbulent).

The flight parameters are $M_\infty = 6.8$ at $H = 31\,\text{km}$, $Re_L = 1.22 \cdot 10^8$ with $L = 55\,\text{m}$, angle of attack $\alpha = 6$.

[9] The decrease of the skin friction of a turbulent boundary layer with increasing wall temperature holds for low-speed flight too [14].

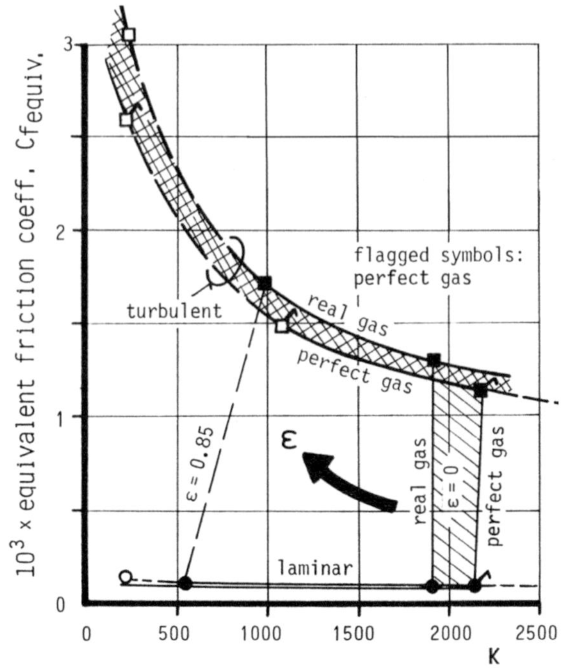

Fig. 3.36 The local skin-friction coefficient as function of the wall temperature, real-gas effect, surface emissivity and boundary-layer state [7]

Regarding laminar flow—curves at the bottom—it does not play a significant role in view of the drag, whether perfect or real gas and high or low emissivity is assumed. The wall temperature, however, is strongly affected by surface radiation cooling, i.e., the emissivity, as was seen above.

With turbulent flow, all four essential parameters play an important role. The skin-friction coefficient is much higher than for laminar flow and depends much more strongly on the wall temperature. For the intended hot primary structure the permissible temperature of 1,000 K is reached for real gas and the emissivity coefficient $\epsilon = 0.85$.

When comparing the curves of the skin-friction coefficient for laminar and turbulent flow, it becomes evident, how important the exact knowledge of the transition location and its extension is. This holds for both the local thermal loads on the airframe and the vehicle's performance in terms of drag, thrust minus drag and aerodynamic performance L/D. This all is much more vital for hypersonic airbreathers with their low thrust/mass and thrust/drag ratios compared to rocket-propelled vehicles with their mass-favorable engines.

3.10.3 Ramp Effect

Now a particular situation occurring on the outer flow path of a hypersonic flight vehicle is considered. It regards both inlet ramps and deflected aerodynamic trim and control surfaces. However, the strong-interaction situation at the foot of a ramp or at the hinge line of a flap, Sect. 3.6.5, is not treated here, only the asymptotic behavior of flow properties along the ramp/control surfaces.

Behind the ramp effect is the change of the unit Reynolds number of the flow across a shock wave. The unit Reynolds number governs the boundary-layer thicknesses as well as the radiation-adiabatic wall temperature and the skin-friction. The effect and its importance, having found not much attention so far in the literature, is discussed in [1].

The basic example is given in Fig. 3.37. The radiation-adiabatic wall temperature at a generic three-ramp inlet of a SÄNGER-type hypersonic airbreather is considered at the flight condition $M_\infty = 7.5$, $H = 35$ km. Turbulent flow is assumed, the emissivity coefficient is $\epsilon = 0.85$.

At each ramp a stepwise reduction of the Mach number occurs and together with that a stepwise increase of the pressure and the unit Reynolds number. This then leads to a decrease of the boundary-layer thickness and hence the radiation-adiabatic wall temperature on each ramp rises stepwise by about 200 K to a higher magnitude. This discrete behavior is due to the employed approximate method. No strong-interaction phenomena such as flow separation, Sect. 3.6.5, are captured. The result must be considered as asymptotic behavior of the flow.

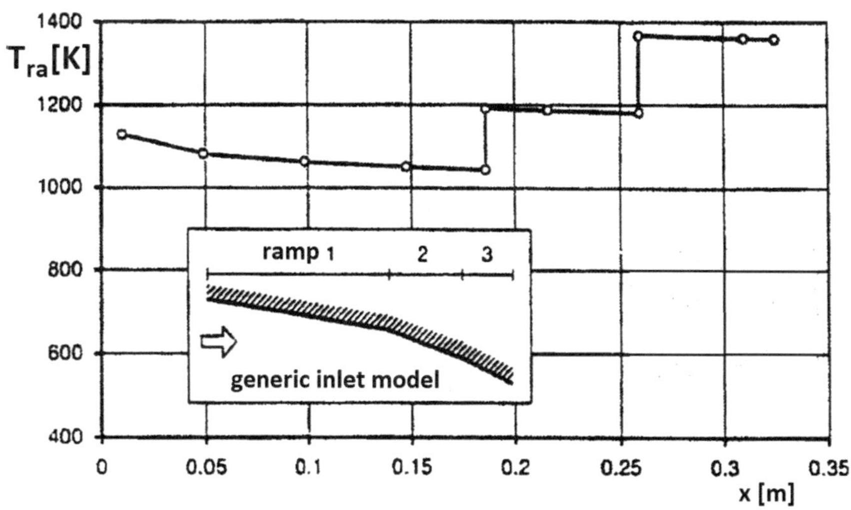

Fig. 3.37 Generic three-ramp inlet model [34]: jump of the radiation-adiabatic wall temperature from ramp to ramp

The result tells us, despite the limitations of the study, that the structure and materials concept of the forebody, based on the radiation-adiabatic wall temperature level present there, is no more adequate once a ramp situation is present. That can be, like in this case, the outer inlet ramps of the airbreathing engine(s) or a deflected flap or any other protruding configuration element.

The unit Reynolds number behavior of the flow across a shock wave hints to another interesting fact, which, however, needs further investigations. In [1] it is shown that beyond a certain flap deflection angle the unit Reynolds number decreases. Hence, there could be a potential to use higher flap angles in order to reduce thermal loads, although this would promote larger regions of strong interaction and flow separation around the hinge line.

3.10.4 Thermal Reversal

We shortly look at a case which resembles thermal reversal [35]. A winged re-entry vehicle of 50.2 m length flies at $M_\infty = 3.2$ with $\alpha = 15°$. The Reynolds number is $Re_{L,\infty} = 2.31 \cdot 10^7$, the radiation-adiabatic wall temperature belonging to this flight condition is $T_w = T_{ra} \approx 500$ K, the emissivity coefficient $\varepsilon = 0.8$.

Fully turbulent flow at the lower side of the configuration is assumed. The inboard wing flap has the deflection $\eta_{iwf} = +20°$ (downward). Figure 3.38 shows at the left the skin-friction distribution for the radiation-adiabatic wall temperature $T_w \approx 500$ K. A high skin-friction coefficient exists along the leading edge. It drops over the wing's

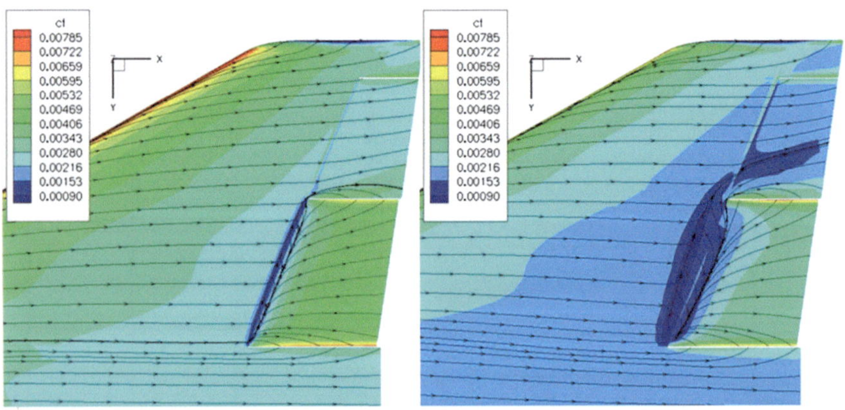

Fig. 3.38 Computed influence of the wall temperature on the distribution of skin-friction coefficient at the left lower-side aft part of the re-entry vehicle configuration HOPPER [35]. The flow is from the left to the right. Left side of the figure: $T_w \approx 500$ K, right side: $T_w \approx 1{,}600$ K, fully turbulent flow in both cases

surface in downstream direction, weak three-dimensionality is present. Along the hinge line of the flap the to be expected zero skin-friction coefficient, Sect. 3.6.5, is evident.

At the right part of Fig. 3.38 the situation for $T_w \approx 1{,}600$ K is shown. This temperature resembles the thermal state of the surface after a preceding hypersonic re-entry flight: the surface is hotter than the actual flight condition (thermal reversal), which is due to the thermal inertia of the thermal protection system.

The figure shows a clearly lower value of the skin-friction coefficient over the whole surface. The skin-friction line pattern has changed slightly, along the flap's hinge line now a large separation region exists. That is due to the influence of the wall temperature on the boundary layer's tangential velocity profile. The wall-near momentum flux is reduced and the separation characteristic is changed, once an adverse pressure gradient is present.

The thermal reversal leads to lower skin-friction drag and lower flap efficiency. For the Space Shuttle Orbiter the subsonic lift-to-drag ratio originally was underestimated by about ten per cent, [36], where probably the decreased skin-friction level due to the heated surface during re-entry was the cause. Another cause could be the condition of the tile surface of the heat protection system with its expansion joints. That would be different in cold and hot status. For BURAN it was reported that the supersonic L/D was underestimated by five to eight per cent. For the hypersonic airbreathers considered here, the effect of thermal reversal must be a matter of attention.

3.10.5 *Transverse Cooling/Heating*

Structure-internal heat transfer mechanisms, in this case transverse radiation cooling or heating, are to be considered [1]. Transverse cooling or heating may happen in the hot structure of airframe elements and also of stability, trim and control surfaces. If for the quantification of a given design-critical thermal-surface effect T_{ra} is too crude an approximation, the true T_w must be determined by means of multidisciplinary simulation. The effect is very important also for the layout of hot primary structures, where anyway in the material a conductive heat transport occurs.

3.11 Laminar-Turbulent Boundary-Layer Transition, Turbulence and Separation

This section is devoted to a short overview of some hypersonic boundary-layer flow issues, which regarding their prediction and simulation—either experimentally or theoretically/numerically—partly still pose very large problems. Particularly the location, the shape and the extent of the laminar-turbulent transition zone is of high importance and at the same time of big concern, because of the shortcomings regarding an exact prediction.

3.11.1 Laminar-Turbulent Transition and Its Implications

Figure 3.31 demonstrates how the wall temperature is affected, and with that the skin-friction coefficient, once laminar-turbulent transition occurs. The location, the shape and the extent of the transition zone—both change (!) along the flight trajectory—hence are of large importance. Its determination is the big problem.

For the National Aerospace Plane(NASP/X-30) it was reported that the take-off mass of the SSTO-vehicle was affected by a factor two or more by the uncertainty of the location of laminar-turbulent transition [37].

The transition location at an airbreathing hypersonic flight vehicle affects directly and indirectly for instance:

- the wall temperature, hence the thermal loads on the structure and with that the structure and materials concept,
- the amount of skin friction drag, hence a large part of the vehicle's drag,
- the amount of the viscosity effects induced pressure drag (form drag),
- the inlet onset flow of the propulsion system, the height of the boundary-layer diverter and the thrust balance,
- the efficiency of trim, control and lift-enhancement surfaces,
- and hence the vehicle's dry mass, the mass of propellant and with that the take-off mass.

These items must be understood definitely, because in the background lurks the vicious snow-ball effect, see Sect. 2.10. Uncertainties for instance in drag prediction lead over a chain of effects finally to a larger take-off mass.

A flight vehicle must be considered to be weight-critical, or mass-sensitive, if the take-off mass grows strongly with the ratio 'empty-vehicle mass' to 'take-off mass'. Large mass-growth factors together with small payload fractions are typical for hypersonic airbreathers. Such vehicles are more or less viscous-effects dominated and especially transition sensitive. Laminar-turbulent transition hence is a key problem in the design of CAVs and ARVs.

In view of the X-43A one sometimes gets the question why not tripping the boundary layer close to the vehicle nose and the problem is gone. For the X-43A indeed boundary-layer tripping was employed [38]. Due to the small size of the flight vehicle, natural transition was unlikely to occur ahead of the inlet and anyway is was—and still is—not possible to predict is location and characteristics. Laminar boundary-layer flow in the engine moreover was seen critical because of possible separation and hence reduced performance.

A tripping device at the lower side not much downstream of the leading edge provided unambiguous turbulent flow over the inlet ramps and in the engine, see, e.g., [39]. The higher thermal loads and the higher skin-friction drag were compensated by a higher mass flow and better fuel mixing.

A CAV or a ARV typically fly in the whole Mach number range from take-off to landing. The two X-43A flights were at $M_\infty \approx 7$ and 10, without the Mach-number span of a CAV or ARV. Hence transition forcing was made for a more or less constant Mach number, Reynolds number and surface condition.

Turbulence tripping for a wide Mach number range from zero to hypersonic speed is unlikely to work. Under- and over-tripping is possible as well as relaminarization, all leading to uncertain flow conditions for the flight vehicle and its propulsion system. Hence the answer to the above question is 'not possible'.

The problems with hypersonic laminar-turbulent transition and with turbulence and turbulent separation are the still insufficient understanding of the involved phenomena on the one hand, and the deficits of the ground-simulation means on the other hand. This holds for both the ground-experimental facility and the computational simulation despite the recent large advances in these fields.

The three key topics regarding laminar-turbulent transition are:

- hypersonic laminar boundary-layer flow,
- phenomena of instability and transition and theoretical and prediction approaches,
- receptivity issues.

3.11.2 Hypersonic Laminar Boundary-Layer Flow

General introductions to boundary-layer theory can be found in [40] and to viscous flow in [41]. Basically low-speed and with some exceptions two-dimensional flow is at the center of [40]. Three-dimensional boundary layers, more in general three-dimensional attached viscous flow, are the topic of [14]. Modeling and computation issues are treated in [42]. In all these monographs laminar and turbulent flow is treated, also laminar-turbulent transition. Hypersonic boundary-layer flow is a major topic in [1, 2]. There also approximate methods are presented, in particular the generalized reference-enthalpy approach by G. Simeonides, [32], which allows fast and quite accurate estimations of laminar and turbulent hypersonic boundary-layer properties.

The following listing gives an overview of features of laminar hypersonic boundary-layer flow in view of laminar-turbulent transition. Assumed are steady or quasi-steady flow fields, see [1].

1. General properties: Depending on the pressure field, which in turn depends on the vehicle's shape, two- and three-dimensional laminar flow develops over the vehicle's surface. At singular points and singular lines hot-spot and cold-spot situations are present [1, 14]. At stagnation points, with the slender configurations only one at the vehicle nose, finite boundary-layer thickness exists. The developing boundary layer "swallows" the entropy layer.
   With a forebody, whose flat lower side functions as pre-compression surface for the inlet, two attachment lines exist (API-type 4 and 5 flight vehicles). Between them the two-dimensional inlet-onset flow develops [14]. At the stagnation point and at the attachment lines hot-spot situations are observed. Cold-spot situations

typically develop along separation lines. The wall temperature there is distinctly lower than in their vicinity.
Depending on the overall configuration non-convex effects can play a role as well as transverse heat transport. On the flight trajectory, in particular during approach and landing, thermal reversal can be an effect, which must be taken into account.

2. Body surface as inner boundary-layer edge: At the body surface, which is the inner edge of the boundary layer, the no-slip condition holds, which is typical for the flight domain of hypersonic airbreathers, Sect. 2.4, as they generally fly in the continuum regime and are free of rarefaction effects. Of deciding importance is the proper consideration of the surface properties, Sect. 3.2.3.
3. Outer boundary-layer edges: Actually at least three different boundary layers are present, each with its own not sharply defined outer edge. These boundary layers are the usually meant flow boundary layer, the thermal boundary layer and the mass-concentration boundary layer [1]. Their thicknesses are defined by the Reynolds number, the Prandtl number and the Schmidt number. The outer edges of these boundary layers are not sharply defined, for the flow boundary layer usually the 99% outer flow velocity is applied as the outer edge.
In numerical simulations with a one-domain approach, i.e., with the Navier-Stokes equations or the RANS equations, the different thicknesses do not pose a problem, with two-domain approaches, Euler plus boundary-layer equations, they must be taken into account [14].
4. Second-order effects: At low Reynolds numbers thick boundary layers develop and surface-curvature effects are present. With a one-domain approach these effects are covered automatically, also the entropy-layer swallowing. With a two-domain approach otherwise a second-order boundary-layer approach is necessary.
5. Thermal state of the body surface: The thermal state means both the wall temperature T_w and the heat flux in the gas at the wall q_{gw}, Fig. 3.30. Both play a role for the generalized wall-compatibility condition, which reads, [1]:

$$\frac{\partial^2 u}{\partial y^2}\Big|_{y=0} = \left\{\frac{1}{\mu}\left[\frac{\partial p}{\partial x} - \frac{\partial \mu}{\partial T}\frac{\partial T}{\partial y}\frac{\partial u}{\partial y}\right]\right\}_{y=0}, \tag{3.19}$$

where q_{gw} is represented by the wall-normal temperature gradient $\partial T/\partial y$.
With laminar flow this condition via the Rayleigh equation, see, e.g., [43], governs the point-of-inflexion instability, which is an important effect also in hypersonic flow.
6. High-temperature real-gas effects: These effects—vibration excitation and de-excitation of molecular oxygen and nitrogen, dissociation and recombination of oxygen, to a degree also of nitrogen, as well as diffusive transport of mass and energy—found in the flight domain, Sect. 2.4, influence the flow-field properties.
7. Ground-facility simulation: In view of hypersonic laminar-turbulent transition, two major problems exist with ground-facility simulation: (1) the disturbance level—free-stream disturbances—in the test section and (2) the simulation of surface radiation cooling. The first problem is tackled with the so-called quiet

tunnel designs, see, e.g., [44]. The second problem to a degree can be overcome with the hot model technique. Viscous thermal surface effects can be studied by heating of the model, for examples see [1]. Very advanced opto-electronic measurement devices are available today.
Demands in view of hypersonic flight vehicle development are

- high model fidelity,
- Mach-number and Reynolds-number similarity for the transition-critical parts of the flight trajectory,
- proper thermal state of the model surface, Fig. 3.30,
- model pre-heating to achieve radiation-adiabatic wall temperature levels,
- elimination of tunnel background radiation (generalized non-convex effects),
- quiet tunnel, but allowing for atmospheric disturbance level.

For the challenges to achieve sufficiently full similitude see, e.g., [32].

8. Numerical simulation of the laminar flow over a flight vehicle demands

- high discretization fidelity of the vehicle shape,
- regard of aerothermoelastic deformations,
- exact surface boundary-condition formulation, [45], instead of $\partial p/\partial y = 0$ at the wall, which is often assumed in codes,
- proper treatment of high-temperature real-gas effects,
- regard of the actual thermal state of the vehicle surface, Fig. 3.30,
- accurate atmospheric properties (air data) as well as fluctuations of all kinds.

3.11.3 Boundary-Layer Instability and Transition

Stability of a laminar boundary layer means that disturbances like free-stream fluctuations as well as surface imperfections do not trigger laminar-turbulent transition. If certain properties of the laminar boundary layer are present—basically the critical Reynolds number and the inflection-point of the stream-wise velocity profile—the boundary layer becomes unstable. Then disturbances are picked up and transition occurs from the laminar into the turbulent state of the boundary layer.

The transition zone—the intermittency region—has a finite length, depending on the kind and magnitude of the disturbances, but also on the flow properties of the boundary layer. Different transition pathways to the turbulent state exist, in particular the Tollmien-Schlichting path and the cross-flow path.

The literature on boundary-layer instability and transition is manifold. Stability and transition in shear flows is treated, for instance, in [43]. Recent results, in particular also regarding high speed and hypersonic cases were presented at the IUTAM symposium on laminar-turbulent transition [46]. Notable are the activities of the NATO STO AVT-240 Task Group on Hypersonic Boundary-Layer Transition Prediction [47].

Relevant flight-vehicle effects, like attachment-line and three-dimensional transition, the thermal state of the surface, shock/boundary-layer interaction, high-temperature real-gas effects etc., are sketched in [1].

A number of theoretical approaches to the prediction of instability and transition encompasses the solutions of the Rayleigh equation, the Orr-Sommerfeld equation, empirical, semi-empirical and non-empirical non-local, non-linear methods.

Empirical and semi-empirical methods are not imaginable due to the many factors involved in the instability and transition processes. These are the flight-vehicle configuration-induced flow field—steady or quasi-steady flight assumed—with two- and three-dimensional portions, attachment lines, the entropy layer, the thermal state of the surface and its properties, Sect. 3.2.3, the boundary layer with possible shock/boundary-layer interactions, surface-curvature effects, Görtler instability, high-temperature real-gas effects and also the possibility of relaminarization.

Finally to be mentioned is the receptivity problem regarding free-stream fluctuations of all kinds and also surface imperfections, for instance expansion joints.[10] Surface imperfections include also vibration—not necessarily flutter—of the surface sheeting, and, on a large scale aerothermoelastic deformations of the forebody, in particular of large API-type 5 flight vehicles. Data of course can be obtained from ground-facility and also flight experiments, but not to the needed degree of fidelity and completeness.

Non-empirical methods, in particular non-linear non-local approaches, simulate the whole transition process. These are the Parabolized Stability Equations (PSE)—with limitation—and the most general Direct Numerical Simulation (DNS). The PSE approach goes back to T. Herbert, [48], and U. Dallmann, [49], see for instance the historical sketch in [50].

Both types of methods have reached a high degree of maturity. DNS even permits to take into account free-stream fluctuations, i.e., the atmospheric disturbance environment. The high computer power needed eventually will pose no obstacle for industrial design purposes, because of the critical dependence of the hypersonic flight-vehicle performance on the proper prediction of the transition location, shape and extent, Sect. 3.11.1.

However, a versatile, exact and reliable method for the prediction of laminar-turbulent transition in hypersonic flow fields past realistic flight-vehicle configurations seems not yet to be available.

---

[10] In Sect. 6.3.2 it is noted that the lower stage of the TSTO-system SÄNGER has a ceramic nose section followed by a titanium skin surface, with an expansion joint between them. Regarding laminar-turbulent transition that can lead to very complex situations. At certain flight conditions the expansion joint may trigger laminar-turbulent transition, which along the lower side of the long forebody may undergo relaminarization and further downstream may undergo plain laminar-turbulent transition. Presently it cannot be imagined how to treat such laminar-turbulent transition mechanism.

3.11.4 Hypersonic Turbulent Boundary-Layer Flow

Once hypersonic attached viscous flow can be considered as turbulent—after location, shape and extent of the laminar-turbulent transition zone somehow have been established—it should be possible to compute the properties of such flow to a reasonable degree of accuracy. The methods generally are transport-equation methods, i.e., Reynolds-averaged Navier-Stokes (RANS) equations with closure expressions. Unsteady flow is treated with the unsteady Reynolds-averaged Navier-Stokes (URANS) equations. The demands on the numerical simulation are the same as for laminar flow, Sect. 3.11.2. Very important is the modeling of the surface properties, Sect. 3.2.3, in particular also with regard to surface roughness, waviness, steps etc., which strongly enhance the wall shear stress, see, e.g., [1].

For boundary-layer edge Mach numbers $M_\infty \lessapprox 5$ Morkovin's hypothesis holds that density fluctuations must not be taken into account. The dynamics of these flows follow the incompressible pattern, see, e.g., [51]. However, a number of flow situations exists, where Morkovin's hypothesis does not hold, for instance in shock/boundary-layer interactions, see, e.g., [14].

At Mach numbers above $M_\infty \gtrapprox 5$ the picture changes. Density fluctuations must be taken into account. They are in principle treatable with the Favre-Averaging Navier-Stokes (FANS) equations. A bridging model are the Partially-Averaged Navier-Stokes (PANS) equations, [52]. The wall-temperature condition, the radiation-adiabatic temperature as conservative condition, needs attention, because it leads to steep temperature and density gradients. Once high-temperature real-gas effects are present, also turbulent heat conduction and turbulent mass diffusion may play a role. Other critical flow issues arise with corner flow and with surface imperfections, see, e.g., [14].

A discussion of basic research regarding turbulent hypersonic flow can be found in [52]. The effect of the wall temperature and that of the Mach number were investigated with direct numerical simulation (DNS) for instance in [53, 54].

An often found but inaccurate consideration is that a turbulent boundary layer has a smooth outer edge. This is because RANS solutions do suggest it. A laminar boundary layer, or shear layer (vortex layer) in general, has a smooth edge—with laminar boundary layers this is chosen to be the 99% of the external flow thickness. Turbulent boundary layers and shear layers, as well as vortices in reality have rugged unsteady edges, which cannot be described with RANS or URANS methods.

In general RANS and URANS methods are well suited to compute attached turbulent boundary layers. But when structural panel flutter of the flight vehicles structure is an issue, or impingement of shear layers or vortices on (downstream located) configuration elements, scale-resolving methods, see, e.g., [55], like large-eddy-simulation (LES) or hybrid RANS-URANS/LES approaches become necessary. These methods permit to determine dynamic loads and/or structural vibration excitation.

For a DNS study of pressure fluctuations in turbulent boundary layers see for instance [56]. Prediction issues in general and closure models for high-speed flow are presented also in [57].

3.11.5 Turbulent Flow Separation

A special problem is separation of turbulent boundary-layer flow with and without vortex shedding. Separation and reattachment happens with formation of recirculation zones. A problem is the determination of the drag increment due to the separation. Another problem is that induced pressure fluctuations can lead to critical fluid-structure interaction, i.e., buffeting, which, like panel flutter, also can be a problem in the hypersonic flight domain.

The experimental determination of such flow and its effects is problematic. Computational simulation with the RANS/URANS approach fails. Scale-resolving methods like large-eddy simulation (LES) in principle permit solutions of such problems. However, because of their need of high computer power and very large storage capacity they are not of practical use up to now.

To overcome these obstacles hybrid RANS-URANS/LES approaches have been developed and applied with good success. We show in Fig. 3.39 as an example the application of such a method to a free-flight configuration of a rocket with a bluff base and the jet of the rocket engine [58].

From the vehicle nose downstream a RANS solution is applied, in a RANS-LES overlapping zone synthetic turbulence is inserted and downstream the LES solution takes over [59]. In this particular study to determine the asymmetrical dynamic loads on the nozzle structure the Dynamic Mode Decomposition (DMD)—a dimensionality reduction algorithm—is applied.

In [60] the reader will find further results of experimental and numerical investigations of the turbulent wake flow of a similar case but at higher Mach numbers $M_\infty = 3$ and $M_\infty = 6$.

The hybrid approach also needs high computer power and large storage capacity but appears to be the ultimate means to overcome the computational simulation problem of turbulent separation.

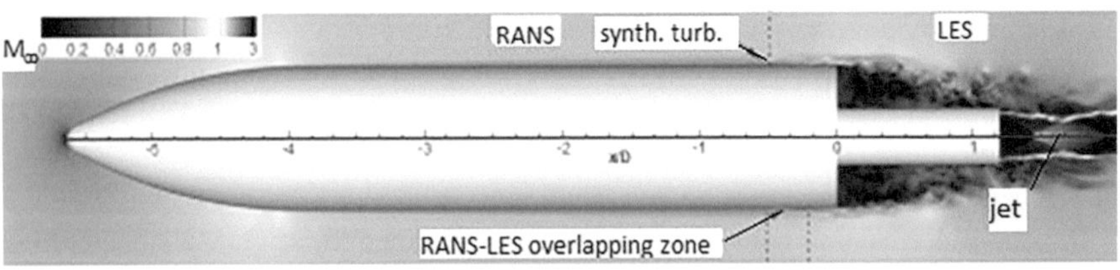

Fig. 3.39 RANS/LES simulation of rocket base flow at $M_\infty = 0.8$ and $Re_D = 0.6 \cdot 10^6$

3.12 Aerothermodynamic Issues of Upper-Stage Integration and Release

In Sect. 2.7 it was shown that with staging—there with an airbreathing lower stage of a TSTO system—the launch-mass problem can be relaxed. Staging is a well-established approach with vertical take-off systems, i.e., rockets. Semi-reusable systems like the Space-Shuttle, the BURAN with only one flight, the not realized HERMES and others were also staged vertical take-off systems.

Horizontal take-off systems as single-stage and two-stage systems were studied in the 1980s and 1990s, Sect. 1.1. Here we present insight and data from our reference system of Chap. 2, the API-type 5 TSTO-system SÄNGER, which was introduced in Sect. 2.8.

The matter of upper-stage integration and separation of that system is discussed in Chap. 4 of [5], a short presentation of the trajectory issues, based on [61], can be found in [2]. Aspects of stage-separation aerothermodynamics are discussed for instance in [62].

The ascent trajectory of the SÄNGER system is sketched in the left part of Fig. 2.13 in Chap. 2. A view of the TSTO-configuration is given in Fig. 3.40.

Shortly considered are five particular aerothermodynamic problems, which must be coped with during the cruise flight of the TSTO-system up to upper-stage release.

1. As was sketched in Sect. 3.2.3, external flight-vehicle surfaces need a coating with a radiation-emissivity coefficient which permits surface radiation cooling to the needed degree. The lower stage has a high-emissivity coating all over. The

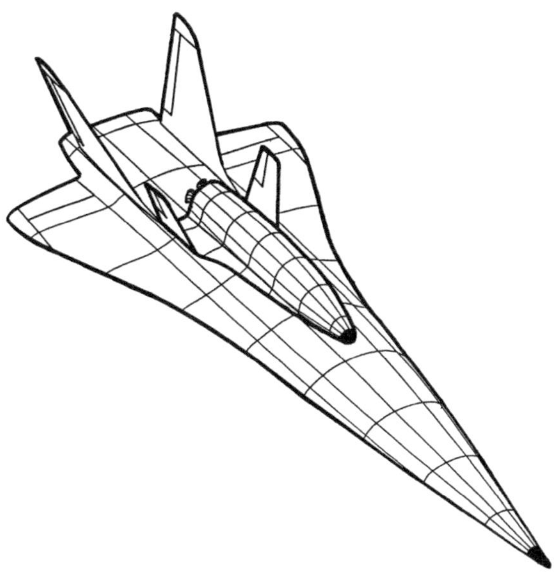

Fig. 3.40 View of the TSTO-configuration SÄNGER [62]

upper stage HORUS should have a low-emissivity coating at the upper side—see the examples in Fig. 3.2—in order to permit in orbit the appropriate thermal management related to Solar radiation heating.
During the transfer flight of about—nominal—40 min to stage separation the upper side of the upper stage is exposed to the same conditions as that of the lower stage. Therefore it should have a high-emissivity coating like that, which contradicts the low-emissivity need in orbit. It is a matter of very accurate and careful determination of the actual situation in order to find the optimum solution to that problem.

2. Fig. 3.40 shows the TSTO-system in its nominal shape, the upper stage without the afterbody fairing, which is necessary to cope with the base-drag problem in the transonic flight regime, Sect. 3.6.4. The afterbody fairing shown in Fig. 3.26 is a possible solution. The problem is how to remove it after the transonic flight regime has been passed. The idea of a simple jettisoning was discarded. A low systems-complexity, low-risk solution is needed.
3. Because of the on-top arrangement of the TSTO-configuration, Fig. 3.40, the interaction of the orbiter fin-flow with the lower-stage fins is a critical issue. Figure 3.41 as an example shows that with increasing flight Mach number an increasing interaction is to be expected.
Even if the shown case has a rather high side-slip angle, the problems of shock/shock interaction, hot-spot situations and buffeting phenomena at the fins of the lower stage are present and must be coped with, in particular at the high flight Mach numbers.
4. Non-convex effects reduce the radiation-cooling effectiveness, in particular at the feet of vertical fins as discussed in Sect. 3.9. The effect of course is present also at the fins of the upper stage during the cruise flight up to its release into orbit. The

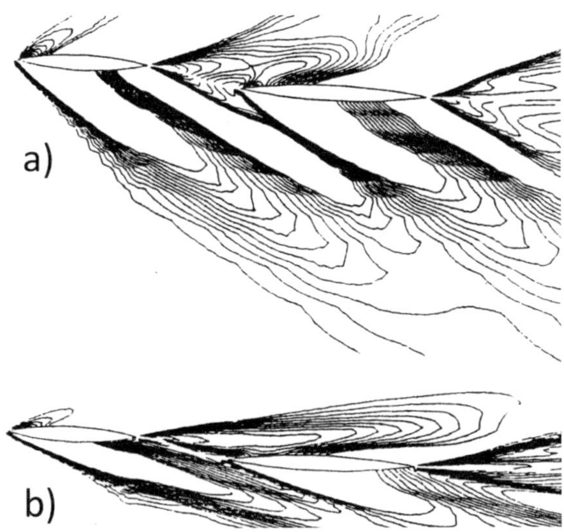

Fig. 3.41 Results of a numerical study of fin-fin flow interaction at the left-hand side of the TSTO-configuration, Fig. 3.40, [62]. Euler solution, **a** $M_\infty = 2$, **b** $M_\infty = 4.5$, side-slip angle in both cases $\beta = 6$

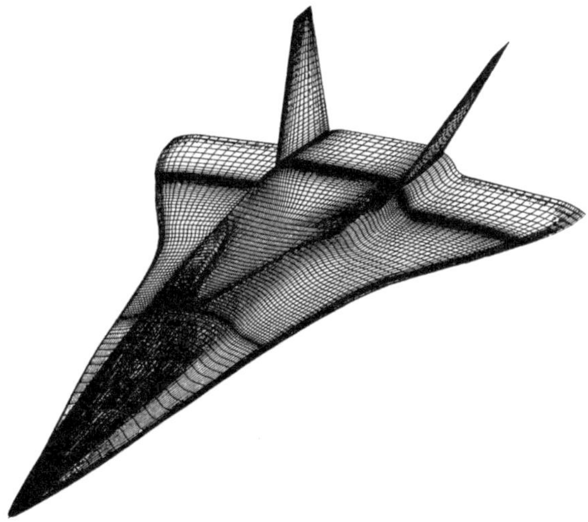

Fig. 3.42 View of the upper side of the lower stage with the trough [62]

ensuing thermal loads must be determined in detail and regarded in the structural layout of the upper stage.

5. The upper stage HORUS during the cruise flight of the TSTO-system sits in a trough at the upper side of the lower stage, Fig. 3.42. The shape of this trough is determined by the shape of the lower side of HORUS. After the release of HORUS the trough represents a part of the upper side of the lower stage.
   The flow over the trough is characterized by a number of interaction phenomena, which must be regarded in view of the structural layout of the lower stage, see, e.g., [62].

The separation process of the upper stage HORUS consists of three stages, Fig. 3.43: 1. Transition from the cruise flight to a pull-up condition, 2. Elevation of the upper stage and preparation for separation, 3. Release of the upper stage.

The pull-up as initiation of the stage separation maneuver is made with a flight-path angle of $\gamma \approx 6°$ and an angle of attack α up to 15° of the lower stage. The upper stage then is mechanically elevated to an angle of 8° related to the lower stage, Fig. 3.44.

The ensuing flow field between the two stages is complex. The configuration is an open channel between the upper side of the lower stage, the trough, and the lower side of the upper stage. Figure 3.45 shows how the bow shock of the upper stage is reflected at the trough's surface and then hits the lower side of the upper stage. There it is reflected again and a complex, partly three-dimensional inviscid/viscous flow field results. The flow field in Fig. 3.45 was calculated with a RANS method with a wind-tunnel flow situation $M_\infty = 6$, $Re_{L,\infty} = 1.13 \cdot 10^6$ with the model length $L = 0.5$ m, and $T_\infty = 242$ K. The angle of attack of the lower stage was $\alpha = 0°$, that of the upper stage $\Delta\alpha = 4°$. For the detailed discussion see [65] and [66].

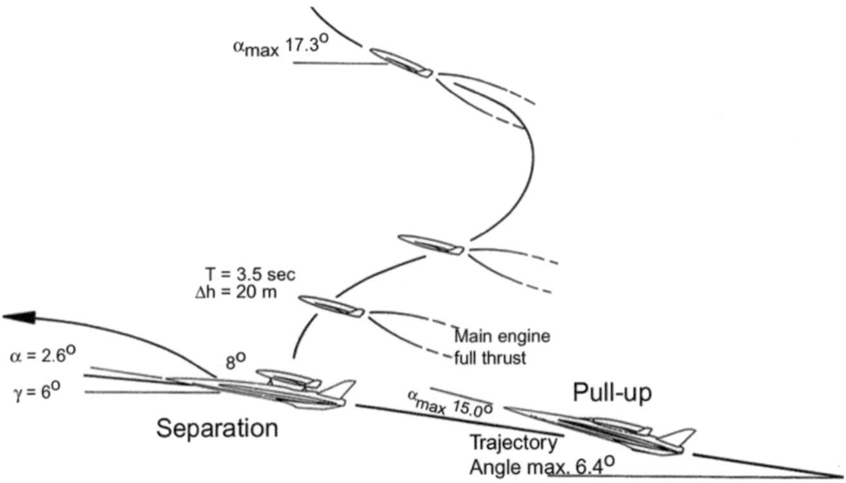

Fig. 3.43 The separation process [63]. Δh is the distance after separation

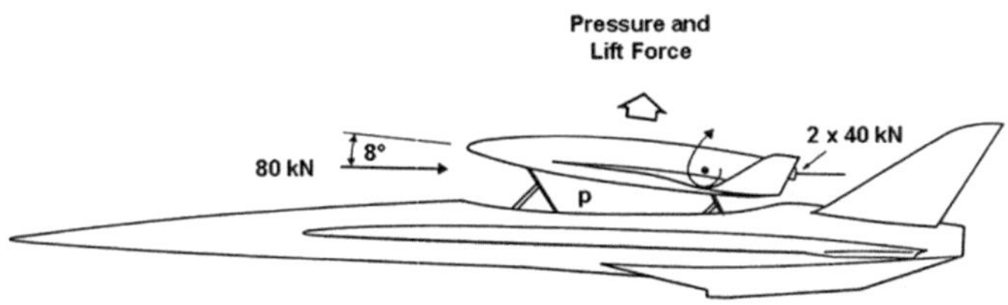

Fig. 3.44 Sketch of the elevation of the upper stage in preparation for release [64]

Fig. 3.45 Flow field in the center plane of the lower stage with the elevated upper stage [62]

The problem then is the induced high pressure field between the two vehicles, which leads to a lift force on the upper stage and a down force on the lower stage. This is in addition to the aerodynamic drag which acts on the elevated upper stage. That is to be compensated by the ignition of the engines of the orbital maneuvering system (OMS), Fig. 3.44. The repercussions on the lower stage due to the jets of the OMS are an induced pitching moment and locally increased thermal loads.

The separation process of the upper stage with its different elements is the result of a general optimization strategy. That had the following elements [5], see also the discussion of an example in [2]:

1. Upper stage release should be at $M_\infty \approx 6.8$ and an altitude $H \approx 35$ km. The release at a smaller Mach number would increase the mass and the size of the orbiter, now being 115 Mg. The Mach number is at the upper limit of ramjet propulsion, a higher Mach number would need a scramjet propulsion system for the lower stage, see also Sect. 2.10.3.
2. Ahead of the separation process the trajectory angle γ of the whole system should be around 6°.
3. The normal load factor n_z during stage separation should be zero. This means a flight along a parabolic arc and an angle of attack of the lower stage $\alpha \approx 0°$ until the start of the separation maneuver.
4. The bank, or roll angle μ_a—defined as negative to the left—is zero during stage separation (wing level condition), and then immediately is changed to about −90°. This is necessary in order to counteract the excess lift after the upper stage is released. The danger of a collision of the two stages is avoided, and the normal load factor constraint is not exceeded.

Figure 3.46 finally shows the upper-stage separation process in terms of its angle of attack and the roll angle as function of the flight time. The angle of attack rises at ≈ 2,600 s with the pull-up, then is reduced to nearly zero and for the return flight initially is elevated to 15°. The roll angle is zero up to the release of the upper stage, then is about 90° and finally reduced to about 64° for the return flight.

For the detailed discussion including a mission-abort scenario see [2, 5], for a case study of the whole flight trajectory.

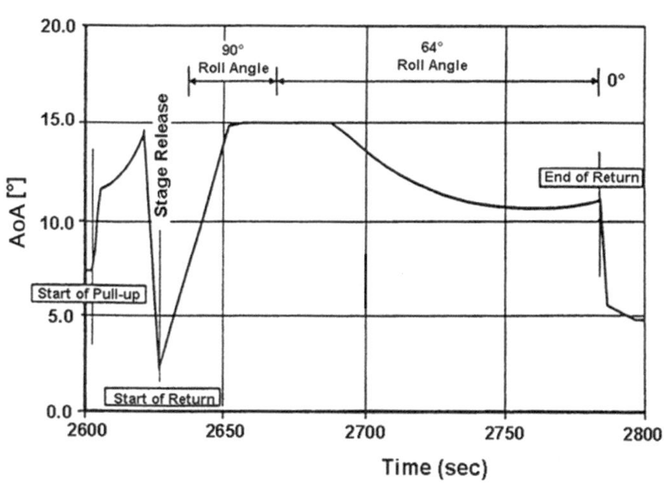

Fig. 3.46 Angle of attack AoA and roll angle during the separation process of the upper stage [67]

3.13 Problems

Problem 3.1 (*Galilean Invariance in Aerothermodynamics*) Explain why the flight situation (aircraft flying through the atmosphere) and the wind-tunnel situation (air flowing past a stationary vehicle) are compatible in the study of aerothermodynamics. Use the concept of Galilean invariance to support your answer.

Problem 3.2 (*Reynolds Number Change with Flight Mach Number Increase*) As flight Mach numbers increase, Reynolds numbers tend to decrease. Explain why this is the case, and describe how this affects the design of hypersonic flight vehicles.

Problem 3.3 (*Thermal Surface Effects on Boundary Layers*) Discuss how the surface temperature of a hypersonic flight vehicle affects the boundary-layer flow. Specifically, explain how the wall temperature influences skin-friction drag.

Problem 3.4 (*Similarity Rules in Hypersonic Aerodynamics*) Use the Oswatitsch Mach-number independence principle to explain why force coefficients, such as drag and pressure coefficients, become independent of Mach number at high speeds (e.g., $M_\infty > 5$).

Problem 3.5 (*Weiland rule*) What does the Weiland mapping rule say? Can it be applied to the airbreathing hypersonic flight vehicles considered here? Where lies its value?

Problem 3.6 (*Role of Surface Emissivity in Thermal Protection*) Given that surface radiation cooling is critical in hypersonic flight, describe how the emissivity of a vehicle's surface influences the thermal loads and the vehicle's overall performance.

Problem 3.7 (*Wave Drag and Thermal Loads at Nose and Leading Edges*) In the design of hypersonic flight vehicles, why must the bluntness of the nose and leading edges be carefully optimized? Discuss the trade-off between wave drag and thermal loads.

Problem 3.8 (*Properties of the surface of the external flow path*) What is demanded regarding the surface properties of the external flow path? What holds for the internal flow path?

Problem 3.9 (*Expansion joints*) What are expansion joints? Why are they needed. What influence can they have on a flow path?

Problem 3.10 (*Flow separation*) Separation of laminar and turbulent flow, the latter in massive form, is present in any flow field. What are the possibilities and problems to compute them with the needed accuracy?

References

1. Hirschel, E.H.: Basics of Aerothermodynamics: Second, Revised. Springer, Berlin (2015); Progress in Astronautics and Aeronautics, AIAA, Reston, Va, vol. 204. Springer, Heidelberg (2004)
2. Hirschel, E.H., Weiland, C.: Selected Aerothermodynamic Design Problems of Hypersonic Flight Vehicles. Progress in Astronautics and Aeronautics, AIAA, Reston, Va, vol. 229, Springer, Heidelberg (2009)
3. Weiland, C.: Aerothermodynamic Data Sets of Space Vehicles. Springer, Heidelberg (2014)
4. Anderson, J.D.: Hypersonic and High-Temperature Gas Dynamics. AIAA, Reston, Va (2002)
5. Kuczera, H., Sacher, P.W.: Reusable Space Transportation Systems. Springer, Berlin Heidelberg (2011)
6. Hirschel, E.H.: Aerothermodynamic phenomena and the design of atmospheric hypersonic airplanes. In: Bertin, J.J., Periaux, J., Ballmann, J. (eds.) Advances in Hypersonics: Defining the Hypersonic Environment, vol. 1, pp. 1–39. Birkh user, Boston (1992)
7. Staudacher, W.: Entwurfsproblematik luftatmender Raumtransportsysteme. Space Course Stuttgart (1995)
8. Perrier, P.C., Hirschel, E.H.: Vehicle Configurations and Aerothermodynamic Challenges. In: AGARD-CP–600, vol. 3, pp. C2-1–C2-16 (1997)
9. Oswatitsch, K.: Ähnlichkeitsgesetze für Hyperschallströmungen. ZAMP, vol. 2, pp. 249–264 (1951). Also: Similarity Laws for Hypersonic Flow. Royal Institute of Technology, Stockholm, Sweden, KTH-AERO TN 16 (1950)
10. Tsien, H.S.: Similarity laws of hypersonic flows. J. Math. Phys. **25**, 247–251 (1946)
11. Weiland, C.: Mechanics of Flow Similarities. Springer Nature (2020)
12. Weiland, C.: The aerodynamics of real space vehicles in the light of supersonic and hypersonic approximate theories. CEAS Space J. **12**(1), 85–96 (2020)
13. Weiland, C.: Aerodynamic similarity of winged and lifting space vehicles. J. Aircr. **58**(1), 17–29 (2021)
14. Hirschel, E.H., Cousteix, J., Kordulla, W.: Three-Dimensional Attached Viscous Flow. Springer, Heidelberg (2013)
15. Peebles, C.: Road to Mach 10–Lessons Learned from the X-43A Flight Research Program. Library of Flight, AIAA, Reston, Va (2008)
16. Detra, R.W., Kemp, N.H., Riddell, F.R.: Addendum to heat transfer to satellite vehicles reentering the atmosphere. Jet Propuls. **27**(12), 1256–1257 (1957)
17. Riley, C.J., DeJearnette, F.R.: Engineering aerodynamic heating method for hypersonic flow. J. Spacecr. Rocket. **29**(3), (1992)
18. Staff, A.R.: Equations, tables, and charts for compressible flow. NACA R-1135 (1953)
19. Hayes, W.D., Probstein, R.F.: Hypersonic Flow Theory: Inviscid Flows, vol. 1. Academic Press, New York, London (1966)
20. Liepmann, H.W., Roshko, A.: Elements of Gasdynamics. Dover Publications, Mineola, NY (2002)
21. Hirschel, E.H., Rizzi, A., Breitsamter, C., Staudacher, W.: Separated and Vortical Flow in Aircraft Wing Aerodynamics. Springer, Heidelberg (2021)
22. Nicolai, L.M.: Fundamentals of Aircraft Design. METS Inc, San Jose, Cal (1975)
23. Hauck, H.: Leitkonzept SÄNGER—Referenz-Daten-Buch. Issue 1, Revision 2. Dasa, Munchen/Ottobrunn, Germany (1993)
24. Wüthrich, S., Sawley, M.L., Perruchoud, G.: The coupled Euler/boundary-layer method as a design tool for hypersonic re-entry vehicles. Zeitschrift für Flugwissenschaften und Weltraumforschung (ZFW) **20**(3), 137–144 (1996)
25. Eggers, T.: Projekt highlift—high hypersonic L/D preliminary reference configurations. DLR-IB 124-2009/901 (2009)
26. Schmatz, M.A., Hold, R.K., Monnoyer, F., Mundt, C., Rieger, H., Wanie, K.M.: Numerical Methods for Aerodynamic Design, II. Space Course 1991, RWTH Aachen, pp. 62-1–62-40 (1991)

27. Hirschel, E.H.: Thermal surface effects in aerothermodynamics. In: Proceedings of the Third European Symposium on Aerothermodynamics for Space Vehicles, Noordwijk, The Netherlands, November 24–26, 1998. ESA SP-426, pp. 17–31 (1999)
28. Weihs, H.: Personal Communication. DLR Stuttgart (2022)
29. Hirschel, E.H., Koc, A., Riedelbauch, S.: Hypersonic flow past radiation-cooled surfaces. In: AIAA-Paper 91-5031 (1991)
30. Hold, R.K., Fornasier, L.: Investigation of thermal loads of hypersonic vehicles with emphasis on surface radiation effects. In: ICAS 94-4.4.1 (1994)
31. Hold, R.K.: Modeling of surface radiation effects by a fictitious emissivity coefficient. In: Kudriavtsev, V.V., Kleijn, C.R. (eds.) Computational Technologies for Fluid/Thermal/Chemical Systems with Industrial Applications, ASME PVP, vol. 397-1, pp. 201–208 (1999)
32. Simeonides, G.: Generalized reference-enthalpy formulation and simulation of viscous effects in hypersonic flow. Shock Waves **8**(3), 161–172 (1998)
33. Simeonides, G.: Extrapolation-to-flight of aerodynamic heating measurements and determination of in-flight radiation-equilibrium surface temperature in hypersonic/high enthalpy flow conditions. Shock Waves **16**, 25–34 (2006)
34. Georg, H.-U.: Personal communication. München (1996)
35. Häberle, J.: Einfluss heisser Oberflächen auf aerothermodynamische Flugeigenschaften von HOPPER/PHOENIX (Influence of hot surfaces on aerothermodynamic flight properties of HOPPER/PHOENIX). Diploma thesis, Institut für Aerodynamik und Gasdynamik, Universität Stuttgart, Germany (2004)
36. Romere, P.O., Whitnah, A.M.: Space shuttle entry longitudinal aerodynamic comparisons of flight 1–4 with preflight predictions. In: Arrington, J.P., Jones, J.J. (eds.) Shuttle Performance: Lessons Learned. NASA CP-2283, pp. 283–307 (1983)
37. Shea, J.F.: Report of the defense science board task force on the national aerospace plane (NASP). Office of the Under Secretary of Defense for Acquisition, Washington, D.C. (1988)
38. Berry, S., Nowak, R., Horvath, T.: Boundary-layer control for hypersonic airbreathing vehicles. In: AIAA-Paper 2004-2246 (2004)
39. Berry, S.A., DiFulvio, M., Kowalkowski, M.K.: Forced boundary-layer transition on X-43 (Hyper-X) in NASA LaRC 31-inch Mach 10 air tunnel. In: NASA/TM-2000-210315 (2000)
40. Schlichting, H., Gersten, K.: Boundary Layer Theory, 9th edn. Springer, Berlin, Heidelberg, New York (2006)
41. White, F.: Viscous Fluid Flow, 3rd edn. McGraw Hill Higher Education (2005)
42. Cebeci, T., Cousteix, J.: Modeling and Computation of Boundary-Layer Flows, 2nd edn. Horizons Publ, Long Beach, Springer, Berlin Heidelberg New York (2005)
43. Schmid, P.J., Henningson, D.S.: Stability and Transition in Shear Flows. Springer, New York, Berlin, Heidelberg (2001)
44. Schneider, S.P.: The development of hypersonic quiet tunnels. J. Spacecr. Rocket. **45**(4), 641–664 (2008)
45. Hirschel, E.H., Groh, A.: Wall compatibility condition for the solution of the Navier-Stokes equations. J. Comput. Phys. **53**(2), 346–350 (1984)
46. Sherwin, S., Schmid, P., Wu, X. (eds.): IUTAM laminar-turbulent transition. In: Proceedings of the 9th IUTAM Symposium, London, UK, September 2–6, 2019. Springer Nature, Cham (2022)
47. N.N.: Summary report of the NATO STO AVT-240 task group on hypersonic boundary-layer transition prediction. TR-AVT-240 (2020)
48. Herbert, T., Bertolotti, F.P.: Stability of nonparallel boundary layers. Bull. Am. Phys. Soc. **32**, 2079 (1987)
49. Simen, M., Dallmann, U.: Nonlocal vs. local instability of compressible flows including body metric, flow divergence and 3d-wave propagation. In: AIAA-Paper 93-2982 (1993)
50. Hein, S.: Nonlinear nonlocal transition analysis. Doctoral thesis, University Stuttgart, Germany (2004)
51. Smits, A.J., Dussauge, J.-P.: Turbulent Shear Layers in Supersonic Flow, 2nd edn. Springer, New York (2010)

52. Smits, A.J., Martin, M.P., Girimaji, S.: Current status of basic research in hypersonic turbulence. In: 39th AIAA Fluid Dynamics Conference and Exhibit, January 5–8, 2009, Orlando FL (2009)
53. Duan, L., Beekman, I., Martin, M.P.: Direct numerical simulation of hypersonic turbulent boundary layers. Part 2. Effect of wall temperature. J. Fluid Mech. **655**, 419–445 (2010)
54. Duan, L., Beekman, I., Martin, M.P.: Direct numerical simulation of hypersonic turbulent boundary layers. Part 3. Effect of Mach number. J. Fluid Mech. **672**, 245–267 (2011)
55. Spalart, P.R.: Detached-Eddy Simulation. Annu. Rev. Fluid Mech. **41**, 181–202 (2009). Palo Alto
56. Duan, L., Choudhari, M.M., Zhang, C.: Pressure fluctuations induced by a hypersonic turbulent boundary layer. J. Fluid Mech. **804**, 578–607 (2016)
57. Gatski, T.B., Bonnet, J.-P.: Compressibility, Turbulence and High Speed Flow. Elsevier (2009)
58. Schröder, W.: Personal communication (2022)
59. Statnikov, V., Meinke, M., Schröder, W.: Reduced-order analysis of buffet flow of space launchers. J. Fluid Mech. **815**, 1–25 (2017)
60. Statnikov, V., Stephan, S., Pausch, K., Meinke, M., Radespiel, R., Schröder, W.: Experimental and numerical investigations of the turbulent wake flow of a generic space launcher at M_∞ = 3 and $M_\infty = 6$. In: 6th European Conference for Aeronautics and Space Sciences (EUCASS), Krakow, July 2015; CEAS Space J. **8**(2), 101–116 (2016)
61. Bayer, R., Sachs, G.: Optimal return-to-base cruise of hypersonic carrier vehicles. Zeitschrift für Flugwissenschaften und Weltraumforschung (ZFW) **19**(1), 47–54 (1992)
62. Weiland, C.: Stage separation aerothermodynamics. In: AGARD-R-813, Aerothermodynamics and Propulsion Integration for Hypersonic Vehicles, pp. 11-1–11-28 (1996)
63. Kuczera, H.: Konzeptstudie Sänger und Fortführung. Workshop Hypersonics Technology, Spitzingsee, Germany (1992)
64. Koelle, D.E., Kuczera, H.: Sänger space transportation system. Progress Report 1989. IAF-89-217, October 1989, Torremolino, Spain (1989)
65. Schröder, W., Behr, R., Menne, S.: Analysis of hypersonic flows around space transportation systems with CFD methods. In: AIAA-Paper 93-5067 (1993)
66. Schröder, W., Hartmann, G.: Analysis of inviscid and viscous hypersonic flow past a two-stage spacecraft. J. Spacecr. Rocket. **30**, 8–13 (1993)
67. Kuczera, H. (ed.): Sänger space transportation system, concept study and system definition. Final Report (Phase Ia, January 1989 until August 1990). BMFT, Referat 513, Reference No. HK 8900-8. Dezember 1991, Bonn, Germany (1991)

Chapter 4
Basics of Ramjet and Scramjet Propulsion

This chapter is devoted to a consideration of basics of ramjet and in particular scramjet propulsion. In the background mainly is the propulsion of airframe-propulsion integration (API-) type 4 and 5 flight vehicles.

The determining parameters of hypersonic airbreathers are flight speed, respective flight Mach number, flight altitude, Reynolds number, dynamic pressure, flight vehicle shape, size, propulsion mode, fuel-air equivalence ratio and others.

Ramjet propulsion takes over from turbojet propulsion at flight Mach numbers around $M_\infty = 3.5$ to 4. The Mach number $M_\infty \approx 7$ today appears to be the upper limit of this propulsion means. Then scramjet propulsion is the means of airbreathing propulsion.

We first have a look at the operational regimes. Then ramjet and scramjet propulsion are characterized. A discussion of general design issues follows.

Results of a numerical scramjet investigation with a simplified configuration then give insight into major issues of this propulsion means. The topic of propulsion integration, i.e., the all-deciding linkage of the propulsion system and the airframe is the topic of Chap. 5. There a large API-type 5 airbreather is at the center of the considerations.

4.1 Operational Regimes: Flight Mach Number and Flight Altitude

The propulsion modes of airbreathing hypersonic flight vehicles of CAV and ARV type are turbojets, ramjets and scramjets. Figure 4.1 schematically gives the range of application of these propulsion means together with the dynamic pressure limits.[1]

[1] See in this regard also Fig. 2.7, because with constant Mach number over altitude the flight speed is not constant.

E. H. Hirschel et al., *Elements of Hypersonic Airbreather Design and Development*,
https://doi.org/10.1007/978-3-031-94219-8_4

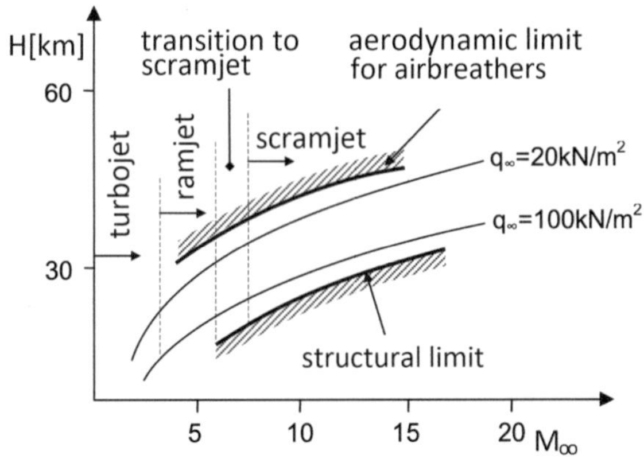

Fig. 4.1 The altitude-Mach number diagram with the flight corridor of the propulsion means [1]

The dynamic pressure

$$q_\infty = \frac{1}{2}\rho_\infty v_\infty^2 = \frac{1}{2}\gamma p_\infty M_\infty^2, \tag{4.1}$$

is a measure of the aerodynamic forces coefficients for any flight vehicle and also for the pressure loads on the vehicle structure and the performance of the propulsion system. With q_∞ too high, structural load limits are exceeded. With q_∞ too low, disproportional large aerodynamic surfaces for lift and control are needed. Further affected is the pressure in the combustion chamber of the engine, such that the efficiency of combustion is impaired.

The turbojet domain extends up to $M_\infty \approx 3.5$, because then the thermal and the pressure loads in the engine become untenable. Total pressure issues to a degree can be alleviated by flying at lower dynamic pressure, i.e., at higher altitude, but total temperature increases with increasing flight speed.

Above $M_\infty \approx 3.5$ the operation domain of the ramjet begins, which reaches up to $M_\infty \approx 7$. Because the ramjet, like the turbojet, reduces the internal flow velocity to subsonic velocity, again the thermal and the pressure loads become high with high flight speeds, and oxygen dissociation begins to occur, which hampers the combustion process.

For flight Mach numbers higher than $M_\infty \approx 7$ and up to $M_\infty \approx 12$, see Sect. 2.1.2, the way out is the scramjet, which features supersonic combustion. The flow through the engine is completely supersonic, with very short residence time. This puts tremendous demands on fuel injection, vaporization, mixing, ignition, and combustion.

A number of variants and combinations of the three propulsion modes has been investigated and partly is in service. Examples are the Air Turbo Rocket (ATR), also called turboramjet, and the Dual Mode Ramjet (DMR), also called Dual Combustion Ramjet (DCR). The latter is a ramjet engine operating in both subsonic and supersonic combustion mode. Necessary is a variable engine geometry, which in the ram mode provides subsonic flow in the combustor, and in the scram mode supersonic flow, see, e.g., [2].

The interested reader is referred to the significant literature on the subject, see, e.g., [3–9]. Everywhere basic considerations can be found, also references to further reading. Interesting is that in [8] the Mach number domains of the propulsion means strongly differ from the usually accepted ones.

As a trajectory example we take the ascent trajectory of the design reference mission of the TSTO-system SÄNGER, Sect. 2.8, whose lower stage is an API-type 5 system. Even if the scramjet domain is not covered, that example shows the high demands on the propulsion system and its components. The critical components are the inlet, the turbo engine, the ram combustor and the nozzle of SERN type, see the compact discussion in [10]. We do not consider the airframe-propulsion integration problem.

The inlet design has to take into account the highest flight Mach number, in our case the upper stage release Mach number $M_\infty = 6.8$ with ramjet propulsion. The transonic flight domain is determining the design of the turbojet engine. This operational pinch-point needs thrust augmentation with afterburner.[2] The design of the afterburner and the ramjet combustion chamber however is driven by the Mach number $M_\infty \approx 3.5$, at which the transition from turbo- to ramjet-operation is to happen.

Critical engine design parameters are shown in Fig. 4.2. They concern the noise and the sonic boom at the beginning of the mission—mission start was foreseen in

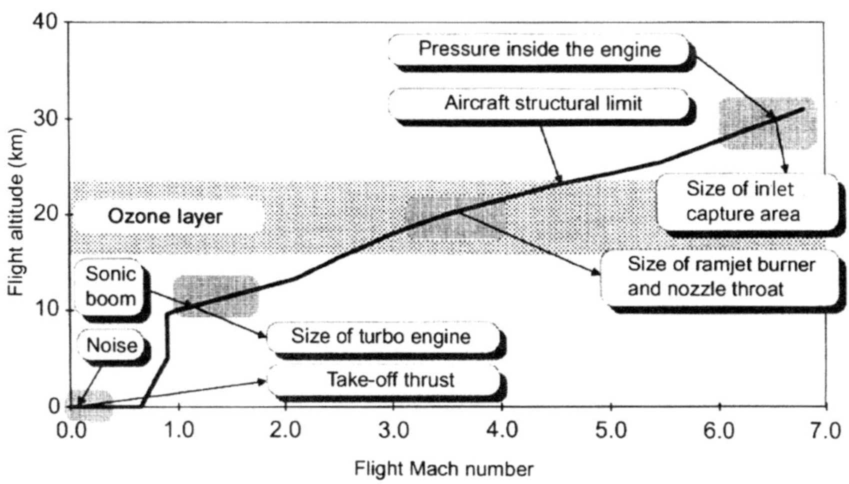

Fig. 4.2 Critical design parameters during the ascent trajectory of the two-stage-to-orbit space transportation system SÄNGER [11]

[2] For the three acceleration pinch-points of the reference trajectory see Fig. 5.36.

the south of Europe—, both determining the take-off thrust, hence the size of the turbo engine. In addition, the vulnerable ozone layer, see Sect. 2.4, has to be crossed with minimum harm to it.

Around $M_\infty \approx 3.5$ the switch from the turbojet to the ramjet is to happen. Critical are the size of the ramjet burner, the nozzle throat and geometry, and the switching mechanism. At the higher Mach numbers the structural limits—pressure and thermal loads—become critical as well as the size of the inlet capture area.

4.2 Brief Characterization of Ramjet and Scramjet Propulsion

Ramjet and scramjet propulsion are intriguing propulsion means. They work without any moving parts, in contrast to reciprocating engines and turbo-jet engines. We give short characterizations following [6]. The literature on this topic was cited above. A recent overview of propulsion cycles can be found in [9].

Details of design issues are treated in the following Sect. 4.3. Basic information about the flight regimes and the flight environment in view of the propulsion means is given in Sects. 2.1.2 and 2.4. Airframe-propulsion integration is the topic of Chap. 5.

4.2.1 Ramjet Propulsion

The ramjet engine simply is a flow channel, consisting of the inlet, the isolator, the combustion chamber with the fuel injector, and the nozzle, Fig. 4.3 in API-type 4 configuration like the next two figures.

In the inlet—we take the usual view with the study object fixed—the flow is slowed down to subsonic speed. In the combustion chamber fuel is added, combustion occurs, the air is heated up and then expanded through a convergent-divergent nozzle to supersonic speed. In this way engine thrust is generated.

At take-off and low flight speeds the pile-up of the air is not sufficient for the engine to generate thrust. Hence it needs to be boosted by, for instance, a turbojet or a rocket to at least high subsonic speed. The optimum performance of the ramjet is found at flight Mach numbers $M_\infty \approx 3$ to 7.

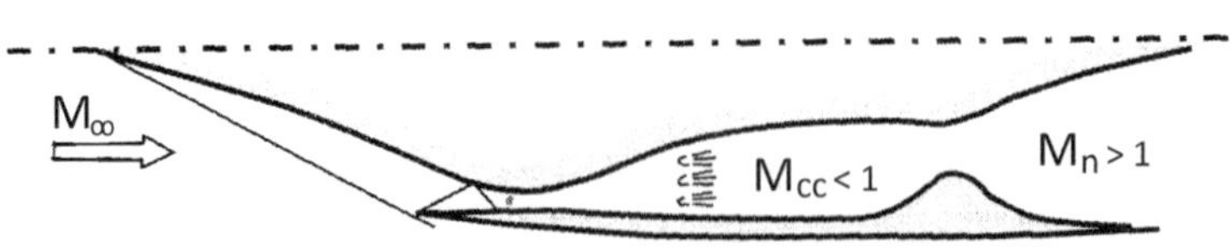

Fig. 4.3 Basic shape of a ramjet engine [6]

At the higher flight Mach numbers the temperature of the air in the engine becomes so high that dissociation of the oxygen molecules happens, when the flow is decelerated to subsonic speed in the combustion chamber. That limits the combustion performance. For the operational limits in terms of the dynamic pressure q_∞ see Fig. 4.1. The thermal and mechanical loads are limiting factors, too.

The way out at the high flight Mach numbers is the scramjet engine.

4.2.2 Scramjet Propulsion

Basically scramjet engines are similar to ramjet engines, Fig. 4.4. The deciding difference is that the air flow is not decelerated to subsonic speed. The Mach number in the combustion chamber is $M_{cc} \approx 2$ to 4. The resulting temperatures are below $T_{cc} \approx 2{,}500$ K and the pressures below $p_{cc} \approx 8$ bar.

The main challenge is the insertion of the fuel into the high speed air stream, its mixing with the air and the supersonic combustion. For the fuel implications see Sect. 2.6. Different injection concepts are characterized in Sect. 4.3.7.

The nozzle now is a purely divergent, and therefore supersonic one. The resulting high speed flow leads to a strong decrease of pressure, density and the static temperature. Because of the very small residence time the corresponding recombination of dissociated gas components does not happen along the nozzle. This must be regarded when dealing with engine performance.

4.2.3 Detonation Wave Scramjet Engine

A particular type of engine is the detonation wave scramjet engine, Fig. 4.5. Fuel is injected at a station upstream of the inlet lip. It is mixed with the air and when the mixing passes the shock wave of the cowl lip it is ignited and burned.

The shock-wave induced combustion is at a very short distance behind the cowl-lip shock wave. Hence one speaks of a detonation wave, see, e.g., [7]. The advantage is that the usual combustor issues like fuel injection and mixing are eliminated and weight, cost etc. are reduced.

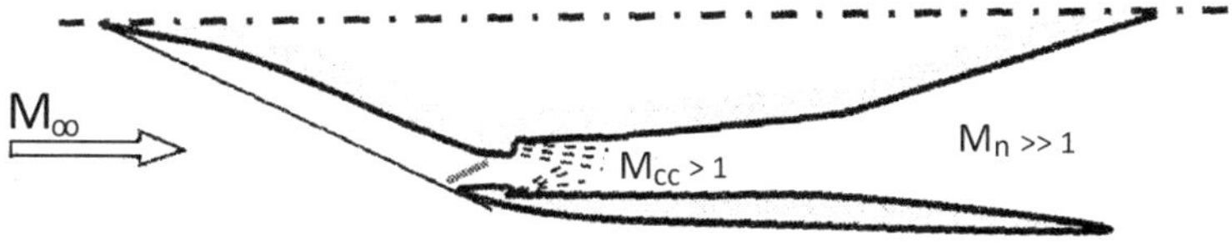

Fig. 4.4 Basic shape of a scramjet engine [6]

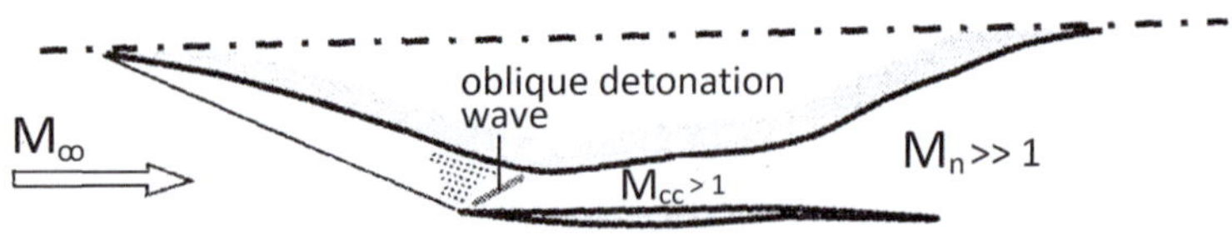

Fig. 4.5 Basic shape of a detonation wave scramjet engine [6]

4.3 General Engine Design Issues

General design issues are considered regarding both ramjet and scramjet propulsion. Figure 4.6 shows the major configuration components of an API-type 4 or 5 hypersonic flight vehicle. This integration type is determined by a forebody, whose lower—windward—side acts as a pre-compression surface. It follow the ramps of the external inlet flow path up to location C. The internal inlet flow path, together with the isolator, extends up to the beginning of the combustion chamber. The nozzle finally, here of SERN type, completes the configuration.

In the following sub-sections short characterizations of general design issues are given. They mainly concern API-type 4 and 5 flight vehicles, but to a high degree also any other airframe-propulsion integration type flight vehicle, in particular those of API-type 3.

4.3.1 Thermal Loads and Cooling Demands

Due to the high flight speeds of hypersonic flight extreme thermal loads occur. This regards the vehicle structure, see Chap. 6, as well as the propulsion system, in particular also the combustion chamber.

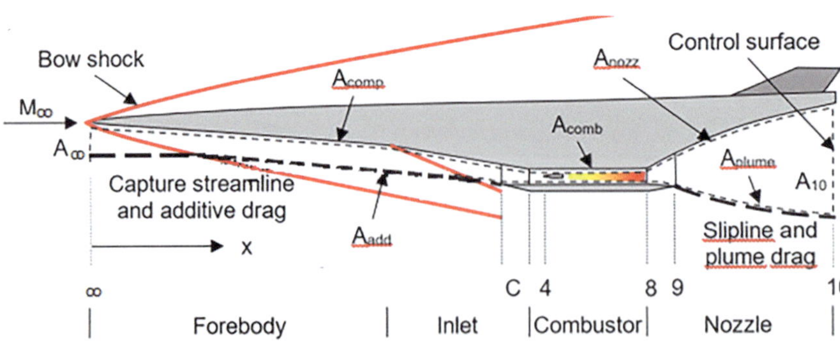

Fig. 4.6 The configuration components of a simplified API-type 4/5 hypersonic flight vehicle [1]

At the vehicle structure the forward stagnation point and its vicinity as well as the attachment lines at the wing and at the stabilization and trim surfaces are the critical areas with the highest thermal loads. Although all surfaces of the external flow path are radiation cooled, in these areas special materials are needed and in the extreme case active cooling measures. Materials which are able to withstand temperatures up to 2,500 K are carbon fibers in a silicon carbide matrix and also ultra-high temperature ceramics.

The surfaces of the internal flow path, beginning at the inlet—where radiation cooling is inhibited—are actively cooled. This of course also holds for the combustion chamber and partly also for the nozzle.

4.3.2 Forebody, External and Internal Compression

The lower side of the forebody of API-type 4 and 5 flight vehicles has the function to channel the needed air stream through the inlet into the combustion chamber of the engine. That is to happen with a compression of the air flow via shock waves and with the smallest possible total-pressure loss and the smallest aerodynamic drag regarding the flight vehicle.

At the same time the forebody function must work in a large operating range and must provide uniform and steady flow at the entrance of the combustion chamber. Aerothermoelasticity of the forebody and gust control play an important role, see Sect. 5.5.

Scramjet propulsion systems (supersonic combustion) due to the higher flight Mach number are able to achieve the needed compression with very slender inlet configurations. They further as a rule work with higher internal compression [12]. This is in contrast to ramjet propulsion systems (subsonic combustion). The external inlet parts of scramjet propulsion systems are radiation cooled and usually do not have suction (bleed) for boundary-layer control.

Figure 4.7 shows a typical two-dimensional configuration. The flow past the forebody and the compression ramps up to the inlet lip (location C) represents the external compression. The internal compression happens between the locations C and 4'. Basically all inlet configurations have the same boundary conditions and flow phenomena, see Sects. 5.2 and 5.5.

With hypersonic airbreathers it is necessary to find a compromise between an effective forebody compression and thermal loads. The latter in particular regards all leading edges, in our case the unswept leading edge of the X43-A, see Sect. 3.5. Small aerodynamic drag demands small nose radii, which leads to very high thermal loads. Hence special requirements arise regarding the material of the nose, see Sect. 6.4.2.

With long flight duration it might be necessary to employ active cooling. The thermal state of the surface is also of interest regarding the infrared signature of the flight vehicle in terms of its stealth capability in view of military applications.

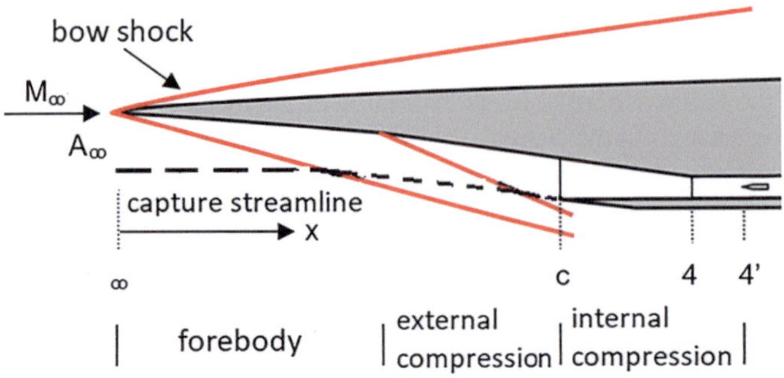

Fig. 4.7 Schematic of a typical API-type 4 forebody with external inlet ramps and internal compression [1]

Regarding the ramp flow, shock-boundary-layer interaction, also too large positive pressure gradients can lead to boundary-layer separation, Sect. 3.6.5. Locally the thermal loads can increase distinctly. Unsteady behavior may lead to acoustic loads, which pose special problems.

All that must be seen in view of the fact, that location, shape and extent of the laminar-turbulent transition zone of the boundary layer plays a deciding role. Today it is still not possible to predict that with the needed accuracy and reliability, Sect. 3.11.

A special boundary-layer phenomenon is the upstream propagation of disturbances, which can unstart the inlet, Sect. 4.3.5. That by all means must be avoided.

It is essential for the modeling and the simulation of an inlet that the shock waves induced by the flight vehicle‘s nose and the ramps are properly characterized. The influence of any viscous phenomenon must be taken into account. All that defines the capture characteristic of the inlet and hence the total-pressure recovery.

This too holds for the properties of the onset flow into the combustion chamber. If in the simulation uncertainties are present of the onset flow—due to an improper description of the bow shock and the ramp shocks including the boundary layer influence—all results regarding the combustion process are questionable. This regards both ramjet and scramjet combustion chambers.

Besides the influence of the location, shape and extent of the laminar-turbulent transition zone also that of high-temperature real-gas effects present in the flow past the forebody and the inlet must be regarded. Between flight speeds of 1 km/s and and about 4.5 km/s, equivalent to Mach numbers about 4–12, oxygen and nitrogen vibration excitation and also oxygen dissociation play a role, Sect. 2.4.

Above these flight speeds dissociation of nitrogen and also ionization become important as well. All these phenomena can occur in equilibrium or non-equilibrium. The reader should remember that ramjet propulsion has its Mach-number limit also due to the occurrence of oxygen dissociation at flight Mach numbers around six to seven.

The problem of a proper prediction of location, shape and extent of the laminar-turbulent transition zone was mentioned in Sect. 3.11. The transition phenomenon regards the wall shear stress, the displacement properties of the boundary layers, the separation phenomena and the thermal loads on the vehicle structure. Turbulent boundary-layer flow, however, withstands higher adverse pressure gradients without separation than laminar flow. Hence turbulent flow can increase the operational range of inlets.

Turbulent flow on the other hand leads to higher wall shear stress and to higher thermal loads than laminar flow. Wall cooling is positive insofar as it allows the boundary layer to cope with a higher adverse pressure gradient before it separates. It also reduces the size of separation regions.

Hence, not only a proper viscous flow modeling is important, but also that of the thermal state of the surface. The surfaces of the external flow path are—passively—radiation cooled, those of the internal flow path are actively cooled. For the radiation-cooled external flow path surfaces the radiation-adiabatic wall temperature T_{ra} can be used as conservative estimation of T_w.

The wall temperature governs the thermal surface effects [13]. These must be identified and treated accordingly, because they locally can lead to high thermal and pressure loads, Chap. 3. Cases are possible, where instead of the radiation-adiabatic wall temperature a more exact wall temperature must be determined with a coupled flow-structure cooling concept consideration.

Above a seemingly simple two-dimensional case was considered. The reality looks very different, see the discussion in Sect. 5.2. Especially complex shock-wave/boundary-layer interactions, side walls with corner flow, bleed, thermal loads, structural deformations etc. make the forebody analysis and design of in particular large hypersonic airbreathers of API-type 5 a formidable challenge [3, 5, 12, 14, 15].

4.3.3 Shock-Wave/Boundary-Layer Interactions

At the external inlet ramps, but also at trim and control surfaces, different types of shock-wave/boundary-layer interactions are present, Sect. 3.6.5. They are classified as Edney-type interactions, see, e.g., [13]. They lead to high pressure and thermal loads and need special attention from the structural side. Regarding the inlets of propulsion systems the shock-on-lip situation, Sect. 5.2, locally leads to high pressure and thermal loads and therefore should be avoided. It also can lead to boundary-layer separation in the internal flow path and to blockage of the inlet.

4.3.4 Inlet Isolator

Between the internal inlet part and the combustion chamber a constant area diffusor is placed, the inlet isolator, see, e.g., [3].

In the ramjet case a train of normal shock waves rises the pressure and reduces the Mach number to a subsonic value. In the scramjet case a train of oblique shocks leads to the pressure rise, but now with a supersonic Mach number at the entry to the combustion chamber, Fig. 4.8. In both cases shock/boundary-layer interactions at the walls play an important role.

The major roles of the inlet isolator are to increase static pressure ahead of the combustor and to prevent upstream disturbances from the combustion chamber to reach the internal inlet and cause inlet unstart.

4.3.5 Inlet Unstart

The inlet is being considered as "started", if it is in function as intended and if the inner flow field does not influence the—upstream—outer flow field as long as that is not strongly disturbed. That concerns any of the airframe-propulsion integration type flight vehicles. In [3] three cases of unstart are noted:

1. The free-stream Mach number is reduced to a sufficient degree below the starting Mach number. If that is the case, the inner flow field of the inlet with or without a shock wave is disgorged by the inlet, unstart is present. In [3] the example is given that an inlet starting at $M_\infty = 2$ with reducing free-stream Mach number will not unstart until $M_\infty = 1.57$ is reached.
   This difference means that an operating margin exists at the instant of inlet starting. However, also hypersonic airbreathers—unless being missiles—must decelerate at the end of their mission and land, which demands an adaptive inlet. Otherwise inlet unstart is expected to happen.

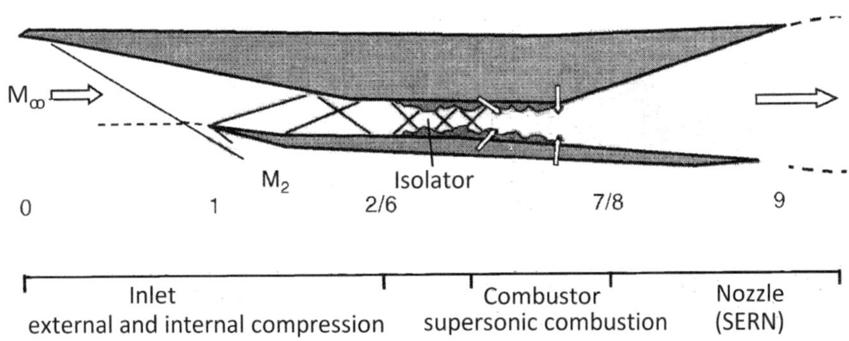

Fig. 4.8 Schematic of a scramjet inlet isolator [6]

2. The external inlet flow is disturbed by upstream effects such that it no longer can enter the inner passage of the inlet. Upstream effects can be unanticipated flow separation at the outer inlet ramps, detrimental changes of the flight attitude (slip, sudden change of angle of attack), exhaust gases from other aircraft. Effects like those demand design margins of the propulsion system.
3. The back pressure is increased. That can be due to excess heat release in the combustion chamber or if the exhaust-nozzle exit area is too small. It does not matter, whether such changes are deliberate or accidental, steady-state or transient, the effect will be the same.

Unstart in any case must be avoided, because not only for the loss of engine thrust, but also because large unsteady mechanical and thermal loads on the structure are occurring. In [3] a factor of ten for such loads is cited. The fuel system moreover will be unable to cope with the situation and overheating or even flame-out may occur.

4.3.6 *Thermal Blockage*

To avoid thermal and mechanical loads being too high, in scramjet propulsion systems supersonic flow speed is present all over. From Fanno or Rayleigh curves it is obvious that wall skin friction or heat addition due to combustion reduces supersonic flow towards the Mach number $M = 1$. Hence it is possible that thermal blockage can happen in the combustion chamber. The flow Mach number can drop below $M = 1$. Structural damage is possible and the inlet characteristic can be changed up to inlet unstart.

4.3.7 *Combustion Chamber and Combustion*

The task of the combustion chamber is to mix the fuel with the incoming air and to burn it up completely. Turbulence plays a large role in the combustion process [9]. The resulting hot gas then is expanded through the nozzle in order to create the thrust of the flight vehicle.

High flight velocities, however, cause a large increase of total-pressure loss once subsonic flow is demanded in the combustion chamber, as is the case in ramjet engines. Increased dissociation processes further lead to thrust losses, since full recovery of energy due to recombination is not possible in the nozzle because of the high flow velocity there and hence short residence times. This and the thermal loads issue limit ramjet operation to about $M_\infty \approx 7$.

In scramjet engines the flow in the engine is slowed down only to a degree that supersonic speed is present everywhere. The increase of energy by combustion of for instance, hydrogen, however is limited. For high flight Mach numbers it can be estimated [16]

$$\frac{\text{stoichiometric heat release } (\Phi = 1)}{\text{kinetic energy of the airstream}} \approx \frac{69}{M_\infty^2}. \tag{4.2}$$

The fuel-air equivalence ratio Φ can be considered in the following way:

- $\Phi = 1$—stoichiometric—means that fuel and air amounts are exactly present for a full combustion.
- $\Phi > 1$—over-stoichiometric—indicates that more fuel is present than needed (rich mixture). That in general leads to increased thrust.
- $\Phi < 1$—under-stoichiometric—indicates that less fuel is present than needed (lean mixture). That leads to decreased fuel consumption, i.e., increased fuel economy.

In [3] it is noted that obviously fuel-air ratios below one are likely to be encountered in practice.

Besides these theoretical, however idealized values other issues are of importance. These are the very high thermal loads present in a scramjet combustion chamber, which lead to limitations. The high flow speeds in the combustion chamber pose additional demands on fuel injection and mixing and further demand a fast and complete combustion.

The major challenge hence is to find for a given scramjet engine an injector concept, which permits to inject the needed fuel, its mixing with the air stream, its ignition and finally its combustion, see, e.g. [17].

The following concepts are the subject of studies and applications:

1. Wall jet, in HyShot, see, e.g., [18] and also Sect. 1.1,
2. Hypermixer, see, e.g., [19],
3. Strut in-stream injection, see, e.g., [20],
4. Cavities, step combustor, see, e.g., [21, 22],
5. Staged injection, see, e.g., [23].

The wall jet, i.e., wall injection, in contrast to concepts with installations in the combustion chamber, has the benefit of small disturbance of the general flow field. Total pressure losses in comparison are small. However, mixing effectiveness is rather low. Thermal loads on the wall of the combustion chamber on the other hand are high, because of the high fuel concentration near the wall.

Better fuel insertion into the flow field is given with ramp or with strut injection. The installed components, however, lead to extra total-pressure losses and if necessary demand active cooling of the injection elements.

To prevent thermal blockage step combustors can be used. Due to the flow separation a recirculation area is present behind the step, which leads to longer flow residence times. Fuel is inserted into this area, mixed and ignited. Resulting are high structural thermal loads. In order to increase the fuel part without the danger of thermal blockage multi-step injection is a possibility.

A combination of strut injection and subsequent wall injection for instance has the benefit of homogenous mixing both in the central flow part and in wall proximity. This benefit, however, requires a longer combustion chamber.

Injector concepts like ramp and strut injection, which of course demand additional built-in components, lead to better fuel/air mixing. However, with them come high total pressure losses and possibly also cooling demands of the injector components and possible blocking.

The fuel injection and the reaction kinetics of course are of highest importance [24]. But the same holds for viscous effects in the whole flow path. These are wall shear stresses, separation and displacement effects as well as heat transfer effects. All must be taken into account.

4.3.8 *Nozzle*

The nozzle, Fig. 4.9, generates the expansion of the combustion gases and produces the thrust of the engine.

Recombination effects in the combustion gases are of high importance. Freezing of reactions play a role as well as friction drag and over- or under expansion. In principle it holds that the longer the nozzle, the higher is the yield rate of thrust.

API-type 4 and 5 flight vehicles usually have a single expansion ramp nozzle (SERN). The flow boundaries are the nozzle contour on the one hand and the flow properties of the external flow field on the other hand. In practice the situation is very complex with high impact on the flight properties of the vehicle. See in this regard Chap. 5, and in particular the aspect of the thrust-vector angle, Sect. 5.7. For details of the nozzle topic the reader is referred to, for instance, [3–5].

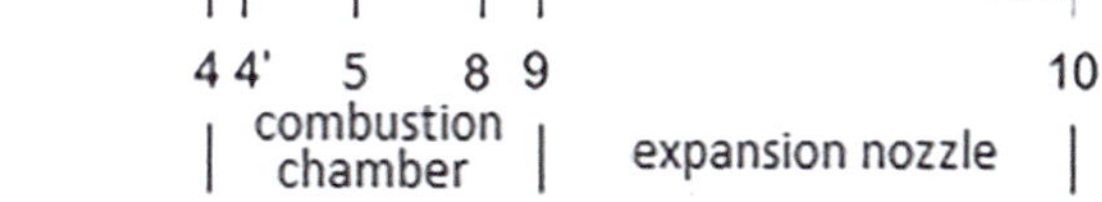

Fig. 4.9 Schematic of a typical API-type 4 or 5 nozzle of SERN type [1]

4.4 Example: Scramjet Demonstration

4.4.1 The Study Case

We give as an example results of an analysis and assessment of scramjet propulsion [1]. The object of the study was the API-type 4 flight vehicle X-43A with a simplified configuration, Fig. 4.10.

The study was performed in the frame of a research cooperation between the Industrieanlagen-Betriebsgesellschaft mbH (IABG) and the University of the Federal Armed Forces Munich (UniBw M), and in association with the Research Training Group (Graduiertenkolleg) 1095 of the German Research Foundation (Deutsche Forschungsgemeinschaft) DFG [20].

Notable are two other studies in that frame and its range. In [25] too the aerothermodynamic design of scramjet propulsion systems for hypersonic space transportation flight systems was investigated. A simulation tool was developed for the performance synthesis of such propulsion systems. In [26] a numerical investigation was performed of the experimental outcome of a HyShot II scramjet study. The latter was made in the High Enthalpy Shock Tunnel, HEG, of the German Aerospace Center (DLR) at Göttingen.

To examine flow, combustion details and performance of a scramjet configuration of course demands the treatment of the whole external and internal flow path. Hence we look at the flow over the forebody, at that through the inlet and the combustion chamber, and finally at that through the nozzle. All is made with a two-dimensional/quasi one-dimensional approach, which sufficiently permits to illustrate the basic phenomena and the related problems.

Figure 4.10 shows the simplified flight vehicle with the basic configuration elements. The lower side flow path as well as that over the upper side—note the small kink of the upper surface at 2,270 mm—of the flight vehicle was recreated by means of a CAD model. The lower side as a pre-compression surface up to the inlet was

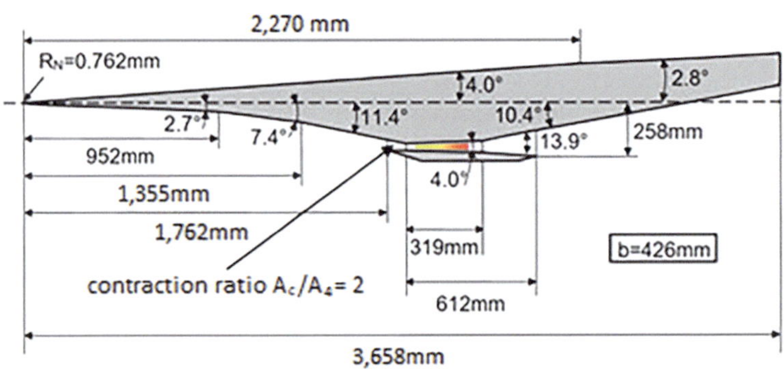

Fig. 4.10 Simplified X-43A configuration [1]

considered as a three-ramp forebody. The external pre-compression hence was performed with the angles 2.7°, 7.4° and 11.4° to the free-stream direction. The total length up to the cowl lip was 1,762 mm.

No information is available regarding the internal flow path, i.e., the isolator and the combustion chamber. The exception is the width of the flow path, which is $b = 426$ mm. For the present study hence an inlet contraction ratio $A_c/A_{4'} = 2$, see Figs. 4.7 and 4.10, was assumed. The cross section of the combustion chamber downstream of location $4^{'}$ over a length of 319 mm opens with a constant half angle of 2°. Downstream the cross section passes over to the single expansion ramp nozzle with the angle 13.9°.

The total length of the combustion chamber is 612 mm, and that of the whole configuration 3,658 mm.

4.4.2 The Computational Approach

The investigation of the external flow path over the forebody and part of the inlet was performed with a coupled Euler/second-order boundary-layer code. The used Euler code is well suited for both shock-fitting and shock-capturing. The boundary-layer code permits to simulate the radiation-adiabatic wall situation. Both the Euler and the boundary-layer code allow to simulate the high-temperature real-gas effects, which are to be expected in the pertinent flight velocity and altitude regime.

The modeling of the flow through the isolator, the combustion chamber and the single expansion ramp nozzle was made with a quasi-one-dimensional Euler code. That permits a fully coupled time-dependent simulation of the combustion mechanisms with an implicit discretization of the unsteady conservation equations.

Investigated were flight situations at altitudes $H = 28$ to 34 km and flight Mach numbers $M_\infty = 6.5$ to 10. The angle of attack always was $\alpha = 2°$, the maximum permissible dynamic pressure was $q_\infty = 100$ kPa. The parameters of the two actual flights of the X-43A at $M_\infty = 6.83$ and $H = 24$ km, and $M_\infty = 9.68$ and $H = 33.5$ km hence are included. Used throughout were the data of the standard atmosphere, Appendix A.

At the lower side of the configuration the boundary layer was tripped at $x = 0.46$ m and turbulent boundary-layer flow was assumed downstream. The transition tripping was the same as that with the X-43A, see Sect. 3.11.1. The upper side boundary layer was considered to be laminar throughout. All surfaces of the external flow path were considered to be radiation-adiabatic with the emissivity coefficient being $\epsilon = 0.8$.

Figure 4.11 shows as an example the computed flow field past the forebody of the simplified X-43A configuration. The flight Mach number is $M_\infty = 10$, the altitude $H = 29$ km, and the angle of attack $\alpha = 2°$.

The bow shock and also the embedded ramp-shock waves are well captured. High-temperature real-gas effects, although modeled in the solution scheme, are not directly apparent. Well discernable is the boundary-layer development along the surfaces.

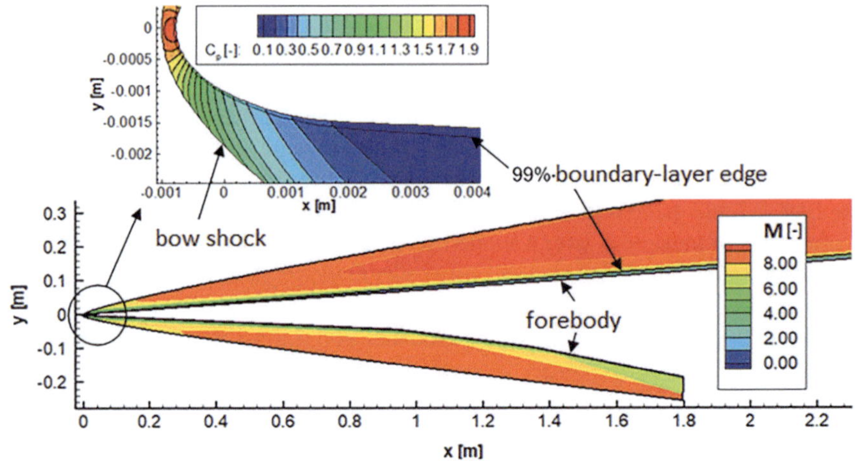

Fig. 4.11 Computed flow field past the forebody of the simplified X-43A configuration at $M_\infty = 10$ and $H = 29\,\text{km}$, $\alpha = 2°$ [1]

Table 4.1 Combustion chamber entry conditions [1]

Velocity	$u_{4'} = 2{,}213\,\text{m/s}$
Density	$\rho_{4'} = 0.1878\,\text{kg/m}^3$
Temperature	$T_{4'} = 661.7\,\text{K}$
Pressure	$p_{4'} = 35.81\,\text{kPa}$
Mach number	$M_{4'} = 4.334$
Air mass flow rate	$m_{4'} = 4.445\,\text{kg/s}$

The basis for the determination of the mixing and ignition characteristic is the flight condition with the flight Mach number $M_\infty = 8$ at the altitude $H = 29\,\text{km}$. This is equivalent to the middle part of the flight envelope. The flow parameters for this case at the entry of the combustion chamber—location 4' in Fig. 4.7—are given in Table 4.1 .

In the one-dimensional computation scheme the hydrogen is injected with a temperature of 300 K and a velocity of 2,000 m/s. The mixing efficiency was assumed to be 100%. In order to simulate the pressure and the temperature rise by means of a central injector and to ensure the ignition, 15% of the fuel—related to a stoichiometric hydrogen-air mixture—is used. The reaction model is a modified Jachimowski scheme. It describes—being optimized for the simulation of supersonic combustion chambers—the combustion of hydrogen with air by means of 13 species and 32 reactions.

In the SERN finally it is assumed that 0.88 m behind the end of the combustion chamber the flow is frozen chemically, that is, chemical reactions do no more occur.

In the computation scheme the boundary-layer displacement and skin-friction effects are determined by means of the reference-temperature concept, see, e.g., [13, 27]. Turbulent boundary-layer flow is assumed in the combustion chamber, whereas the boundary-layer flow in the SERN is assumed to be laminar due to the high acceleration there. The wall of the combustion chamber is considered to be adiabatic. The wall temperature always is the temperature of the combustion gases.

The interaction of the mixing and the ignition model leads to an instantaneous rise of the temperature at the entry of the combustion chamber. The temperature then drops due to the divergent contour of the combustion chamber. At about 0.3 m the temperature again rises due to the hydrogen-air combustion up to about 1,600 K at the end of the combustion chamber. In the SERN finally a steady drop of the temperature is observed. The installed thrust is found to be $F = 140\,\text{N}$.

4.4.3 *Presentation of Results*

The sketched approach and the parameters of the combustion chamber and the combustion processes are the basis for the discussed results of the performance synthesis of the simplified two-dimensional X-43A configuration in the following sections. The flight envelope considered is 28–34 km altitude with Mach numbers $M_\infty = 6.5$ to 10. The angle of attack is considered to be constant with $\alpha = 2°$. The fuel-air equivalence ratio is varied from $\Phi = 0.4$ to $\Phi = 1.4$.

The results in the following figures are given in terms of contours of equal performance parameters in the considered altitude-Mach number domain. The air compression is characterized by the captured air stream, the recovery of total pressure and the conditions at the entry of the combustion chamber. The characterization of the supersonic combustion chamber is made with the flow properties just downstream of the flame stabilizer and those at maximum present pressure and temperature.

The installed thrust is a measure of the performance of the whole flight vehicle. The mechanical and thermal loads on the flight vehicle are given in terms of the maximum pressures and temperatures at the forebody and in the combustion chamber. The data are presented in color-coded contours. The figures always are completed in the flight envelope with lines of constant dynamic pressure q_∞.

4.4.4 *Forebody and External Inlet*

4.4.4.1 Thermal and Pressure Loads at Two Locations of the Forebody

Figure 4.12 illustrates thermal loads in terms of the radiation-adiabatic wall temperature as well as pressure loads at the stagnation point of the forebody of the simplified X-43A configuration. The same is made in Fig. 4.13 for the foot of the first ramp, see Fig. 4.10.

The wall temperature at the stagnation point more or less is a function only of the Mach number M_∞. Nevertheless, a slight reduction is observed with increasing

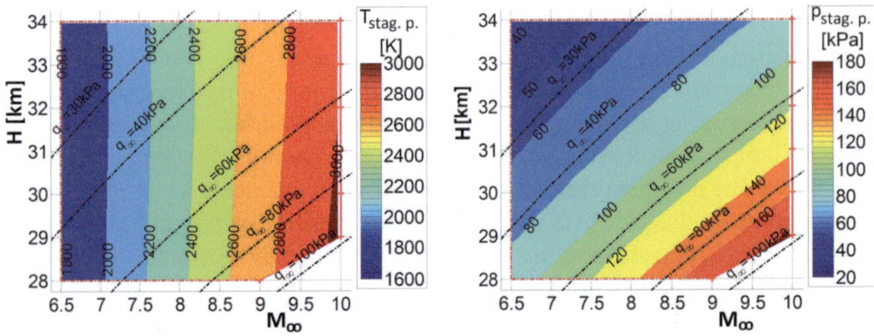

Fig. 4.12 Simplified X-43A configuration, stagnation point: contours of the maximum temperature T (left) and the maximum pressure p (right) [1]

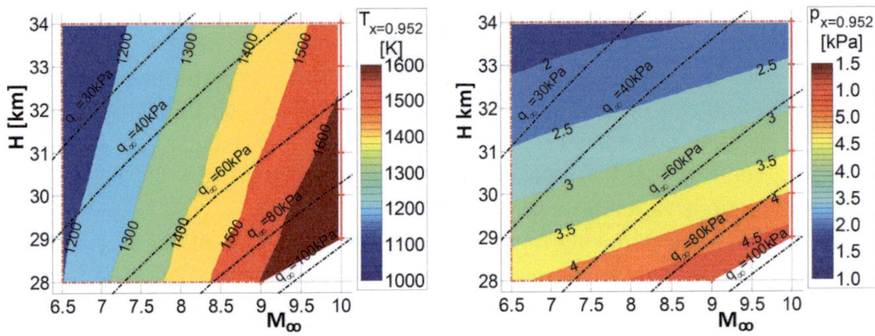

Fig. 4.13 Simplified X-43A configuration, first ramp foot $x = 0.952$ m: contours of the maximum temperature T (left) and the maximum pressure p (right) [1]

altitude. At the foot of the first ramp a strong decrease of the temperature T is present with altitude. Mind the different color coding in the two figures.

The reason for the different behavior is the following. At the forward—nose—stagnation point a laminar boundary layer is present. It has a finite but very small thickness [28]. At the foot of the first ramp—$x = 952$ mm, Fig. 4.10—a fully developed turbulent boundary layer exists. Remember that laminar-turbulent transition was enforced at $x = 0.2$.

The radiation-adiabatic wall temperature of a turbulent boundary layer depends more strongly on the Reynolds number than that of a laminar boundary layer [13]. This is the reason for the different behavior of the temperature with altitude for the two locations. Compare also with Fig. 3.31.

For the pressure it holds that it decreases with increasing flight altitude. Figure 4.12 right—stagnation-point data—shows the iso-pressure lines to be parallel to the lines of constant dynamic pressure. This result is to be expected, however, because at the stagnation point this is the total pressure, with higher data than the latter. At the foot of the first ramp, the pressure is much lower than the dynamic pressure.

4.4.4.2 Compression Effectiveness

Figure 4.14 shows the effectiveness of the compression of the three-ramp external inlet and Fig. 4.15 that including the internal compression up to the location 4', Fig. 4.7.

Contours of the efficiency factors $\eta_{pt,c} = p_{t,c}/p_{t,\infty}$ and $\eta_{ke,c} = u_c^2/u_\infty^2$ in Fig. 4.14 are a measure of the total pressure behavior and the degree of efficiency of the recovery of the kinetic energy of the external inlet. $\eta_{pt,4'}$ and $\eta_{ke,4'}$ in Fig. 4.15 give the effectiveness of the full inlet up to location 4'.

The total-pressure recovery to first approximation is a function of the flight Mach number, such that $\eta_{pt,c}$ decreases with increasing M_∞. The efficiency factors $\eta_{ke,c}$ and $\eta_{pt,4'}$ at the right of the two figures behave the other way around and are slightly increasing with increasing flight Mach number.

They are to first order a function only of M_∞. With stronger deceleration of the flow the behavior of the atmospheric temperature profile, Fig. A.1 in Appendix A, plays a role. The efficiency factor $\eta_{ke,4'}$ therefore rises slightly with increasing flight altitude H.

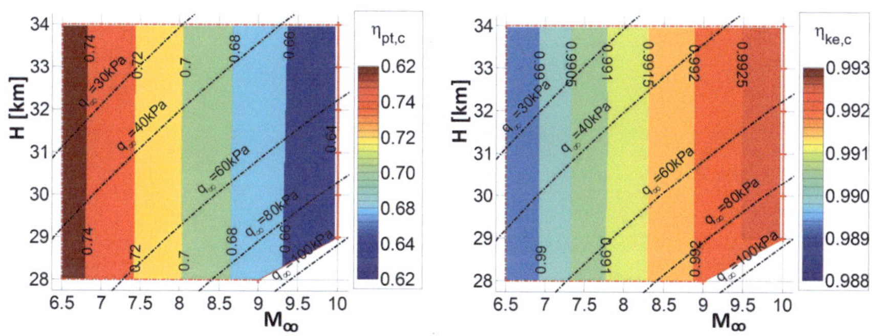

Fig. 4.14 Efficiency factors $\eta_{pt,c}$ and $\eta_{ke,c}$ of the external compression of the simplified X-43A forebody [1]

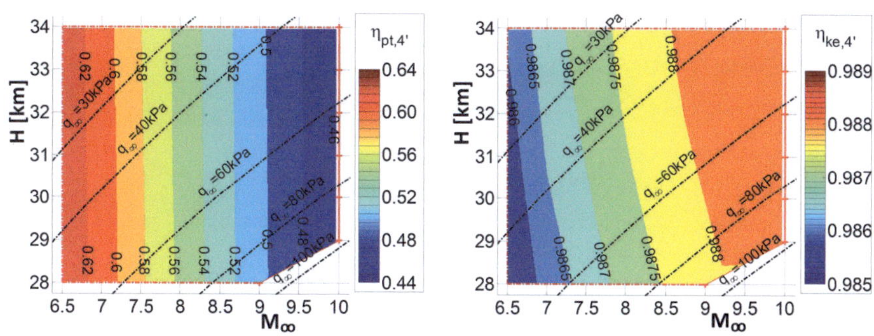

Fig. 4.15 Efficiency factors $\eta_{pt,4'}$ and $\eta_{ke,4'}$ of the whole compression of the inlet of the simplified X-43A forebody [1]

4.4.4.3 Captured Mass Flux and Combustion-Chamber Entry Data

The captured mass flux $\dot{m}_{air}$ is shown at the upper left of Fig. 4.16 together with the combustion-chamber entry data pressure p, temperature T and Mach number M at location 4'.

The captured mass flux depends on the free-stream density ρ_∞ and the capture cross section, which depends on the location of the oblique shock waves of the external inlet and hence on the flight Mach number M_∞.

It is further to be observed that the static pressure $p_{4'}$ and the temperature $T_{4'}$ additionally depend on the free-stream conditions in contrast to the total pressures $p_{t,4'}$ and $p_{t,\infty}$. That is due to the fact that in the considered flight domain the pressure p_∞ drops, whereas the temperature T_∞ rises, Fig. A.1 of Appendix A.

Hence we observe that with increasing flight altitude the pressure $p_{4'}$ slightly drops—it is lower than the dynamic pressure—and the temperature $T_{4'}$ rises. The Mach number $M_{4'}$ first of all is a function of the flight Mach number M_∞.

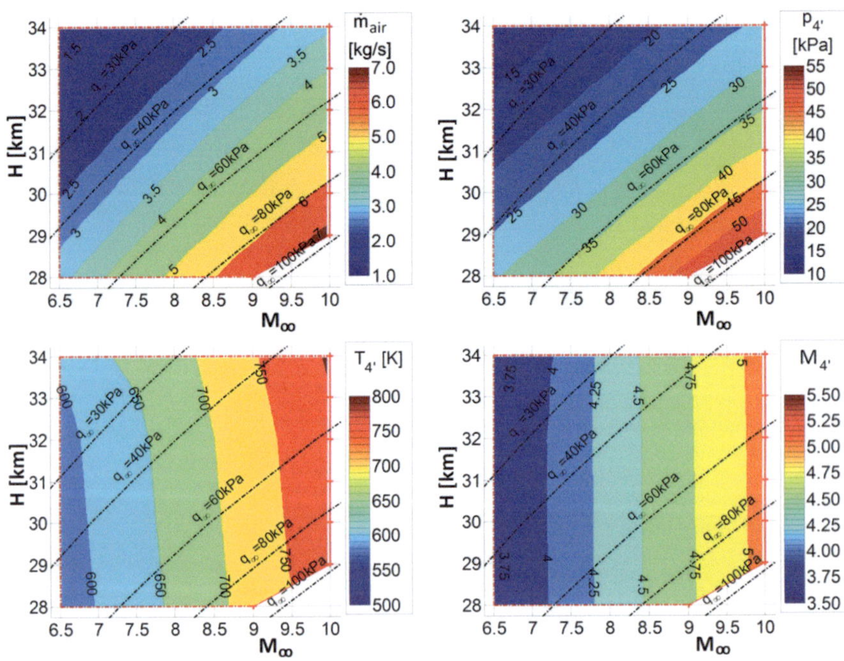

Fig. 4.16 Simplified X-43A configuration: contours of captured mass flux $\dot{m}_{air}$ and of pressure p, temperature T and Mach number M at location 4' [1]

4.4.5 Combustion Chamber

4.4.5.1 Conditions Downstream of the Flame Holder

We show in Fig. 4.17 for the comparison with the data of Fig. 4.16 the combustion chamber conditions just downstream of the flame holder, location 5, Fig. 4.9.

That is made together with the fuel mass flux $\dot{m}_{fuel,\Phi=1}$, i.e., with stoichiometric conditions. The trend of the fuel mass flux is proportional to the captured mass flux $\dot{m}_{air}$, Fig. 4.16, upper left. It rises with increasing flight Mach number M_∞, but drops with increasing flight altitude H.

Pressure p_5, temperature T_5 are increased due to the energy addition whereas the Mach number M_5 is reduced. Qualitatively the trends of p_5, T_5 and M_5 are the same as those of them at location 4'.

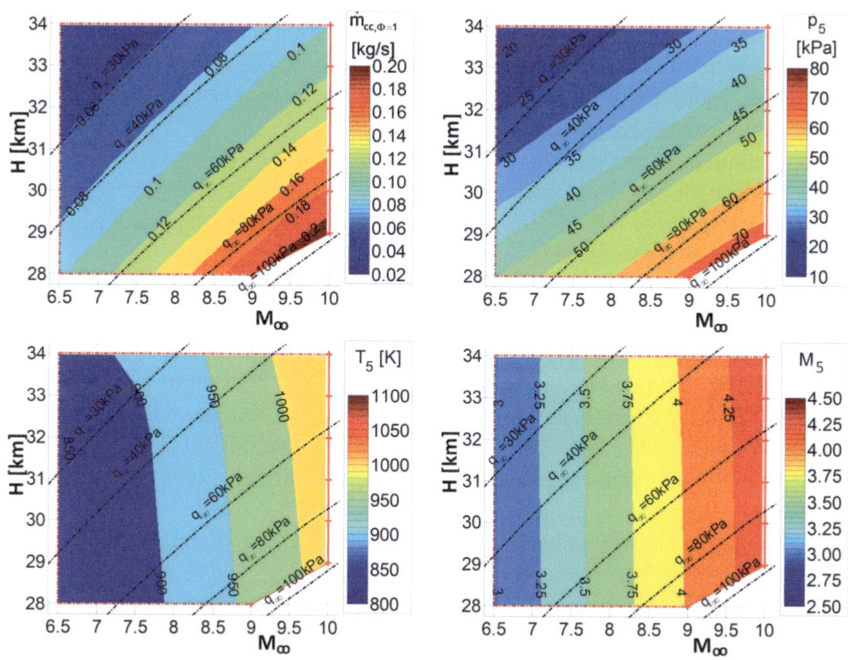

Fig. 4.17 Simplified X-43A configuration: contours of the conditions downstream of the flame holder: fuel mass flux $\dot{m}_{fuel,\Phi=1}$, pressure p, temperature T and Mach number M at location 5 [1]

4.4.5.2 Influence of the Fuel-Air Equivalence Ratio Φ on the Flow Properties

The following graphs show in the considered altitude-Mach number domain the maximum temperature $T_{CC,max}$, the maximum pressure $p_{CC,max}$ and the minimum Mach number $M_{CC,min}$ for each of six fuel-air equivalence ratios in the range $0.4 \leqq \Phi \leqq 1.4$.

The maximum temperatures and pressures characterize the thermal and the pressure loads on the combustion chamber, Figs. 4.18 and 4.19.

The increase of the fuel-air equivalence ratio for a given flight condition leads to a rise of the maximum temperature and the maximum pressure, but to a reduction of the minimum Mach number. The maxima in the combustion chamber are the higher, the earlier the ignition of the hydrogen-air mixture happens. The ignition on the other hand depends on the mixing characteristic and the flow entry conditions of the combustion chamber, Fig. 4.16.

Maximum temperature and pressure in the combustion chamber rise with increasing flight Mach number, but decrease with flight altitude. Both increase, once sufficient fuel is available—increasing fuel-air equivalence ratio—and sufficient time to allow a significant combustion.

Up to an fuel-air equivalence ratio $\Phi = 0.8$ virtually no combustion occurs. The maximum temperatures and pressures in the combustion chamber are rising only marginally. Beginning with $\Phi = 1$ and for the flight domain with dynamic pressure $q_\infty \gtrapprox 60$ to 80 kPa they begin to increase distinctly.

This domain is enlarged with $\Phi = 1.2$, such that already for dynamic pressures $q_\infty \gtrapprox 40$ temperatures higher than 2,000 K and pressures higher than 60 kPa are present. But at the low flight altitudes 28–30 km and low flight Mach numbers up to $M_\infty \approx$ 7 to 7.5 already thermal blockage, Sect. 4.3.6, happens in the combustion chamber.

For the fuel-air equivalence ratio $\Phi = 1.4$ maximum temperatures in the combustion chamber higher than 2,000 K and pressures higher than 60 kPa are present almost in the whole considered flight domain. Hence for all flight Mach numbers below $M_\infty \approx 7$ and all flight altitudes below $H = 29$ km thermal blockage happens.

The minimum Mach numbers behave in accordance with the maximum temperature and pressures, Fig. 4.20.

Up to the fuel-air equivalence ratio $\Phi = 0.8$ only a marginal reduction of the minimum Mach numbers is evident. With increasing fuel-air equivalence ratio $\Phi \geqq 1$ a reduction of the minimum Mach number is seen. That is due to the increasing heat addition which happens owing to the combustion reactions in the range of dynamic pressure $q_\infty = 60$ to 80 kPa.

Noticeable in this regard becomes the dependence of the rate of the H_2-air combustion reaction on the temperature and the pressure, see Fig. 4.17. With increasing flight altitude and decreasing flight Mach number a drop of the pressure p_5 happens, whereas the temperature T_5 slightly rises with increasing flight altitude and Mach number. The Mach number M_5 rises with M_∞, such that in the combustion chamber the minimum Mach numbers also increase and become higher with increasing flight altitude in dependence on T_5 and p_5.

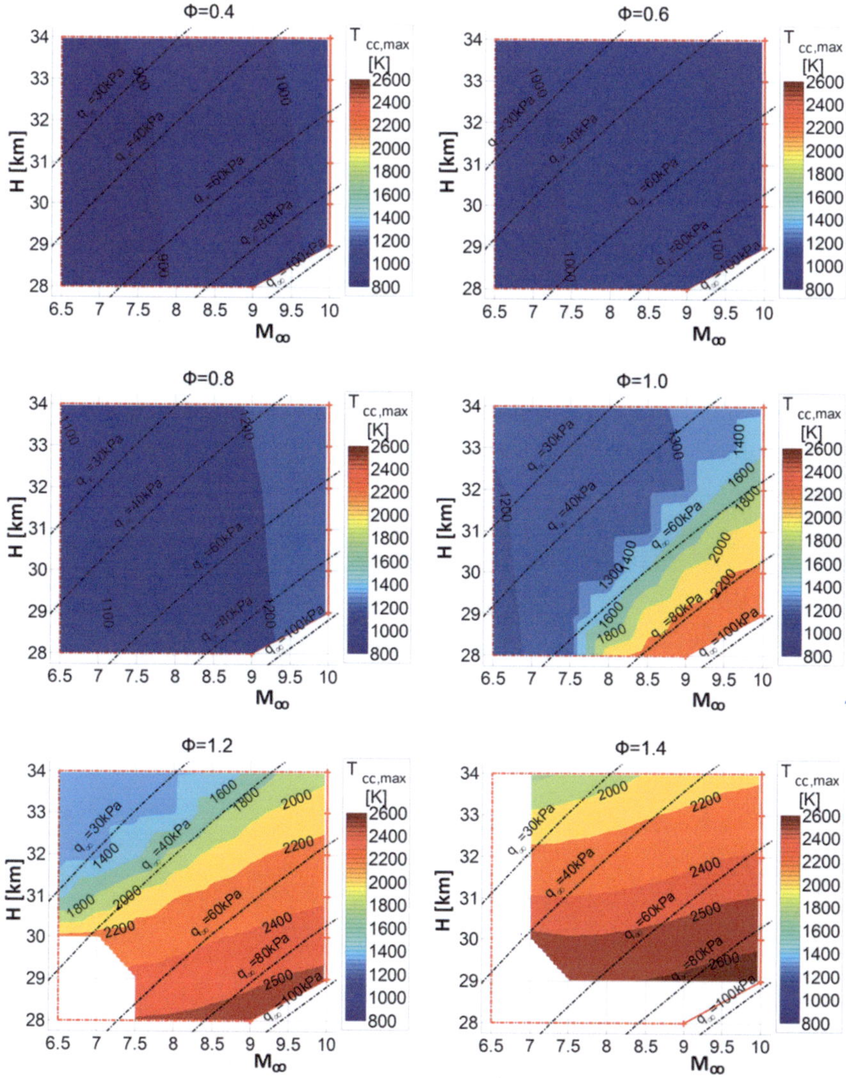

Fig. 4.18 Simplified X-43A configuration, combustion chamber: contours of the maximum temperature T of the combustion gases for different fuel-air equivalence ratios Φ [1]

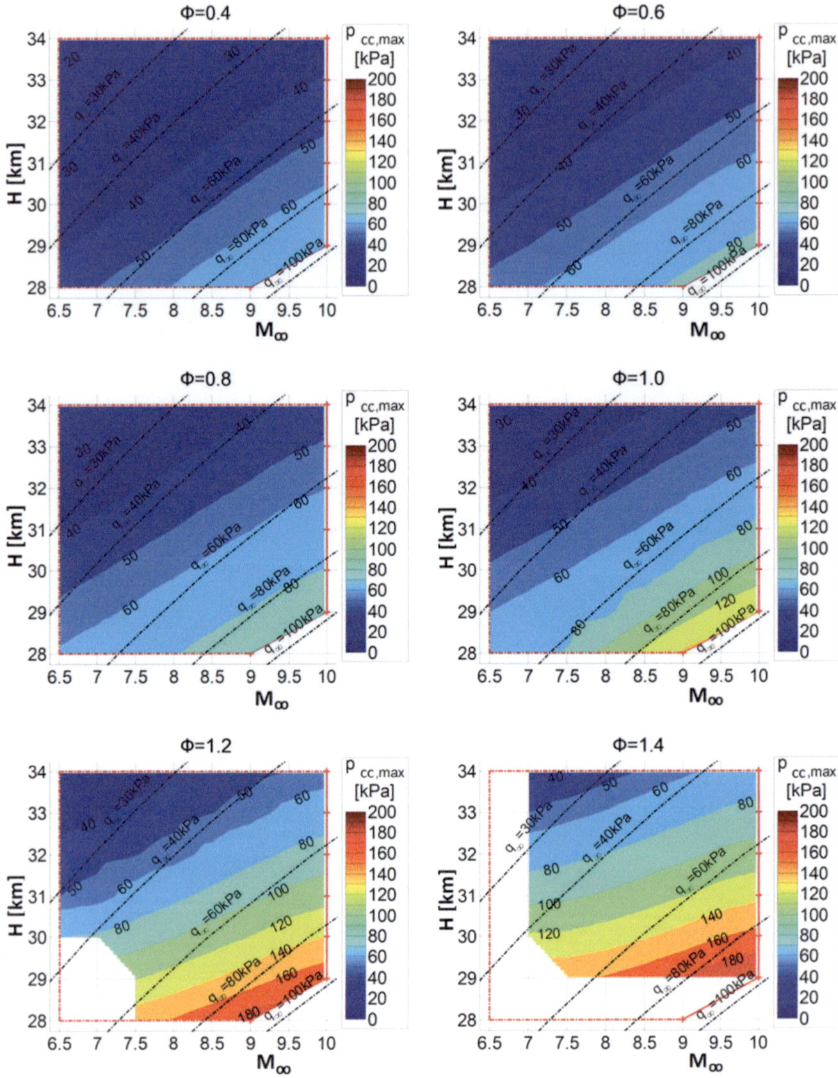

Fig. 4.19 Simplified X-43A configuration, combustion chamber: contours of the maximum pressure p of the combustion gases for different fuel-air equivalence ratios Φ [1]

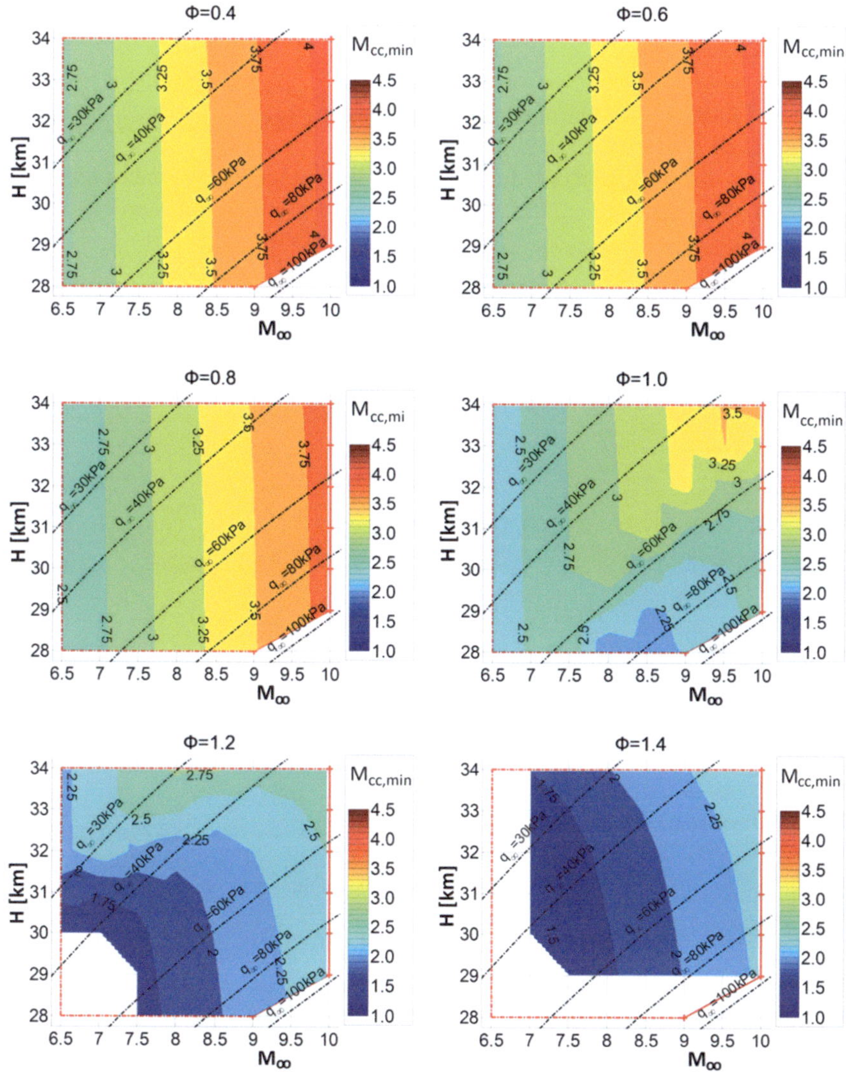

Fig. 4.20 Simplified X-43A configuration, combustion chamber: contours of the minimum Mach numbers M of the combustion gases for different fuel-air equivalence ratios Φ [1]

4.4.5.3 Influence of the Fuel-Air Equivalence Ratio Φ on the Installed Thrust

Figure 4.21 finally shows the contours of the constant installed thrust F_x. It is given for the fuel-air equivalence ratios $\Phi = 0.4$ to 1.4.

A positive thrust is given only for $\Phi \geqq 1$. For this fuel-air equivalence ratio also a significant rise of the temperature and the pressure is present for dynamic pressures

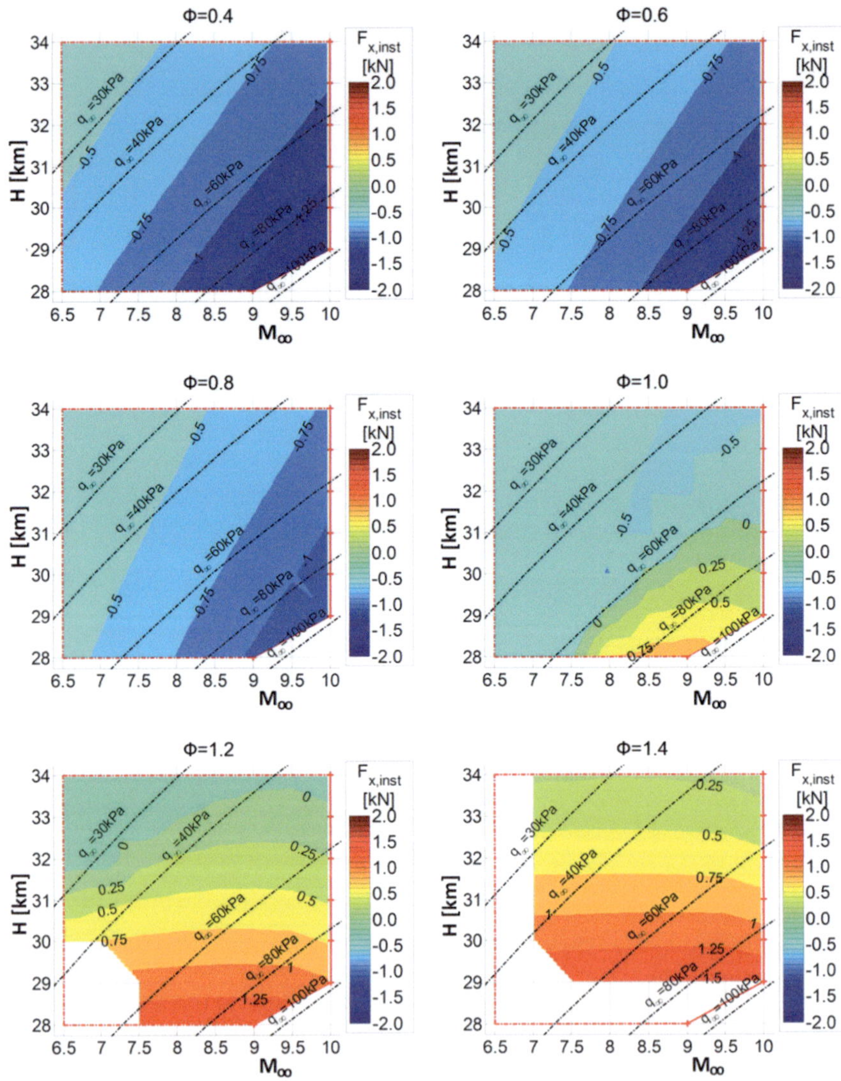

Fig. 4.21 Simplified X-43A configuration: contours of the installed thrust F_x for different fuel-air equivalence ratios Φ [1]

$q_\infty \gtreqqless 60$ to 80 kPa. With the higher fuel-air equivalence ratios a positive thrust up to $F_x = 2$ kN is generated. This does not hold for the flight conditions with thermal blockage.

4.5 Discussion and Assessment

This section is devoted to a discussion and an assessment of the results presented in the previous sections. Considered are the performance and the critical performance parameters of the simplified X-43A scramjet configuration. The possible capabilities of future scramjet propulsion systems are deduced in general.

Of interest besides the thrust are the thermal and mechanical loads, drag components and the temperature rise in the combustion chamber. Not performed were sensitivity considerations, parameter variations and the effects of different modeling approaches.

The flight envelope is bounded by the specification of the flight altitude—$H =$ 28 to 34 km and the flight Mach number—$M_\infty = 6.5$ to 10. The specification of a strength limit in terms of a maximum permissible dynamic pressure of $q_\infty = 100$ kPa limits the flight domain at $H = 28$ km to $M_\infty = 9$.

Not explicitly evaluated has been the maximum permissible temperature at the forebody. Hence the thermal loads over the whole flight domain have not been quantified. Nevertheless, some data were obtained.

At $M_\infty = 10$ temperatures up to 3,000 K have been found at the nose of the configuration and at other stagnation point locations. If with longer flight periods the maximum admissible temperatures are less than 2,500 K, the operational flight Mach number reduces to approximately $M_\infty = 8.5$, Fig. 4.12.

With a larger nose radius, like that for the $M_\infty = 9.6$ configuration of the X-43A with $R_{Nose} = 1.27$ mm, it is possible to reduce distinctly the thermal loads and hence the wall temperatures, because, [13, 27],

$$T^4_{rad,nose} \propto q_{rad,nose} \propto \frac{1}{\sqrt{R_{Nose}}}. \tag{4.3}$$

With the here considered case flight Mach numbers up to $M_\infty \approx 9$ can be reached with the radiation-adiabatic wall temperature not exceeding $T = 2{,}500$ K. However, the aerodynamic drag increases, in particular with slender API-type 4 and 5 configurations. Hence the nose radius always is a compromise between aerodynamic drag and thermal loads, see the discussion in Sect. 3.5.

The static temperatures of the combustion gases in the combustion chamber reach up to 2,500 K, too. Hence with longer flight times active cooling of the chamber's structure becomes necessary. Thermal blockage moreover can result in high mechanical loads. The resulting drop of thrust does not only reduce flight speed, but also, due to the single expansion ramp nozzle, massively disturbs the pitching moment behavior of the fight vehicle.

Operation points with thermal blockage due to boundary-layer effects must be identified during any performance investigation. In the cases considered here, boundary-layer separation and thermal blockage does not happen.

The inlet in all cases was considered as being "started", such that no restrictions were present in the operation domain of the simplified X-43A configuration. Further it is to be noted, that the bow shock in all cases was outside the lip of the inlet cowl, see also case (b) in Fig. 5.13. Hence the shock-on-lip problem was not present.

The possible flight envelope of the simplified X-43A configuration with thrust surplus is restricted to the flight conditions shown in Fig. 4.22 with the fuel-air equivalence ratio $\Phi = 1$.

Shown too are the thermal-load limits as function of flight altitude and Mach number (see also Fig. 4.12). The nose radius is $R_N = 0.762$ mm. In addition visualized is the case with $R_N = 1.27$ mm.

With increasing fuel-air equivalence ratio at low flight altitudes and Mach numbers thermal blockage happens in the combustion chamber. However, the operational regime with thrust surplus is extended to higher flight altitudes. At high altitudes a limitation is found regarding the maximum fuel mass and its mixing with the air.

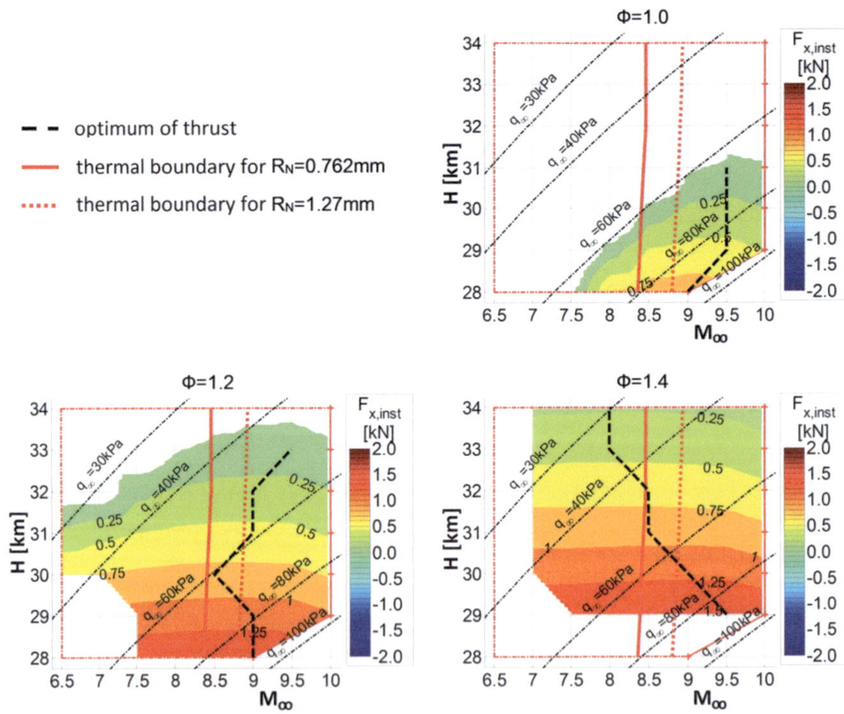

Fig. 4.22 Possible flight envelope of the simplified X-43A configuration with thrust surplus as function of different fuel-air equivalence ratios [1]

Another limitation is due to the dynamic pressure limit with $q_\infty \leqq 20\,\text{kN/m}^2$. The sizes of aerodynamic surfaces for lift generation, stabilization, trim and control grow strongly below this limit. Figure 4.22 further shows those flight altitudes for the respective fuel-air equivalence ratios in dependence on the flight Mach number in which the installed thrusts have their optima.

These thrust optima increasingly flatten out with increasing Φ, however, that becomes more distinct with increasing flight altitude. Hence, it is possible to reach an optimum of the specific impulse of the propulsion system with a clever control of the fuel-mass stream. In addition one sees that with increasing fuel-air equivalence ratio a trend of the thrust optima toward smaller Mach numbers exists.

For different fuel-air equivalence ratios contours of percentage temperature rise at the simplified X-43A configuration are given in Fig. 4.23. The main factors for the restriction of the operational flight regime at the high flight Mach numbers are the reaction kinetics, the thermal loads limit at $M_\infty \approx 8.5$, respectively $M_\infty \approx 9$, and the disproportionate reduction of the thrust at $M_\infty \gtrapprox 9$.

The thermal loads can be reduced by means of active cooling, using the available fuel. The reduction of the thrust is due to the caloric value of the fuel and hence to the effective rise of the temperature of the fuel-air mixture due to the combustion process.

The effective rise of the temperature in the combustion chamber also depends on the total enthalpy of the free stream. The temperature at the entry of the combustion chamber is a function of the flight speed. The effective temperature rise in the combustion chamber hence decreases with this temperature. A continual rise of the thrust hence cannot be expected. Not to be forgotten is that the aerodynamic drag rises with increasing flight speed.

Another factor regarding the limitation of the flight envelope at high Mach numbers and altitudes are the friction forces $F_{x,\tau}$ at the surfaces of the combustion chamber. They act opposite to the pressure force $F_{x,p}$ on the surfaces. Figure 4.24 shows this for the here considered fuel-air equivalence ratios as function of flight Mach number and altitude.

The friction forces are reciprocal to a power of the Reynolds number for a given dynamic pressure, see, e.g., [13, 27], such that they increase with increasing flight altitude, if small fuel-air equivalence ratios are present with no significant temperature rise in the combustion chamber. The increase is also due to the proportionality to the dynamic pressure, which increases with increasing Mach number for a given Reynolds number, Eq. (4.1).

The pressure rises with increasing fuel-air equivalence ratios and hence increasing temperature due to the combustion process, but at the same time the friction forces are dropping. They are reciprocally proportional to the temperature of the combustion gases.

The ratios $F_{x,\tau}/F_{x,p} = 0.5$ to 0.6 hence lead to a positive installed thrust of the simplified X-43A configuration. But $F_{x,\tau}/F_{x,p}$ still rises with increasing flight Mach number and flight altitude. However, due to the limitation of the effective temperature rise with increasing Mach number it is no more possible to compensate the rise of the friction forces with the temperature rise.

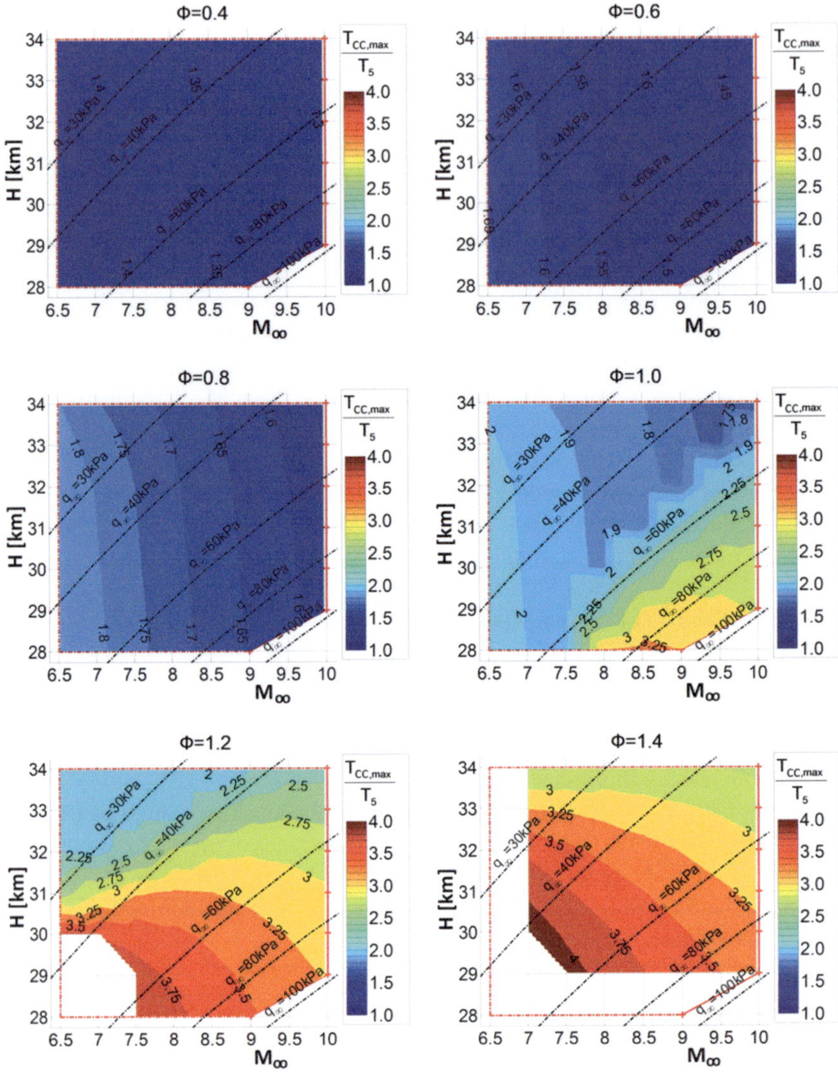

Fig. 4.23 Contours of percentaged temperature rise of the simplified X-43A configuration as function of flight Mach number and altitude for different fuel-air equivalence ratios Φ [1]

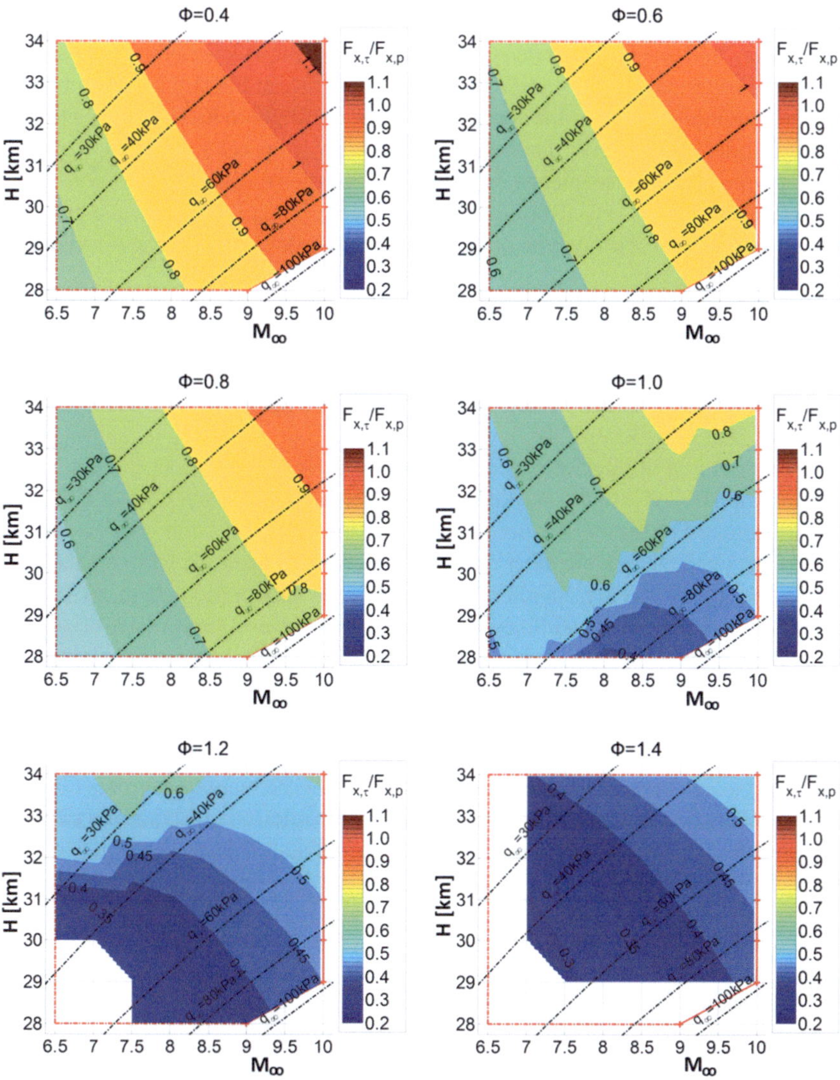

Fig. 4.24 Contours of equal pressure to friction force ratios $F_{x,\tau}/F_{x,p}$ in the combustion chamber of the simplified X-43A configuration as function of flight Mach number and altitude for different fuel-air equivalence ratios Φ [1]

With increasing total free-stream enthalpy not only the possible effective rise of temperature and pressure decreases, but in the combustion chamber the friction drag increases. In the case of a significant temperature rise in the combustion chamber ($\Phi \geqq 1$) an equilibrium in the force balance of the flight vehicle occurs. Hence a constant thrust plateau over the flight Mach number is the consequence.

Beginning with approximately $M_\infty \approx 9.5$ the installed thrust of the here considered configuration is negatively affected by both the decreasing pressure rise and the rising friction drag. Hence it is no more possible to compensate the increasing aerodynamic drag of the flight vehicle.

The drop of the installed thrust at $M_\infty \gtrapprox 9.5$ for all flight altitudes can be observed independently of the maximum temperature in the combustion chamber (compare with Fig. 4.18). Hence the sole cause of the thrust drop due to slow or not reversible dissociation processes can be ruled out. The total enthalpy of the considered free-stream conditions is lower than $h_t = 5\,\text{MJ/kg}$.

In the frame of this performance synthesis investigation of the simplified X-43A configuration no sensitivity analysis was made regarding the inner configuration and its parameters. For the contraction ratio of the inner compression contour and the shape of the combustion chamber only one generic configuration was considered.

Also not made was a variation of the fuel mixing and ignition characteristics, as well as the fuel injection conditions. The influence on the installed thrust of changes of the angle of attack, of the laminar-turbulent transition location, of possible boundary-layer re-laminarisation of the combustion-chamber flow was not considered.

All this can have an substantial influence on the performance and the operating range of the flight vehicle, see, e.g., [29]. Not to be forgotten is that only a highly idealized two-dimensional treatment was performed. The reality comprises complex three-dimensional flow details at the forebody, the inlet, the isolator, the combustion chamber, the SERN, and at the aerodynamic stabilization, trim and control surfaces [30].

4.6 Final Remarks

The present discussion was centered on the basic phenomena and issues in order to give the reader an introduction to the problem of treating scramjet performance. The degree of exactness anyway should be in accordance with the actual knowledge of the configuration details.

Moreover it must be realized that with API-type 4 and 5 scramjet vehicles on the one hand a relatively simple basic geometry is present, but on the other hand very complex aerothermodynamic phenomena are given. Not to mention the actual configurational and performance issues of small (API-type 4) and in particular large (API-type 5) flight vehicles, as they are presented in Chap. 5.

In the case of the X-43A a small operational flight envelope with thrust surplus is combined with sensitive thrust optima and very high thermal loads. This can only be countered with geometrical variations and with demanding cooling concepts of in particular the internal flow path. The benefits of the simple geometry are thus easily lost.

Of course the ratio of vehicle mass to thrust can be varied by additional design measures. Anyway a compromise must be found between acceleration performance and possible range of operation. Airbreathing hypersonic propulsion systems in any case have a higher specific impulse than rocket systems, even if their range of application is limited to flight Mach numbers $M_\infty \lessapprox 12$, Sect. 2.1.2.

4.7 Problems

Problem 4.1 (*Flight Mach Number and Propulsion Mode*) Given the operational range of the propulsion modes:- Turbojet: up to $M_\infty \approx 3.5$-Ramjet: $3.5 \leq M_\infty \leq 7$-Scramjet: $M_\infty > 7$ At what flight Mach number will the transition from ramjet to scramjet propulsion occur? What are the main reasons for switching propulsion modes at this Mach number?

Problem 4.2 (*Dynamic Pressure Calculation*) Calculate the dynamic pressure for a hypersonic flight at Mach 8, assuming an atmospheric pressure $p_\infty = 1.2\,\text{kPa}$ and specific heat ratio $\gamma = 1.4$. Use the formula for dynamic pressure:

$$q_\infty = \frac{1}{2}\gamma p_\infty M_\infty^2.$$

Problem 4.3 (*Scramjet Combustion Chamber Design*) Why must the combustion process in a scramjet engine be completed very quickly? Describe the challenges in fuel injection, mixing, and ignition in supersonic flow.

Problem 4.4 (*Heat Addition and Mach Number*) Explain the phenomenon of thermal blockage in a scramjet engine. How does heat addition affect the Mach number within the combustion chamber, and what are the potential consequences?

Problem 4.5 (*Fuel-Air Equivalence Ratio Impact*) For a scramjet engine, explain the difference in engine performance when the fuel-air equivalence ratio Φ is greater than 1 (rich mixture) versus when Φ is less than 1 (lean mixture). What are the trade-offs between thrust and fuel economy?

Problem 4.6 (*Inlet Isolator Functionality*) Describe the role of the inlet isolator in a scramjet engine. What are the two primary functions it serves, and why is it crucial to prevent upstream disturbances from reaching the inlet?

Problem 4.7 (*Shock-Wave/Boundary-Layer Interaction*) What are the consequences of shock-wave/boundary-layer interaction in a scramjet engine? How can this interaction affect the performance of the engine, particularly at high speeds?

Problem 4.8 (*SERN Boundary Layer*) When can the boundary-layer flow in the SERN assumed to be laminar?

References

1. Kliche, D.: Multidisziplinäre Analyse und Bewertung von Staustrahltriebwerken mit Überschallverbrennung (Multidisciplinary Analysis and Rating of Ramjet Engines with Supersonic Combustion). Doctoral Thesis, Universität der Bundeswehr, Neubiberg. Verlag Dr. Hut, München, Germany (2012)
2. Falempin, F.: Ramjet and dual mode operation. Advances on Propulsion Technology for High-Speed Aircraft. RTO-EN-AVT-150 7-1–7-36 (2008)
3. Heiser, W.H., Pratt, D.T.: Hypersonic Airbreathing Propulsion. Education Series. AIAA, Washington, D.C. (1994). Online edition (2012)
4. Curran, E.T., Murthy, S.N.B. (Eds.): Scramjet Propulsion. Progress in Astronautics and Aeronautics, vol. 189. AIAA, Reston, Va. (2000)
5. Segal, C.: The Scramjet Engine—Processes and Characteristics. Cambridge University Press (2009)
6. Rick, H.: Gasturbinen und Flugantriebe. Springer, Berlin-Heidelberg (2013)
7. Jiang, Z.: Standing oblique detonation for hypersonic propulsion: a review. Progr. Aerosp. Sci. (2024). https://doi.org/10.1016/j.paerosci.2023.100955
8. Babu, V.: Fundamentals of Propulsion. Springer, Cham (2022)
9. Bruno, C.: Airbreathing Hypersonic Propulsion. Springer, Singapure (2023)
10. Kuczera, H., Sacher, P.W.: Reusable Space Transportation Systems. Springer, Berlin Heidelberg (2011)
11. Zellner, B., Sterr, W., Herrmann, O.: Integration of Turbo-Expander and Turbo-Ramjet Engines in Hypersonic Vehicles. ASME-Paper No. 92-GT-204, June 1–4, 1992, Cologne, Germany (1992)
12. Van Wie, D.M.: Scramjet Inlets. In: Curran, E.T., Murthy, S.N.B. (Eds.), Scramjet Propulsion Progress. Astronautics and Aeronautics, vol. 189, pp. 447–511. AIAA (2000)
13. Hirschel, E.H.: Basics of Aerothermodynamics. Springer, Berlin (2015) and Vol. 204, Progress in Astronautics and Aeronautics. AIAA, Reston, Va.; Second, revised Springer, Heidelberg (2004)
14. Seddon, J., Goldsmith, E.L.: Intake Aerodynamics. AIAA Education Series (1989)
15. Reinartz, B.: Performance analysis of a 3D scramjet intake. In: 26th Congress of the International Council of the Aeronautical Sciences (ICAS), Anchorage, USA (2008)
16. Anderson, G.Y., McClinton, C.R., Weidner, J.P.: Scramjet Performance. In: Murthy, S.N.B., Curran, E.T. (Eds.) Scramjet Propulsion. Progress in Astronautics and Aeronautics, vol. 189. AIAA, pp. 369–446 (2000)
17. Gerdroodbary, M.B.: Scramjets—Fuel Mixing and Injection Systems. Elsevier (2020)
18. Smart, M.K., Hass, N.E., Paull, A.: Flight data analysis of the HyShot 2 scramjet flight experiment. AIAA J. **44**(10), 2366–2375 (2006)
19. Sunami, T., Itoh, K., Satoh, K., Komuro, T.: Mach 8 Ground Tests of the Hypermixer Scramjet for HyShot-IV Flight Experiment. AIAA-Paper 2006-8062 (2006)
20. Weigand, B., Gaisbauer, U.: An Overview on the Structure and Work of the DFG Research Training Group GRK 1095: Aerothermodynamic Design of a Scramjet Propulsion System. AIAA-Paper 2009-7276 (2009)

21. Sabelnikov, V.A., Penzin, V.I.: Scramjet research and development in Russia. In: Murthy, S.N.B., Curran, E.T. (Eds.), Scramjet Propulsion. Progress in Astronautics and Aeronautics, vol. 189, pp. 223–367. AIAA (2000)
22. Chinzei, N., Mitani, T., Yatsuyanagi, N.: Scramjet engine research at the national aerospace laboratory in Japan. In: Murthy, S.N.B., Curran, E.T. (Eds.), Progress in Astronautics and Aeronautics, vol. 189, pp. 159–222. AIAA (2000)
23. Kirstein, S., Maier, D., Fuhrmann, T., Hupfer, A., Kau, H.-P.: Experimental Study on Staged Injection in a Supersonic Combustor. AIAA-Paper 2009-7343 (2009)
24. Choubey, G., Tiwari, M.: Scramjet Combustion—Fundamentals and Advances. Elsevier (2022)
25. Fuhrmann, T.: Auslegung und Betriebsverhalten von SCRamjet-Antriebssystemen für Raumtransporter-Hyperschallflugzeuge (Design and Operating Behavior of Scramjet Propulsion Systems of Space-Transportation Hypersonic Aircraft). Doctoral Thesis, Technische Universität München, Germany (2009)
26. Karl, S.: Numerical Investigation of a Generic Scramjet Configuration. Doctoral Thesis, Technische Universität Dresden, Germany (2011)
27. Hirschel, E.H., Weiland, C.: Selected Aerothermodynamic Design Problems of Hypersonic Flight Vehicles. Progress in Astronautics and Aeronautics, AIAA, Reston, Va., vol. 229. Springer, Heidelberg (2009)
28. Hirschel, E.H., Cousteix, J., Kordulla, W.: Three-Dimensional Attached Viscous Flow. Springer, Heidelberg (2013)
29. Kliche, D., Mundt, Ch.: Combuster Modelling for Multidiscipinary Analysis and Evaluatiion of Supersonic Combustion Ranjets. AIAA-Paper 2009-7282 (2009)
30. Peebles, C.: Road to Mach 10-Lessons Learned from the X-43A Flight Research Program. Library of Flight, AIAA, Reston, Va (2008)

Chapter 5
Issues of Airframe-Propulsion Integration

This chapter is devoted to a consideration of airframe-propulsion integration of hypersonic airbreathers. That does not only concern the physical integration—depending on the integration type, Sect. 1.5—but in particular also the implications regarding flyability and controllability of the vehicle.

Mainly the issues of Airframe-Propulsion Integration-type 5 flight vehicles are considered. From the classical subsonic and transonic transport aircraft to the slender API-type 5 hypersonic aircraft, Sect. 1.5—both CAV and ARV—a change of lift distribution on the airframe is evident. The geometric and configurational integration of wing and body brings the constituent parts of the flight vehicle to an almost equal level of lift generation.

At the slender API-type 5 hypersonic aircraft the influence of the propulsion system on the shape of the lower side of the configuration reaches from the fuselage nose to the inlet cowl lip and from the end of the engine to the afterbody and the nozzle. Aerothermoelasticity plays a large role.

First made is a characterization of the propulsion systems of an API-type 4 and an API-type 5 flight vehicle, the X-43A and the lower stage of the TSTO space transportation system SÄNGER, and of one API-type 2 flight vehicle, the MR2/MR3 concept of the STRATOFLY study. They are shortly sketched.

Then more specific issues are discussed briefly: inlet flow, integration aspects, book-keeping, forebody pre-compression, thrust and installation losses, thrust-vector angle, cooling balance, and trim and longitudinal stability.

5.1 Three Flight Vehicles: Short Characterizations of Their Propulsion Systems

We characterize in a compact manner the basic configurations of three hypersonic flight vehicles, one of API-type 4, one of API-type 5 and one of API-type 2. Of the three vehicles, only the first one, the API-type 4 X-43A actually was flown. The

E. H. Hirschel et al., *Elements of Hypersonic Airbreather Design and Development*,
https://doi.org/10.1007/978-3-031-94219-8_5

other two, the lower stage of the TSTO-system SÄNGER (API-type 5), Sect. 2.8, and the STRATOFLY concept MR2 (API-type 2) are studies of different depth, see the overview in Sect. 1.1, however illustrating the severe challenges of hypersonic flight.

5.1.1 The Experimental Vehicle X-43A

The concept of the scramjet engine as 'airframe-integrated scramjet' was developed in the 1970s. Many years later, after several attempts—see also Sect. 1.1—the successful flights of the X-43A happened with $M_\infty = 6.83$ and $M_\infty = 9.68$ [1].[1]

The shape of the API-type 4 experimental flight vehicle X-43A can be seen as the epitome of a hypersonic scramjet configuration, Fig. 5.1. The configuration of the X-43A has an overall length of 3.66 m with all-moving wings and vertical stabilizers with rudders.

The external propulsion flow path at the lower side of the vehicle, Fig. 5.2—which at the same time is also the aerodynamic lift flow path—is beginning at the slightly blunt unswept leading edge, see also Sect. 3.5.

Over three ramps with small ramp angles a flow compression happens, with the oblique ramp shocks at the nominal flight Mach number focussed on the cowl lip

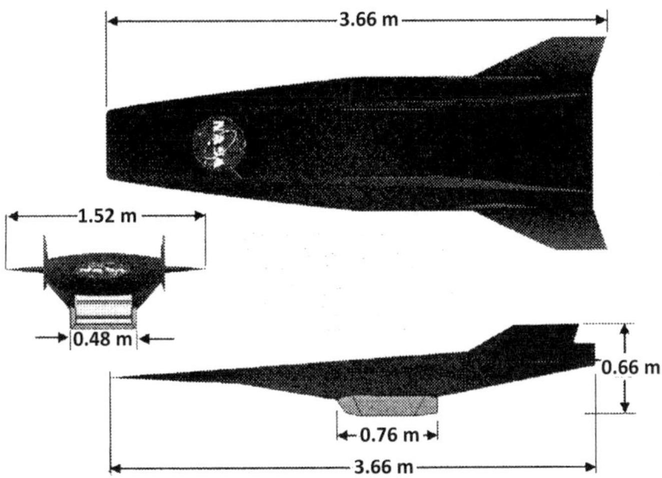

Fig. 5.1 Schematic of the experimental vehicle X-43A [2]

[1] To Peebles' book [1] a CD is attached with twenty-nine AIAA, NASA and other papers on the X-43A project.

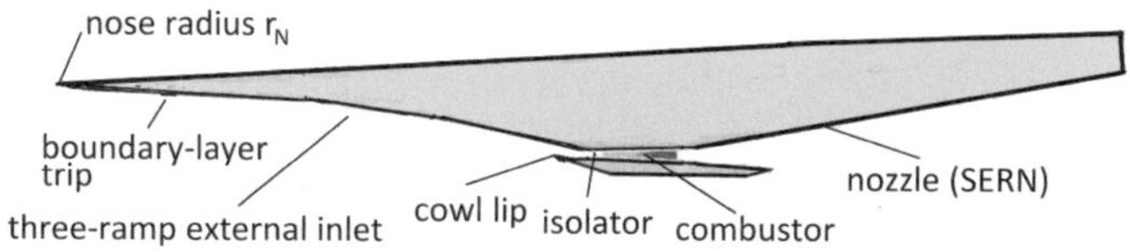

Fig. 5.2 Schematic of the propulsion flow path of the X-43A [3]

at the inlet, Sect. 5.2.[2] Because the forebody is nearly rectangular in planform, the onset flow to the inlet is two-dimensional, as it appears from [4].

Between the cowl and the vehicle body lies the internal propulsion flow path. It is beginning with the isolator in which a train of oblique shock waves further decelerates the flow until a Mach number around $M \approx 3$ to 3.5 is reached. It follows the combustor in which the fuel is injected, mixed and burned. The exhaust gas then is expanded through the single expansion ramp nozzle (SERN) generating in this process the intended thrust.

The thermal loads are very high. For the $M_\infty \approx 7$ flight the equilibrium real-gas totaltemperature was about 2,500 K and for the $M_\infty \approx 9.7$ flight about 3,800 K. The unswept cylindrical nose radius was $r_N = 0.762$ mm of the vehicle for the $M_\infty \approx 7$ configuration and $r_N = 1.27$ mm for the $M_\infty \approx 9.7$ configuration. Surface radiation cooling of the outer flow path led to manageable thermal loads. The inner flow path of course demanded active cooling.

For structural details see Sect. 6.4.2.

5.1.2 The Lower Stage of the TSTO-System SÄNGER

The selection process of the propulsion system for the lower stage of the TSTO-system SÄNGER converged to an integrated turbo/ramjet system with a co-axial internal flow path, Sect. 2.8. Five turboramjet engines were arranged side by side at the lower side of the fuselage, Fig. 5.3, each with 1,300 kN thrust at $M_\infty = 1.2$ [5].

Important is a proper two-dimensional onset flow to each of the five inlets. That is ensured, if the lower side of the forebody is flat. Figure 5.4 shows the basic situation.

If the—in our case—lower surface, inclined against the free-stream, is flat, then the skin-friction lines between the two attachment lines are straight and parallel to each other. This indicates that a two-dimensional flow is present between the attachment lines. The inlet arrangement must be such that the whole package lies between the two attachment lines. If the surface is not fully flat, the flow is only approximately two-dimensional.

The configuration of the lower stage of the TSTO-system, Fig. 5.5, has an overall length of 86.4 m with a wing span of 45.1 m, configuration 12/92 [5].

---

[2] It can be argued that the first ramp is the forebody, and that only two inlet ramps are present. Whether the ramp shocks were focussed on the cowl lip could not be ascertained.

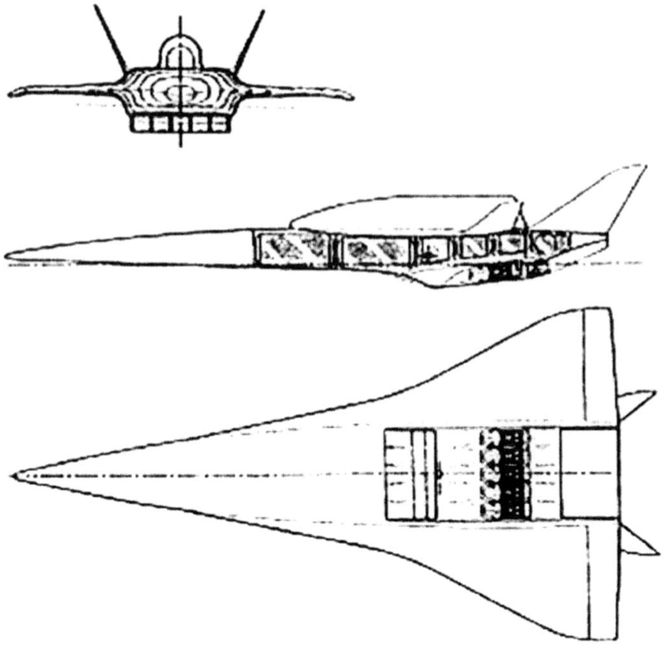

Fig. 5.3 Sketch of the SÄNGER configuration [6]. Lower part: view from below at the propulsion system

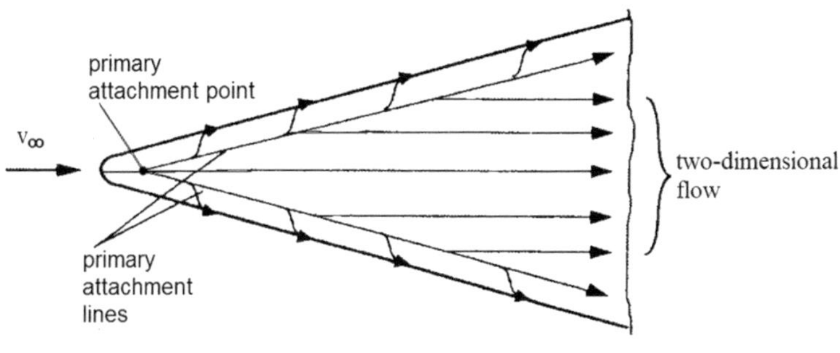

Fig. 5.4 Delta wing: sketch of the skin-friction lines against the free stream inclined flat lower surface [7]

The maximum fuselage width is 14.4 m. The lower side flow path is beginning at the blunt vehicle nose. The lower side up to the five inlets is a pre-compression surface. The five inlets have three external ramps each. The oblique ramp shocks at the nominal flight Mach number are focussed on the cowl lips, Sect. 5.2. Regarding the vehicle's bow shock, the situation is not clear. The internal part of the propulsion system is housed in the engine cowl. The nozzles are of SERN type. Structural details are given in Sect. 6.3.2.

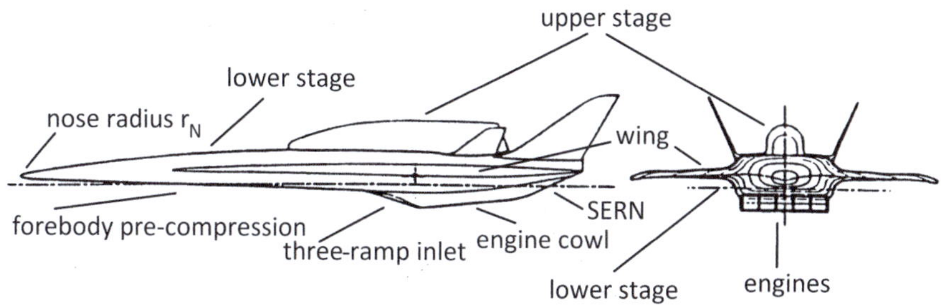

Fig. 5.5 Schematic of the TSTO-system SÄNGER

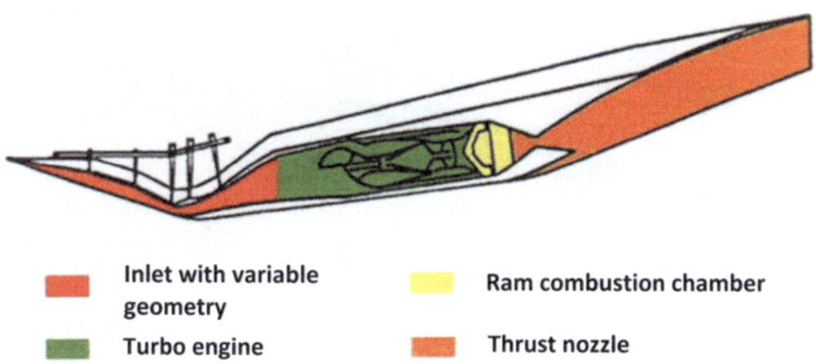

Fig. 5.6 The selected propulsion system of the lower stage of the TSTO-system SÄNGER [8]

Three propulsion systems were studied [5]. The finally selected propulsion system consists of four basic elements, Fig. 5.6.

These are the variable-geometry inlet including the boundary-layer diverter for the turbojet mode, the turbo engine, the ram combustion chamber, and the single ramp expansion nozzle. The structure and materials concept, except for the outer ramps of the inlet, demands actively cooled components. Auxiliary systems are the air cooler, the hydrogen pump, and so on, and finally a high-temperature measurement system for the monitoring and control of the system.

The operation modes of the co-axial turbojet-ramjet propulsion system are shown in Fig. 5.7.

A forward/backward moving cone channels the airflow toward the respective operation system. The turbojet operation domain—upper part of the figure—extends up to $M_\infty = 3.5$. In the turbojet mode the boundary layer from the forebody of the lower stage is diverted before it enters the inlet ramps. It is led around the propulsion system into the exhaust nozzle. In the ramjet mode it enters the engine flow path.

Above $M_\infty = 0.9$ the afterburner is active. After reaching $M_\infty = 3.5$, the cone moves forward, closing the flow path through the turbojet—lower part of the figure. The airflow now is led around the turbojet toward the afterburner, which hence acts as ramjet until the maximum flight Mach number $M_\infty = 6.8$ is reached.

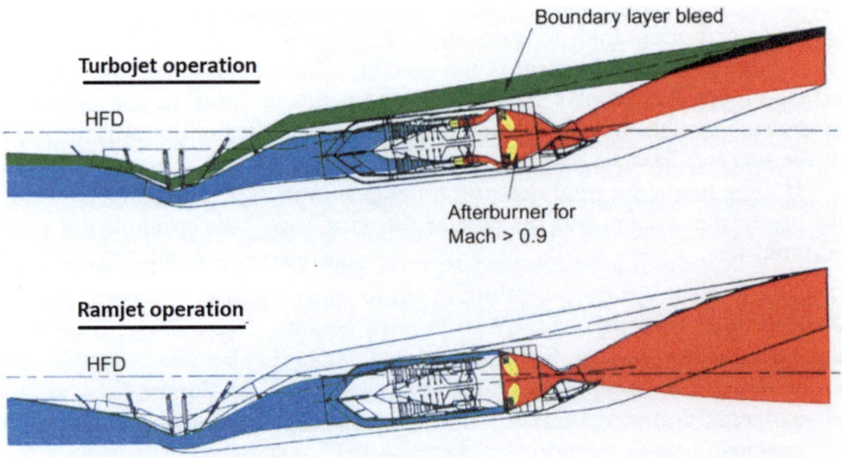

Fig. 5.7 The SÄNGER lower stage turboramjet propulsion system [9]. Above: turbojet operation mode up to $M_\infty = 3.5$, below: ramjet operation mode up to $M_\infty = 6.8$. HFD: Horizontal Fuselage Definition

For the nozzle geometry two concepts were studied, a two-dimensional rectangular nozzle and an axially symmetric plug nozzle. The latter had a smaller mass and an advantage in cooling handling. The former needed a cross-section change from the axial symmetric into a two-dimensional shape with a complex adaption mechanism for the different flight modes. The two-dimensional nozzle finally was selected, because it allowed to reduce the base drag and over the SERN led to a thrust increase at high Mach numbers.

A major problem are the high thermal loads on the internal flow path. The thermal loads in terms of the total temperature, see also Fig. 2.8, are shown in Fig. 5.8 [9].

The total temperatures given there in the turbojet mode at $M_\infty > 2.5$ and $H > 15$ km—below—and the ramjet mode at $M_\infty = 6.8$ and $H = 31$ km—above—of course are constant up to the respective combustion location. The engine feasibility was determined by the ability to cool the internal flow path.

5.1.3 *The MR2 and MR3 Concept of the STRATOFLY Study*

Several concepts were examined in the STRATOFLY study, see, e.g., [10]. We have a short look at details of the MR2 and the MR3 concept. The MR2, Fig. 5.9, is the forerunner of the MR3 concept.

The length of the $M_\infty = 8$ concepts is 94 m, the wing span is 41 m. The total temperatures in the turbojet mode at $M_\infty > 2.5$ and $H > 15$ km—below—and the ramjet mode at $M_\infty = 6.8$ and $H = 31$ km—above—of course are constant up to the respective combustion location. The engine feasibility was determined by the ability to cool the internal flow path.

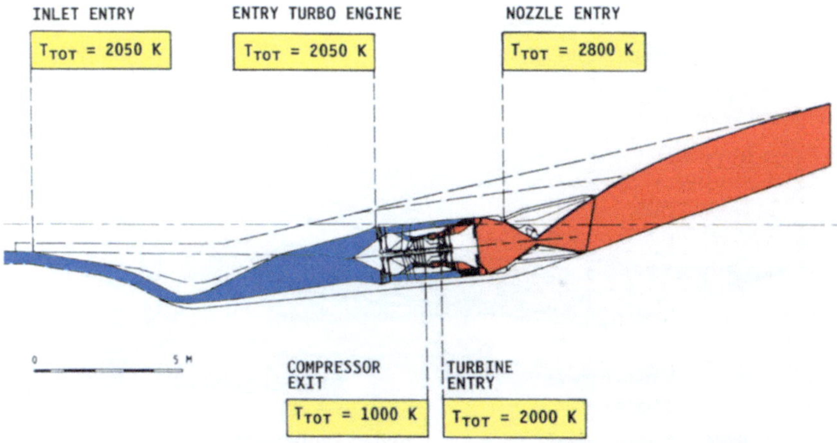

Fig. 5.8 The SÄNGER lower stage propulsion system: thermal loads in terms of the total temperature T_{TOT} [9]. Above: ramjet operation at $M_\infty = 6.8$ and $H = 31$ km, below: turbojet operation at $M_\infty > 2.5$ and $H > 15$ km

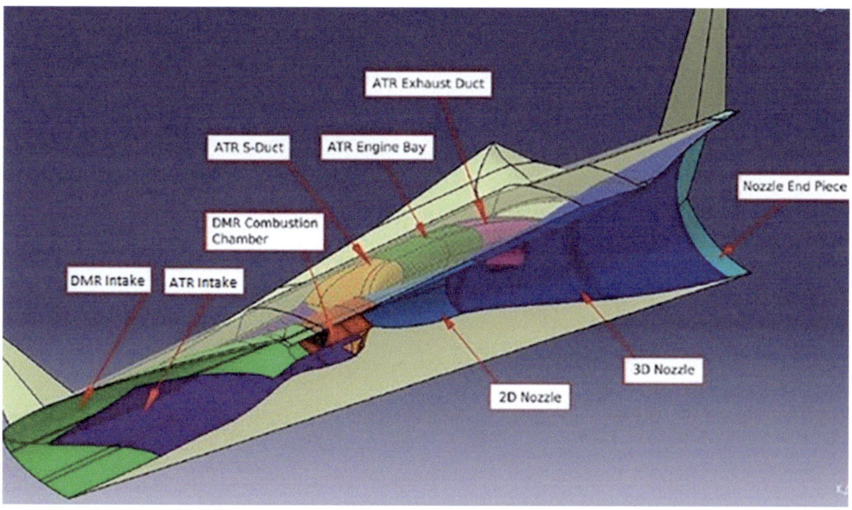

Fig. 5.9 The MR2 concept as half model with the sub-systems [11]

The reference mission is long-haul hypersonic flight, for instance from central Europe to Australia, with $R \approx 16{,}000$ km. With a total flight time of about 8,000 s, hypersonic speed at $M_\infty = 8$ and $H \gtrsim 30$ km was assumed to last for about 3,000 s. Degradation of passenger comfort and physical strain determine the acceleration and deceleration time periods, [12], hence the time-restriction of the $M_\infty = 8$ flight segment.

The fuselage is a multi-bubble one, see, e.g., [10]. Above located is the propulsion system, The cryogenic tanks are located in the front, at the sides and at the rear of the flight vehicle. The passenger cabin, located beneath the propulsion system, has a width of 12 m and a length of 34 m.

MR2 and MR3 are of API-type 2, i.e., the propulsion flow path is completely on top and inside of the fuselage. The inlet is an inward turning one, located at the upper side of the fuselage. The Air Turbo Rocket (ATR) engines and the Dual Mode Ramjet (DMR) have different internal inlets.

The DMR lies between the ATR engines, Fig. 5.10. The fuel is hydrogen for both propulsion means. The ATR engines operate up to $M_\infty = 4$ to 4.5, the DMR then up to $M_\infty = 8$.

The two-dimensional nozzle has a length of 13 m, downstream it changes into the three-dimensional nozzle of 40 m length. An end piece matches the nozzle with the external shape. The exit cross section is nearly quadratic with about 10 m side length.

Figure 5.11 gives the schematic of the DMR propulsion system and Fig. 5.12 that of the ATR system.

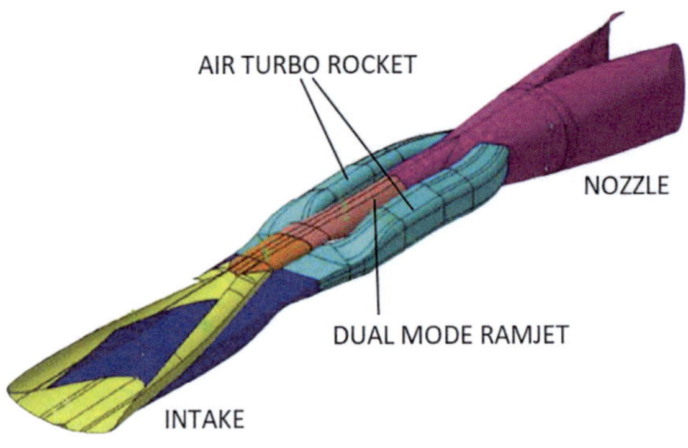

Fig. 5.10 The propulsion system consisting of the ATR engines and between them the DMR engine [11]

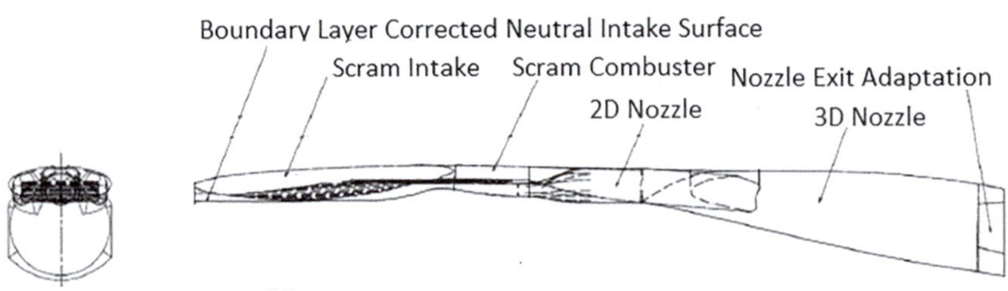

Fig. 5.11 Sketch of the DMR propulsion system [11]. Front view and side view

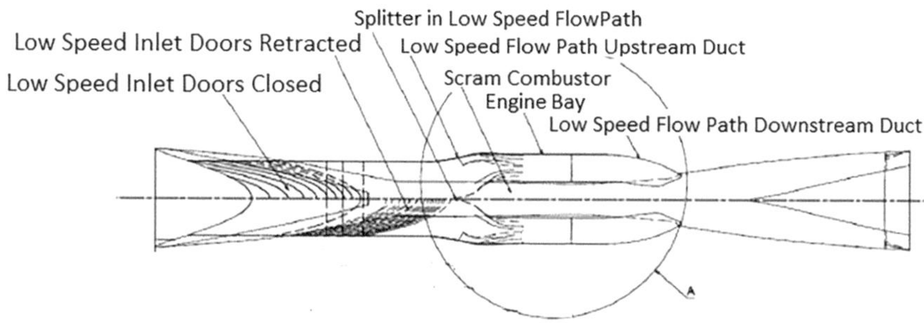

Fig. 5.12 Sketch of the ATR propulsion system [11]. View from above

The problem of aerothermoelasticity of the forebody and its influence on the inlet pre-compression hence appears to be avoided, Sect. 5.5.2. How far aeroelasticity effects play a role for the whole flight vehicle is not evident.

5.2 The Rectangular Ramp Inlet of API-Type 3, 4 and 5 Flight Vehicles

The purpose of the inlet is to deliver with the right local Mach number a steady, uniform and distortion-free airstream to the compressor of the turbojet or to the combustor of the ramjet or scramjet. The local Mach number for the turbojet and the ramjet then is subsonic, for the scramjet supersonic. For the flight vehicles of the considered API-types the inlet usually is a mixed external-internal compression inlet. Figure 5.13 schematically shows a rectangular ramp inlet for the turbojet or ramjet operation, i.e., with finally subsonic Mach number in the diffuser.

The figure also shows with situation (a) the ramp shocks, and the bow shock too, focussed on the cowl lip: shock-on-lip situation. That is the case at the nominal design flight Mach number. The captured mass flux is maximum. For any lower supersonic flight Mach number the situation (b) is present with a lower captured mass flux. Situation (c) finally represents the situation at a flight Mach number higher than the nominal one, with a lower captured mass flux, too. The reader should note, that for the different lower/higher Mach-number cases the ramp-shock locations are different, which is not indicated in the figure.

In the literature these cases usually are discussed as critical, sub-critical and super-critical cases in view of the mass-flow rate, which is required by the engine. The reader should consult [13–15]. In [16] the discussion is in view of the reference concept SÄNGER of the German Hypersonics Technology Programme at that time.

It is necessary to consider the tacit assumptions behind these discussions. They are (a) flight through an undisturbed atmosphere at constant altitude, (b) a fixed angle of attack α and (c) with API-type 4 and 5 flight vehicles a fixed inclination angle $\overline{\alpha}$

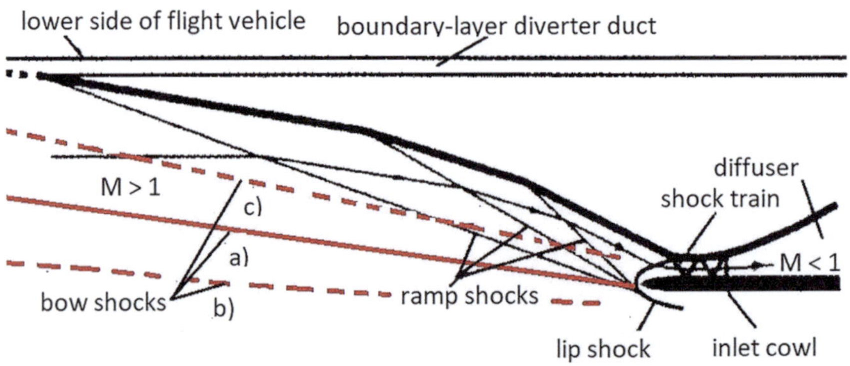

Fig. 5.13 API-type 4 and 5 three-ramp inlet with mixed external-internal compression. Red lines: **a** flight-vehicle bow shock at nominal—maximal—flight Mach number, **b** bow shock at lower flight Mach number, **c** bow shock at higher flight Mach number

of the lower side of the forebody, which acts as a pre-compression surface, Sect. 5.5. For all flight vehicles it is assumed (d) that the forebody is an entirely rigid one.

All that does not necessarily hold in the reality of hypersonic flight vehicles in particular of API-type 4 and 5. The API-type 4 X-planes X-43A and X-51A, especially the first of them, probably had a stiff forebody, but that does not hold for the forebody of the API-type 5 SÄNGER configuration, see in this regard also Sect. 2.10.6.

Particular issues are shortly considered:

1. **Boundary-layer diverter** Figure 5.13 shows on the upper side the boundary-layer diverter duct. That is necessary in order to remove the forebody boundary layer in the turbojet propulsion mode. In that mode the boundary-layer material would lead to flow distortion, which at any rate must be avoided. In the ramjet or scramjet mode the duct is closed—the broken line—because then boundary-layer ingestion is not of concern. Where in the latter two modes the limit is for boundary-layer ingestion remains an open question.
   The X-43A at the lower forebody side had a boundary-layer trip about 46 cm downstream of the leading edge. Forcing turbulent flow was deemed to be necessary in order to have unambiguous flow conditions at the inlet and in the engine [17].[3] Higher thermal loads and friction drag were counterbalanced by a higher air mass flux in the engine, reduced susceptibility of flow separation and improved fuel mixing.
   For a long forebody like that of the SÄNGER lower stage, a boundary-layer trip approach appears not to be feasible. Over-tripping as well as relaminarization, see, e.g., [18], would make questionable a reliable estimation of the properties

[3] The upper Mach-number limit of the possibility of effective turbulence tripping usually is seen to be around a **local** boundary-layer edge Mach number $M_e \approx 5$ to 8, see, e.g., [18].

of the boundary-layer reaching the external ramps and hence also of the diverter height.

2. **Bleed** At the foot of the oblique ramp shocks Edney type VI interaction is present, see also Sect. 3.6.5. However, boundary-layer separation must be avoided. That holds also for the shock train in the inlet cowl. Boundary-layer bleed is a possibility to prevent it. In [16] a detailed discussion is given.
   We note for the turbojet mode that bleed is necessary in order to achieve high pressure recovery and low flow distortion. Bleed momentum must be recovered by injection into the nozzle flow, see Sect. 5.6. That adds weight and structural complexity. Bleed holes generally pose surface roughness, which leads to drag increments.
   In the ramjet mode bleed is not possible because of the high thermal loads in the ducts. Moreover, thrust is dominated by the mass flow, if the pressure recovery is high enough. Hence mass should not be removed. To close bleed holes adds complexity and mass.
   The conclusion is to avoid bleed or to minimize its application needs by a suitable layout of the flow path.
3. **Side walls** With several engines side by side, as with the lower stage of SÄNGER, Fig. 5.5, side walls become an issue. Side walls between adjacent inlet ramps and inlets would lead to corner flow, non-convex surface radiation cooling effects, Sect. 3.9, additional friction drag, boundary-layer/shock-wave interaction and, last but not least, to weight increments.
   Hence side-walls, at least their extension between inlet ramps, should be minimized. A special issue are the outer inlets, where a side-spillage into the free-stream happens. In [16] side walls only from the foot of the second ramps to the inlet cowl are seen as a possibly optimal compromise solution.
4. **Wall temperature jump over ramps** The inlet ramps are part of the external flow path. At the lower side of the lower stage of SÄNGER at the highest Mach number the radiation-adiabatic wall temperature was determined to be around 1,000 K, Fig. 3.31. At the inlet ramps, however, different temperature levels are present. In Sect. 3.10.3 with Fig. 3.37 it is shown that a stepwise wall-temperature jump happens over the ramps of the inlet. That is due to the rise of the pressure and the unit Reynolds number [18]. Hence the surface radiation cooling effect is reduced.
   Consequently the structure and materials concept of the forebody is not adequate for the inlet ramps. Also for SÄNGER this was an issue of the structure and materials concept of the inlet ramps. For them CSiC ceramic was foreseen and specimens were manufactured and tested.
   We observe further that the Mach number and the unit Reynolds number changes influence the general inlet boundary-layer and flow behavior and properties. The Edney type VI interaction around the ramp corner line was mentioned above.
5. **Shock-on-lip** At the nominal flight Mach number, case (a), Fig. 5.13, the bow shock and the ramp shocks are centered on the lip of the cowl. The cowl has its own bow shock, Fig. 5.14. Here an Edney interaction type IV is present, with locally very high pressure and thermal loads, see, e.g., [18]. Such an interaction

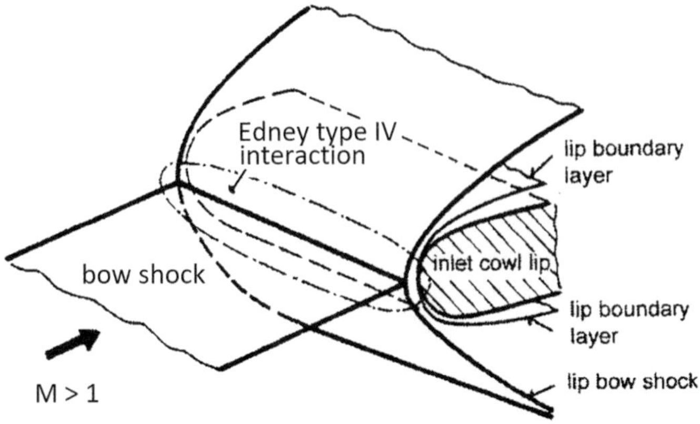

Fig. 5.14 Schematic of the shock-on-lip situation of the inlet. The ramp shocks are not indicated

was present at the pylon of a ram engine on the X-15, Sect. 2.9.4. A similar interaction happens at double-delta wings, see Fig. 3.22.
For the flight vehicle design it is mandatory to identify such interactions and to take care that they are avoided at all or treated accordingly. The problem in this case is that the cowl lip must have a small thickness in order to reduce the wave drag. With the small thickness, radiation cooling as well as active cooling become problematic. A way out is the use of very heat resistant material for the cowl.

6. **Gap flow** Gap flow present for instance at a ramp foot with variable angle or at flap hinges is very problematic. High thermal loads are present, because in the gap radiation cooling is not possible, see, e.g., [12]. Such situations are the most challenging in hypersonic flight vehicle design. Their simulation, either experimentally or computationally is very demanding.

5.3 Aspects of Airframe-Propulsion Integration

From the large aspect ratio subsonic/transonic transport aircraft configuration to the slender hypersonic vehicle shape of API-type 5 a change of lift distribution is evident. The geometric and configurational integration of wing and body of the latter type brings these configurational elements on almost equal levels of lift generation.

The same trend is found regarding the propulsion system of such vehicle types. The area of influence—regarding the shape of the lower side of such configurations—is reaching from the nose of the forebody to the end of the afterbody, including the inlet, the engine and the nozzle.

5.3.1 Balance of Thrust

The background of the increasing necessity and complexity of airframe-propulsion integration with the flight Mach number is given in Fig. 5.15. Comparing three high-speed configurations, the external forces on the inlet (inlet-drag force) and the gross thrust vector are shown. For the surfaces impinged by the propulsion system see the three configurations at the right of Fig. 2.5.

Before we consider the balance of thrust and drag in detail, we need to have a look at the forces, which act on the flight vehicle, Fig. 5.16. Besides the lift and the drag of the "clean" configuration it is the inlet-drag force, also called ram-drag force and the gross thrust force of the engines. With high flight velocities the centrifugal force plays a role.[4]

The centrifugal force, also called g-reduction, approximately reads

$$\frac{g_{eff}}{g} = (1 - (v_\infty / v_0)^2), \tag{5.1}$$

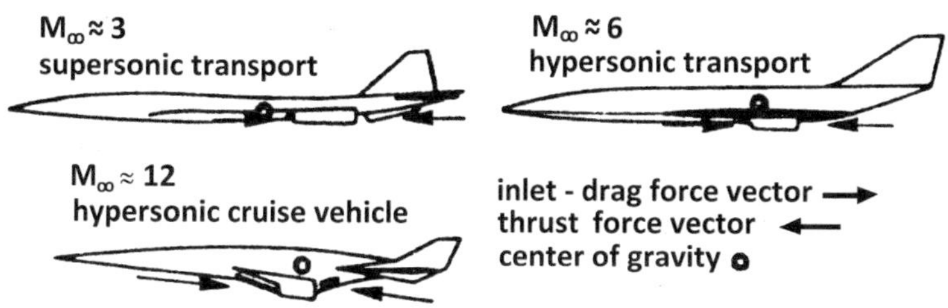

Fig. 5.15 Examples of balance of thrust [19]

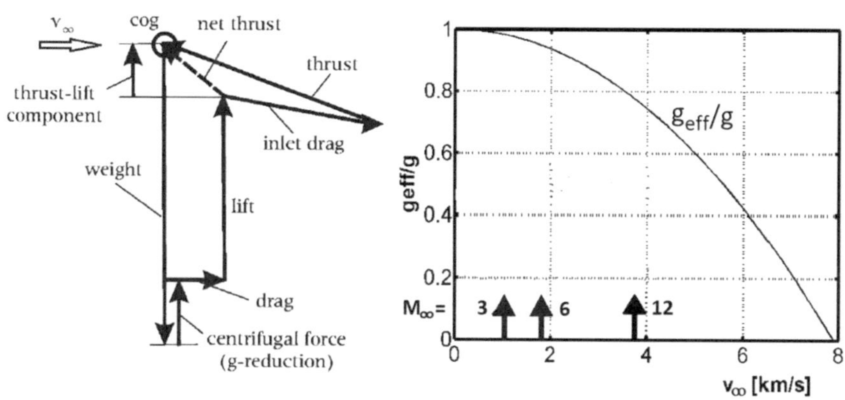

Fig. 5.16 About the forces acting on the hypersonic aircraft. At the left shown is the schematic of the mass-force polygon, at the right the g-reduction with flight speed [20]

[4] What commonly is called horizontal flight actually is a circular flight over the surface of Earth.

Table 5.1 Comparison of the magnitude of the vectors of inlet drag, gross thrust and their differences of the three flight vehicles shown in Fig. 5.15 [19]

Flight vehicle	Inlet drag (kN)	Gross thrust (kN)	Net thrust (kN)	Per cent of gross thrust
Supersonic transport turbojet, $M_\infty \approx 3$	242	547	305	56
Hypersonic transport ramjet, $M_\infty \approx 6$	979	1,468	489	33
Hypersonic cruise scramjet, $M_\infty \approx 12$	8,674	9,341	667	7

with $v_0 = 7.9087$ km/s being the mean circular orbit velocity around Earth. The g-reduction is of particular interest for hypersonic passenger transport [12].

The actual engine net thrust is indicated by the broken line at the left of Fig. 5.16. Note that it also has a thrust-lift component.

Coming back to Fig. 5.15 we compare in a simplified manner—i.e., only the magnitude of the vectors—the inlet drag, the gross thrust and the net thrust, Table 5.1.

The useful net thrust of the engines is a measure of the total efficiency of the propulsion system. For the supersonic transport vehicle the ratio of net thrust to gross thrust—the direction of the vectors being neglected—is about 56%. For the hypersonic transport vehicle at $M_\infty \approx 6$ the ratio is down to 33% and for the hypersonic cruise vehicle at $M_\infty \approx 12$ down to only 7%.

Hence if the inlet drag is estimated 1% too low—and the gross thrust 1% too high—the correction of this error reduces in the $M_\infty \approx 12$ case the net thrust from 7 to 5.27%, say by one fourth. The net thrust thus is hit with a 25 fold effect!

This is an example of an extremely high sensitivity of design with respect to engine performance (thrust and drag). It also shows the need of a highly efficient airframe-propulsion integration.

For the $M_\infty \approx 6$ case only a loss of approximately 3.3% is found with (33 − 32)/33. The sensitivity hence did decline from 14% for the $M_\infty \approx 12$ case to this low number. For the lower stage of SÄNGER with the upper-stage separation Mach number of $M_\infty = 6.8$ this meant a drastic reduction of the design risk.

5.3.2 *Comparison of Two Types of Airframe-Propulsion Integration*

We compare the two basically different API-types shown in Fig. 5.17. The SR-71—although not being exactly a hypersonic flight vehicle with $M_\infty < 4$—with podded engines is an API-type 1 flight vehicle, the SÄNGER lower stage with $M_\infty = 6.8$, is a highly integrated API-type 5 flight vehicle.

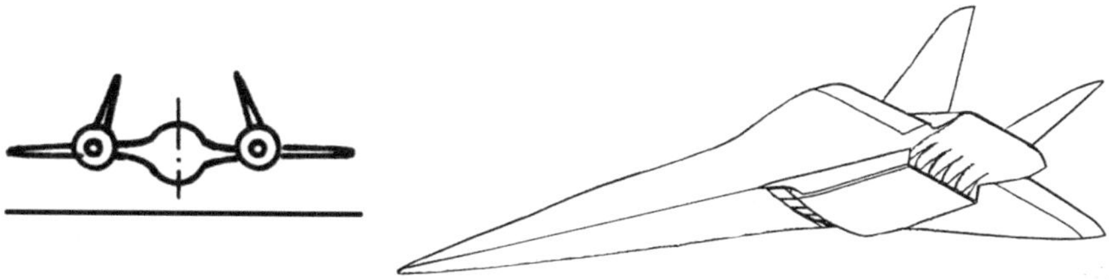

Fig. 5.17 Two API-types: left the SR-71 as API-type 1 flight vehicle, right the lower stage of the TSTO-system SÄNGER as API-type 5 flight vehicle

Both API-types have their benefits and disadvantages.

- **API-type 1 flight vehicle SR-71**

  1. Benefits:

     – the weight of the engines reduces the effect of the aerodynamic loads on the wing (root bending moment, weight advantages),
     – the engine inlets are in undisturbed flow, no boundary-layer diverter is needed.

  2. Disadvantages:

     – the pre-compression happens in the inlet,
     – the inlet is exposed to asymmetric flow conditions with angle of attack and sideslip,
     – high moments of inertia around the vehicle's longitudinal (x-) and vertical (z-) axes,
     – extreme yawing moment in the case of loss of thrust of one engine (safety aspects, the constructive solution of this problem of the SR-71 was found with the inwardly inclination of the two vertical tails to avoid induced yawing moments).

- **API-type 5 flight vehicle SÄNGER lower stage**

  1. Benefits:

     – synergetic effects of forebody pre-compression: reduction of inlet capture area, Sect. 5.5, generation of lift,
     – two-dimensional inlet-onset flow, Sect. 5.1.2,
     – compact construction,
     – smaller engine system due to pre-compression, ground inspection possible.

2. Disadvantages:

   – thick boundary-layer at the inlet location, boundary-layer diverter necessary in turbojet propulsion mode,
   – adjacent engine influencing in case of engine failure,
   – coupling of forebody aerothermoelasticity into engine flow path (change of pre-compression, hence of captured air mass and engine thrust).

5.4 API-Type 5 Flight Vehicles: The Book-Keeping Problem

With API-type 5 flight vehicles the distinction of the aerodynamic and the propulsion forces is extremely demanding. Different book-keeping approaches of forces and moments are used in predesign and design phases, see Chap. 7. Behind this is the fact that in these phases a complete simulation of the whole flight-vehicle system is not possible. With increasing computer power it may become possible in the frame of the Virtual Product approach, Sects. 2.2 and 7.6. So far the participating disciplines aerodynamics and propulsion make available data of forces and moments, which then are combined in a suitable book-keeping procedure.

Obviously the separation of the individual loads becomes the more complicated, the higher the degree of airframe-propulsion integration is. Figure 5.18 presents an overview of the composition of the net thrust vector $\overline{F}_N$.

Details of the forces acting on the propulsion system, beginning at the external inlet ramps and extending up to the single expansion ramp nozzle, are given in Fig. 5.19.

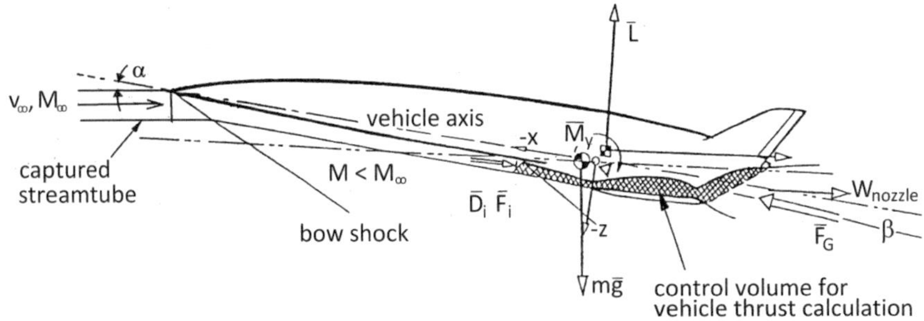

Fig. 5.18 Schematic of the book-keeping of forces and moments at a API-type 5 flight vehicle. Aerodynamic forces act on the airframe, forces of the propulsion system act in the hatched area [21]

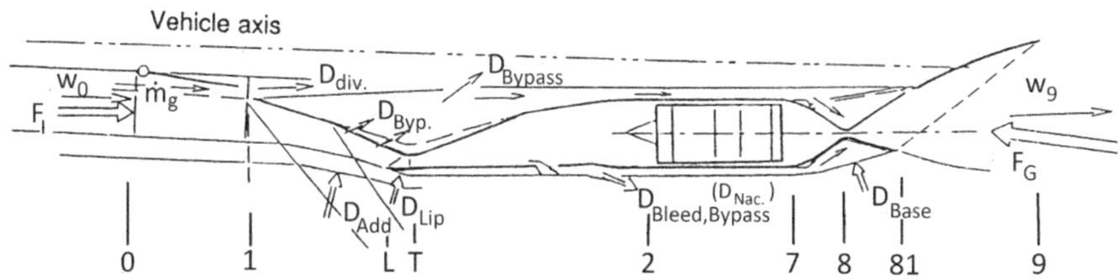

Fig. 5.19 Details of the forces acting on the propulsion system [21]

Equation (5.2) then shows the composition of the net thrust vector $\overline{F}_N$:

$$
\begin{aligned}
\overline{F}_N &= \overline{F}_G - \overline{F}_I - \sum \overline{D}_I + \sum \overline{F}_I \\
&= \overline{F}_G - \overline{F}_I - (\overline{D}_{Add} + \overline{D}_{Lip}) - \overline{D}_{Base} - \overline{D}_{Div} \pm \overline{F}_{I\,Bleed,Byp} \\
&= (\dot{m}_9 \overline{w}_9 + \Delta p_9 \overline{L}_9) - (\dot{m}_0 \overline{w}_0 + \Delta p_0 \overline{L}_0) - \int_{L_{A/L}} \Delta p \, d\overline{L} - \int_{L_B} \Delta p \, d\overline{L} \\
&- \overline{D}_{Div} \pm \overline{F}_{I\,Bleed,Byp}.
\end{aligned}
\tag{5.2}
$$

The vector is composed of 20 partial forces acting at different locations and in different directions, Fig. 5.19. $\overline{L}$ is the aerodynamic lift, $\overline{mg}$ the weight of the flight vehicle, $\overline{D}_I$ are components of drag forces, $\overline{F}_I$ are components of thrust forces, and so on.

5.5 Aspects of Forebody Pre-compression

Forebody pre-compression is the primary feature of API-type 3, 4 and 5 flight vehicles. We give a short overview of this topic and its implications.

5.5.1 The Concept of Forebody Pre-compression

The influence of the shape of the forebody and the inclination of its flat lower side against the free stream $\overline{\alpha}$—angle of attack α plus the angle between the lower side and the reference x-axis of the vehicle—is a matter of stream-tube compression. That is shown schematically in Fig. 5.20 for the flat plate with the inclination angle α against the free stream.

Here M_1 is the free-stream Mach number, h_1 the free-stream stream-tube height, Θ the angle of the oblique shock wave, h_2 the compressed stream-tube hight, and finally M_2 the inlet-onset flow Mach number downstream of the oblique shock wave.

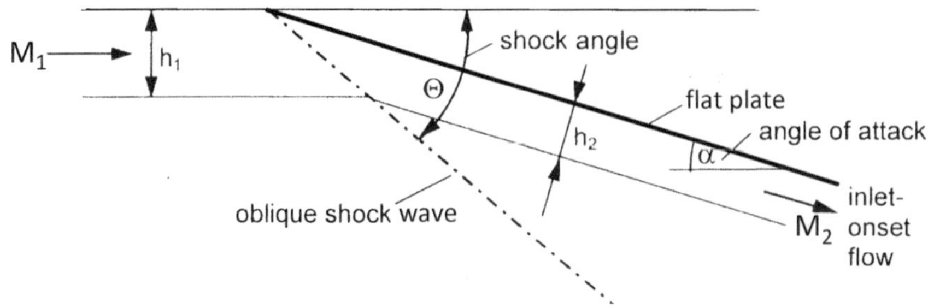

Fig. 5.20 Schematic of the stream-tube compression at a flat plate at angle of attack α

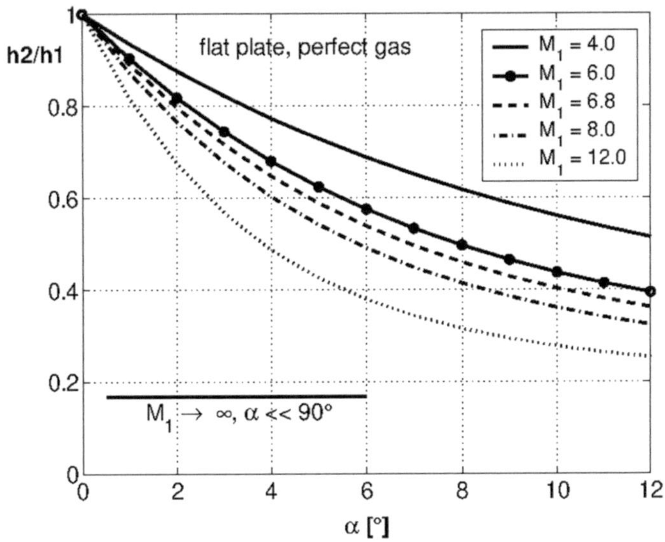

Fig. 5.21 The result of the stream-tube compression: h_2/h_1 as function of α, perfect gas

Figure 5.21 shows the ratio h_2/h_1 of the compressed stream tube as function of the angle of attack α for different free-stream numbers Mach M_1. The effect is strong and increases with increasing M_1. For $M_\infty \to \infty$ and small α we obtain $h_2/h_1 = 0.167$ [12].

The situation is different for a round-nosed forebody like that of the SÄNGER lower stage, Fig. 5.22. There four oblique shock waves are present, the bow shock and the three ramp shocks. Of interest here is only the effect of the bow shock.

Compared to the flat-plate case we find a quasi-linear dependence of A_∞/A_0 on the incidence $\overline{\alpha}$ (h_2/h_1 corresponds to A_0/A_∞). Note that the reciprocal value is given. Compared to Fig. 5.21 a somewhat higher effect of pre-compression is seen.

Not to be forgotten, at the flight vehicle the lower side of the forebody, the pre-compression surface, is inclined against the free stream by the angle $\overline{\alpha}$, which is larger than the nominal angle of attack α of the flight vehicle.

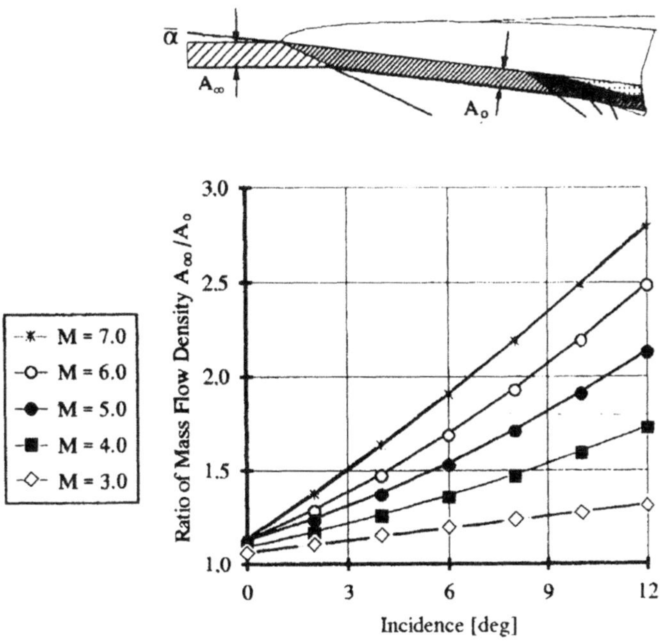

Fig. 5.22 The pre-compression system at a SÄNGER lower-stage type forebody for five flight Mach numbers [22]: A_∞/A_0 as function of $\overline{\alpha}$

5.5.2 *Changes of Flow Properties with Changes of $\overline{\alpha}$*

We consider qualitatively changes of the flow properties along the lower side of the forebody, which are due to changes $\Delta\overline{\alpha}$ of its inclination angle $\overline{\alpha}$. The changes may be due to changes of the flight conditions, due to gusts or to aeroelastic and to aerothermoelastic deformation of the forebody, for the latter see the Banana Effect in Sect. 2.10.6. In any case $\Delta\overline{\alpha}$ is considered to be small, such that the effects are in the linear domain.

- The static pressure increases with positive $\Delta\overline{\alpha}$ and decreases with negative $\Delta\overline{\alpha}$.
- The same is true for the total pressure loss.
- The static temperature increases with positive $\Delta\overline{\alpha}$ and decreases with negative $\Delta\overline{\alpha}$. The general temperature distribution along the lower side of the forebody, as shown in Fig. 3.31, does not change.
- The forebody lift part increases with positive $\Delta\overline{\alpha}$ and decreases with negative $\Delta\overline{\alpha}$.
- The same is true for the drag part.
- The pitching moment increases with positive $\Delta\overline{\alpha}$ and decreases with negative $\Delta\overline{\alpha}$.
- Aerodynamic loads, hence aeroelastic effects, increase with positive $\Delta\overline{\alpha}$ and decrease with negative $\Delta\overline{\alpha}$.

- The boundary-layer edge Mach number M_e decreases with positive $\Delta\overline{\alpha}$ and increases with negative $\Delta\overline{\alpha}$.
- The same is true for the unit Reynolds number Re_e^u.
- The net thrust of the propulsion system increases with positive $\Delta\overline{\alpha}$ and decreases with negative $\Delta\overline{\alpha}$, see next sub-section.

This qualitative consideration emphasizes the strong coupling of lift and propulsion of airbreathing hypersonic flight vehicles of API-type 3, 4 and 5. If, as an example, the net thrust is increased by increasing the mass flux via pre-compression, also the drag increases, as well as the lift and the pitching moment, which in turn affects the trim drag.

As indicated, thermal loads are increased, as well as the skin friction, which is a part of the vehicle drag. The change of the net thrust with $\Delta\overline{\alpha}$, i.e., the problem of thrust sensitivity, is considered next.

5.5.3 An Example and the Problem of Net Thrust Sensitivity

For the lower stage of the TSTO-system SÄNGER at $M_\infty = 6.8$, $H = 31$ km and $\alpha = 6°$ the inclination angle of the lower side is $\overline{\alpha} = 9.4°$. That leads to a reduction of the capture area at infinity ahead of the flight vehicle by a factor of two at the five inlets [12], Fig. 5.23.

Figure 5.24—based on the SÄNGER forebody with a constant capture area A_0—presents the relative net thrust of the combined engines—turbojet for $M_\infty < 3.5$ and ramjet for $M_\infty \geqq 3.5$—as function of the angle of attack of the flight vehicle, i.e., of the rate of pre-compression at the inlets. Note that the actual incidence angle $\overline{\alpha}$ of the flat lower side of the fuselage is higher than α (= aircraft incidence).

The thrust is always referenced to $\alpha = 0°$. The design point of the propulsion system including inlet and nozzle in this case is 150% thrust at $M_\infty = 6.8$ and $\alpha = 5°$. The turbojet engine is reacting, as usual, negatively on the excessive supply of air (curve at the bottom, $M_\infty = 3.0$, Turbojet).

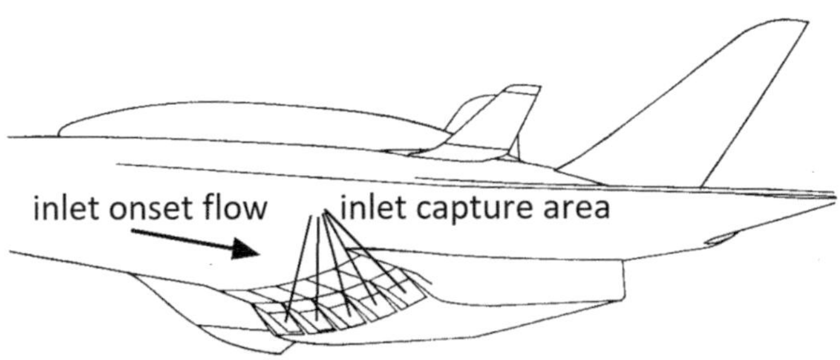

Fig. 5.23 The five engine inlets at the SÄNGER lower-stage

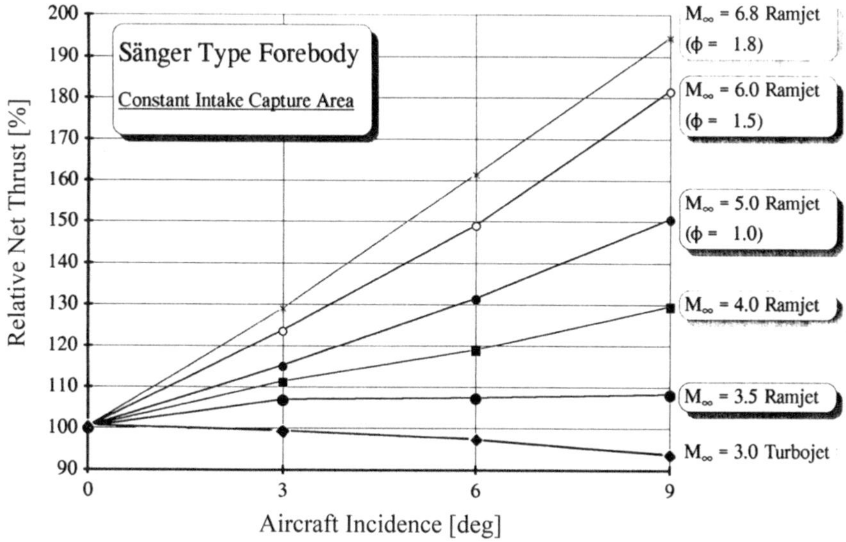

Fig. 5.24 Effect of forebody pre-compression on the net thrust of a turbojet/ramjet propelled hypersonic airbreather [22]

The switch from turbojet to ramjet propulsion is made at moderate pressure, i.e., angle of attack, in order to avoid excessive loads in this critical phase of flight. With increasing flight Mach number $M_\infty = 4$ to 6.8 increases the relative net thrust of the ramjet due to the increasing pre-compression, and—for $M_\infty > 5$—due to the additional over-stoichiometric fuel (LH_2) supply (parameter $\Phi > 1$ in the figure). The combination of both effects virtually leads to a duplication of the reference thrust at $\alpha = 0°$.

Figure 5.24 also indicates a high net-thrust dependence on the angle of attack, which increases with increasing flight Mach number. This is the problem of net thrust sensitivity. As we have seen with Fig. 5.16, the net thrust is the small difference of large forces, which act in different directions. All these forces and in particular the net thrust change with speed, attitude, and engine settings along the trajectory. The result is a large sensitivity of the net thrust.

We give another example from the SÄNGER studies, for details see [12], where results of the study [23] are sketched. The installed thrust on the trajectory is defined as the thrust produced by the nozzle minus the flow momentum entering the inlet, the drag due to forebody boundary layer diversion, spill drag, bypass drag, and bleed drag.

Figure 5.25 shows the sensitivity of the net thrust on changes of the inclination $\overline{\alpha}$ of the lower surface of the forebody. The inclination angle is a measure of the forebody pre-compression.

The result is that a change of $\Delta\,\overline{\alpha}$ results in changes of the net thrust, positive for positive, and negative for negative $\Delta\,\overline{\alpha}$. We see that at $M_\infty = 6.8$—the highest flight

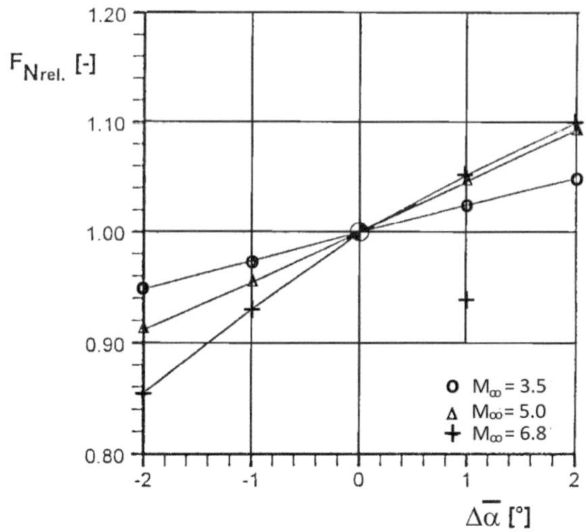

Fig. 5.25 Influence of the incidence $\Delta\,\overline{\alpha}$ against the free stream of the lower side of the forebody on the net thrust at different flight Mach numbers [23]

Mach number—only one degree less than the nominal $\overline{\alpha}$ reduces the net installed thrust by approximately seven per cent! One degree more would give about five percent more thrust. For the lower Mach numbers the effect is smaller. All this is due to the forebody pre-compression, which changes with changing angle of attack. Note that the effect is smaller than that shown in Fig. 5.22, where only positive $\Delta\,\alpha$ data are given.

As sketched above, not only the lift is affected, but also the other flight-mechanical parameters, as well as the fluid-mechanical ones. Hence the inclination angle $\overline{\alpha}$ of the lower side of the forebody must be controlled precisely. Consequently the air data and attitude sensors, the propulsion control system, the flight control system including the aerodynamic control surfaces must react in real time on even small changes of $\overline{\alpha}$. All this is governed by the static and dynamic aerothermoelastic properties of the forebody.[5]

The problem hence is different for the different API-type flight vehicles. A hypersonic cruise missile, API-type 3, would have a rather rigid airframe, like the API-type 4 experimental flight vehicle X-43A. In the case of a large lower stage of a TSTO space transportation system like the API-type 5 lower stage of SÄNGER, however, the aerothermoelastic properties of the 52 m long forebody would pose a huge problem.

When looking at the design approach which is the usual one in aircraft design, see Sect. 7.2, the conclusion is that in view of the very small payload fraction of such space transportation systems another approach must be developed, see also the discussion in [12].

---

[5] Preliminary evaluations during the SÄNGER studies have shown that for the then given structure concept amplitude ranges from $\pm 1°$ to $\pm 3°$ with complicated eigenmodes and frequencies from 1 to 3 Hz were to be excepted [24]. Such structural properties of course are not acceptable.

5.6 About Thrust and Installation Losses

Thrust and installation losses of the propulsion system need a close consideration, too. In this sub-section they are considered for the ascent trajectory of the API-type 5 TSTO-system SÄNGER. Figure 5.26 shows the "market situation" of offer and demand for the inlet and the power plant.

The schematic presents the direction and the magnitude of the different parts of the airflow. For $M_\infty = 1.2$ only 41 per cent of the total captured mass flow is useful for the generation of thrust, the rest has to be rendered drag harmless or to be bypassed. For the design flight Mach number $M_\infty = 6.8$, the inlet makes use of about 98% of the mass flow, diverter flow, inlet bleed and bypass flow are zero.

Besides the inlet also the nozzle is directly interfering with the outer flow. Therefore, the book-keeping problem, as addressed in Sect. 5.4, again is of importance in view of the relevant forces and moments. This is visualized in Fig. 5.27.

Small double-hatched regions at the afterbody are part of the external flow path, hence, belong to the external aerodynamics. Single-hatched surface belong to the propulsion system. Hence their force and moment portions belong to the propulsion system. Therefore, these are found within the thrust data. The sketch also shows the components for the adjustment of the nozzle and the boundary-layer diverter flow, the latter for the turbojet propulsion mode only.

The dimensioning of the nozzle's final cross-section is given in Fig. 5.28. It shows the relative net thrust as function of the relative nozzle exit area. ($A9$ is at the location 9 in Fig. 5.19, whereas CA is the capture area of the inlet.)

Parameters are the design-critical flight Mach number $M_\infty = 1.2$ (turbojet mode) and the maximum flight Mach number $M_\infty = 6.8$ (ramjet mode). The effect of

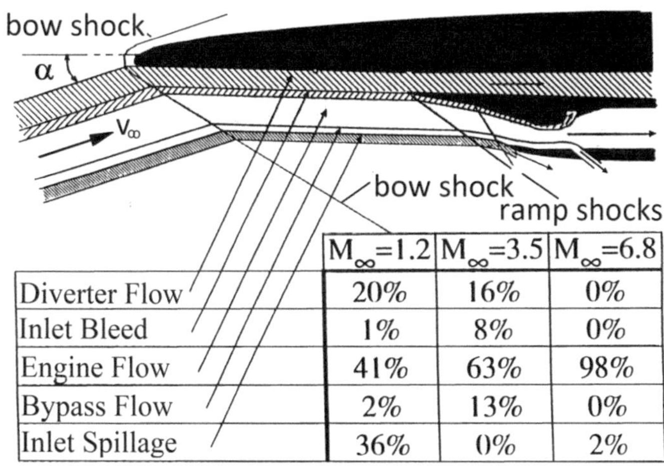

	M_∞=1.2	M_∞=3.5	M_∞=6.8
Diverter Flow	20%	16%	0%
Inlet Bleed	1%	8%	0%
Engine Flow	41%	63%	98%
Bypass Flow	2%	13%	0%
Inlet Spillage	36%	0%	2%

Fig. 5.26 Typical thrust and installation losses at the lower stage of the TSTO-system SÄNGER along the ascent trajectory [25]

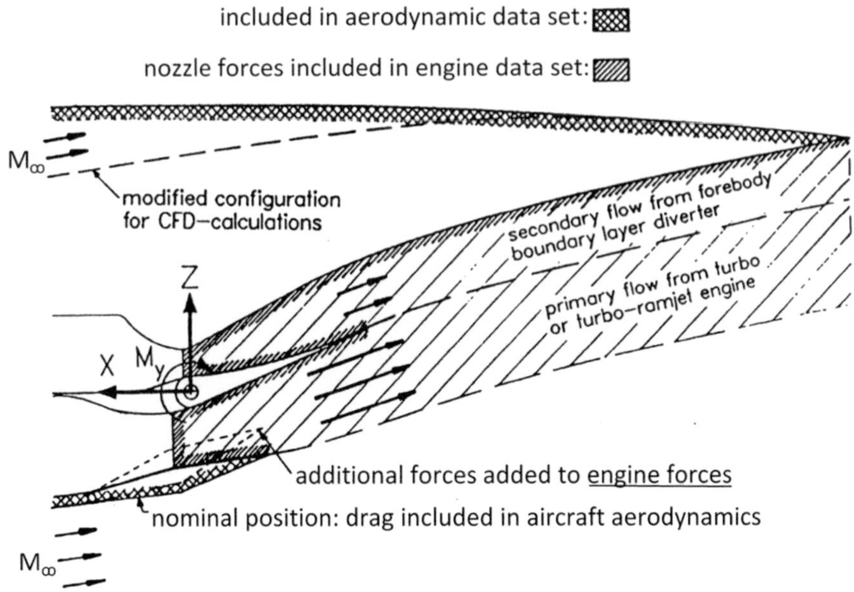

Fig. 5.27 Force and moment book-keeping for the SERN-type nozzle and the afterbody area [26]

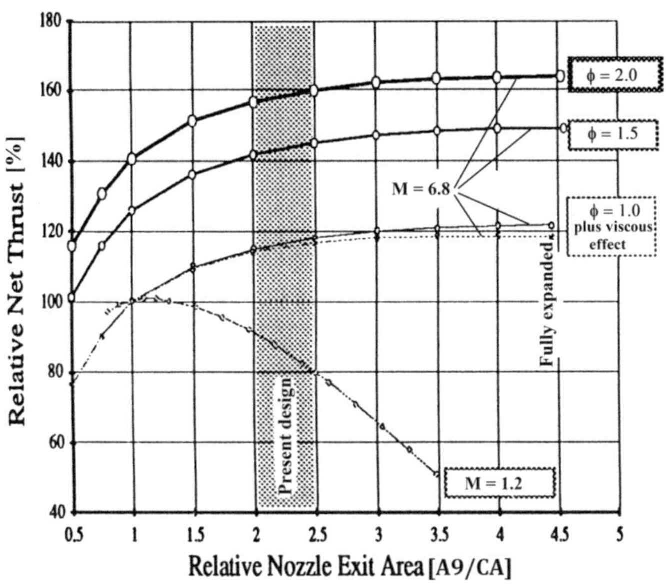

Fig. 5.28 Influence of the nozzle exit area on the achievable net thrust [22]

hyper-stoichiometric modes ($\phi = 1.5$ and 2) is also shown, as well as the very small effect of viscosity—found with a RANS calculation. The choice of the optimum is governed by the compromise between optimum performance (net thrust) at $M_\infty = 6.8$ and $M_\infty = 1.2$ (contradicting tendencies), the resulting contour of the afterbody, respectively the base area (afterbody or base drag), and connected to it, aspects of weight development and also consequences for the usable space within the airframe.

5.7 Thrust-Vector Angle at API-type 4 and 5 Flight Vehicles

API-type 4 and 5 flight vehicles are characterized by single expansion ramp nozzles (SERN). With them the direction of the thrust vector is not fixed as it is with bell nozzles. The contribution of the thrust-vector angle to the characteristics of the aircraft layout is of considerable importance in view of three items:

(a) It influences the **performance balance** (vertical and horizontal component),
(b) if it depends on the angle of attack, it gives a contribution to the **longitudinal stability** (positive or negative, depending on the direction of the vertical component),
(c) if it does not depend on the angle of attack, it anyway will modify the **trim balance**.

Changes of (b) and (c), being primarily stability and trim effects, in turn will result in changes of performance (a). These aspects will be treated in Sect. 5.9.2. Here we solely consider the cause, the magnitude and the direction of the changes of the thrust-vector angle.

This problem was already indicated in Sect. 5.3.1. Figure 5.29 now shows for four typical points on the trajectory the flight Mach number, the flight altitude, the inlet drag, the gross thrust, and the respective vector angles.

Case (1) and (2) are turbojet cases, whereas case (3) and (4) are ramjet cases. Table 5.2 gives the thrust and the thrust-vector angle at the inlet and the nozzle. Note that the thrust-vector angle is counted negative for case (1) and positive for the other cases.

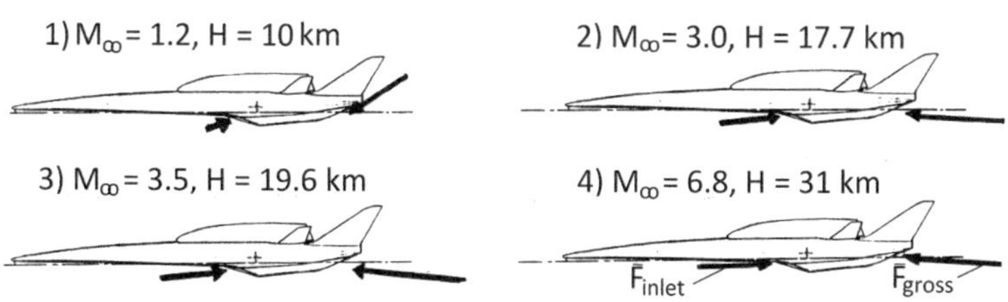

Fig. 5.29 Mach number and altitude of four flight cases along the trajectory of the TSTO-system SÄNGER [26]. Shown too are the vectors of the inlet drag and of the gross thrust

Table 5.2 Thrust and thrust-vector angles at the inlet and the nozzle along the trajectory of the TSTO-system SÄNGER [26]

Case	M_∞	H [km]	F_{Inlet} [kN]	Angle [°]	F_{Gross} [kN]	Angle [°]
(1) turbojet	1.2	10	198	18.6	531	−32.7
(2) turbojet	3.0	17.7	450	5.4	744	3.6
(3) ramjet	3.5	19.6	507	4.8	866	6.5
(4) ramjet	6.8	31	533	0.1	750	3.7

For the cases 2–4 the contribution of the flow momentum entering the inlet (F_{inlet}) can be considered more or less as fixed. Regardless of the flight Mach number this vector always is directed positively against the vehicle axis.

The alternating angle of the gross thrust F_{Gross} cannot be controlled. It depends on the degree of asymmetry of the SERN contour, hence on the nozzle design. A symmetric nozzle would fix the direction of gross thrust. The extremely strong inclination of the gross thrust of case (1) with −32.7°—i.e., with a pitch-up moment—is due to the impossibility to adapt the nozzle contour to the optimum expansion ratio. The final cross section of the nozzle is too large, see also Fig. 5.27.

These tendencies can be found along the trajectory for the Mach number range $0 < M_\infty \leqq 6.8$ in Figs. 5.30 and 5.31. The first of them shows the gross thrust coefficient as function of the flight Mach number. The hatched band represents the error bandwidth (range to be expected).

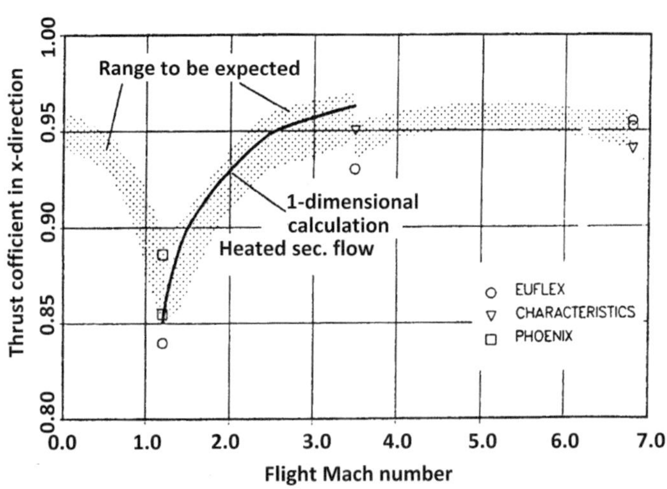

Fig. 5.30 Generic nozzle: thrust coefficient in x-direction as function of the flight Mach number [27]

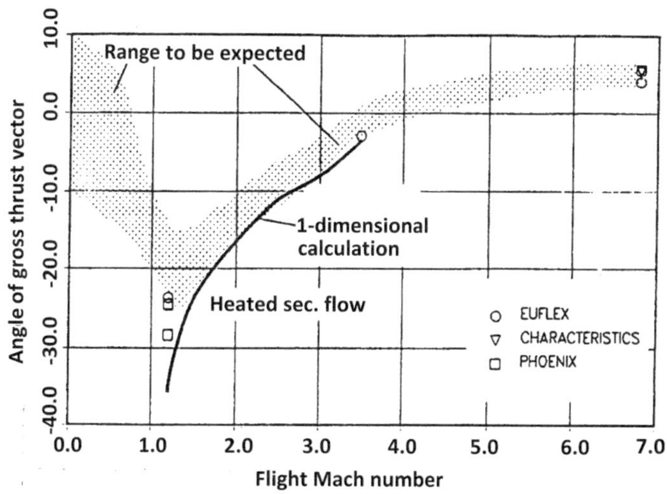

Fig. 5.31 Generic nozzle: gross thrust-vector angle as function of the flight Mach number [27]

The reason for the transonic acceleration pinch point at $M_\infty \approx 1.2$ from the side of the propulsion system is well discernible. From the side of aerodynamics it is the transonic drag divergence. The figure also shows that at the high flight Mach numbers the thrust coefficient in x-direction at most reaches approximately 0.95.

Figure 5.31 shows for the thrust-vector angle below $M_\infty \approx 1$ a broader hatched band. Around $M_\infty \approx 1.2$ the dip is similar to the dip of the thrust coefficient. The zero-crossing is found at $M_\infty \approx 4$, the tendencies are similar in both figures.

5.8 Cooling Balance of the Propulsion System

In view of the cooling demands for a mission it was shown in Sect. 2.6 that LH2 has an approximately nine times higher cooling capacity than JP-7. Figure 5.32 gives a summary of the fuel needed for propulsion and cooling for the complete nominal mission—ascent and return—of the TSTO-system SÄNGER.

The capital letters along the x-axis (flight Mach number) have the following meanings: RH: reheat (afterburner active), C: cruise, D: throttled for return flight only. The consumers of fuel are—from the bottom to the top—the 2-D nozzle, the ram burner, the ram duct, which surrounds the central turbojet, the cooler of the cooling air, which is needed to cool other fuel consumers.

For the whole ascent mission—including cruise, and the separation of the upper stage—, thanks to LH2, no deficit of cooling flow exists. Part of the fuel needed for the propulsion of the flight vehicle (H2-Fuel Flow) first is used for cooling purposes. However, if one would—after the release of the upper stage—too soon change to

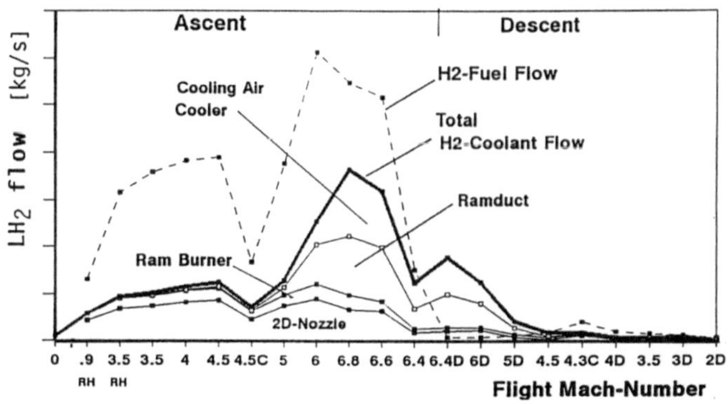

Fig. 5.32 Ascent and return flight of the TSTO-system SÄNGER: cooling demand and fuel need of the different users [28]

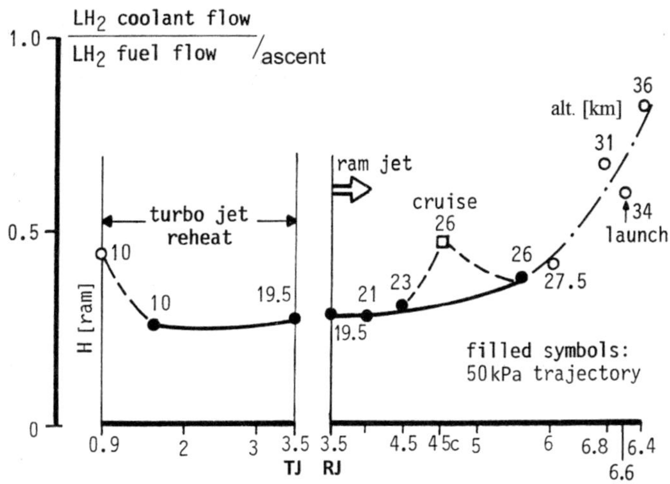

Fig. 5.33 Ratio of LH2 coolant flow to LH2 fuel flow as function of the flight Mach number [28]

the throttling position—compare $M_\infty = 6.4$ with $M_\infty = 6.4$D on the abscissa of Fig. 5.32—the hydrogen for cooling would have to be blown off useless.

In Fig. 5.33 for the ascent flight of the lower stage a comparison is given between availability and demand of LH2—turbojet (TJ) mode at the left, ramjet (RJ) mode at the right. The pull-up maneuver for the launch of the orbiter with $M_\infty = 6.6$ at 34 km altitude can be found at the utmost right part of the figure.

5.9 Trim and Longitudinal Stability

The reader is referred to the Sects. 2.9.2, 2.9.3 and 5.7, which contain preconditions for the following two sub-sections.

5.9.1 *Influence of the Propulsion System*

We first show how and how much the variation of the thrust vector angle influences the effective stability.

Stable in the sense of flight mechanics is a configuration when it automatically returns to the trimmed situation after a disturbance in the atmosphere (gusts), a change of the angle of attack, or a change of the flight speed. In the case of longitudinal stability hence the gradient of the pitching moment coefficient must be negative: $dC_m/d\alpha < 0$, see also Sect. 2.9.2.

With regard to this demand, the effect of the thrust vector direction, which also depends on the angle of attack, has to be examined in view of the stability problem. Indications in this respect are to be found in Figs. 5.22, 5.24, and 5.28 to 5.31.

For a hypersonic experimental flight vehicle of API-type 4 and 5 the effect of the thrust-vector direction is analyzed in Fig. 5.34.

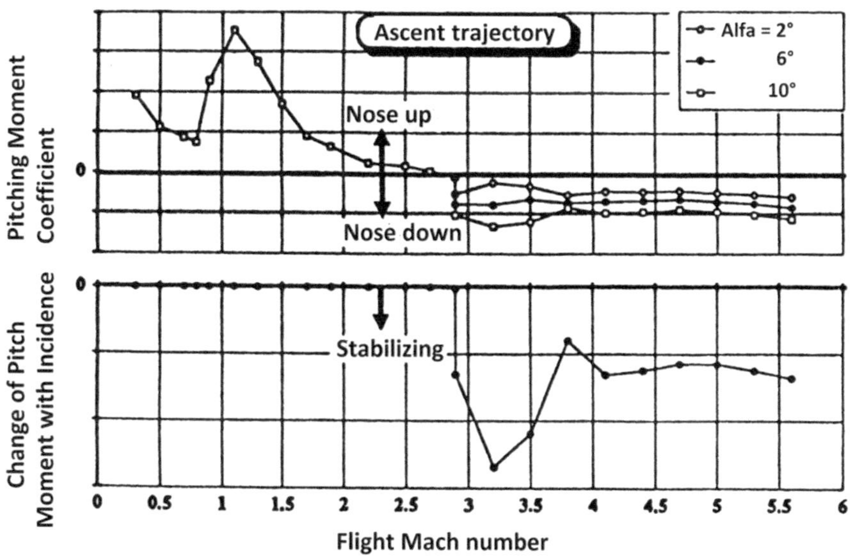

Fig. 5.34 Propulsion influence on trim and stability of an API-type 5 experimental vehicle [28]. Switch to ramjet propulsion at $M_\infty = 2.8$

The upper part illustrates the pitching moments due to a change of the thrust vector. In the turbojet domain pitch-up moments are induced, in particular, as is to be expected, in the transonic flight domain, see Figs. 5.29 and 5.31. No dependence on thc angle of attack is indicated.

Switching to ramjet propulsion at $M_\infty = 2.8$—the thrust vector direction, due to the forebody pre-compression, now depending on the angle of attack, Figs. 5.24 and 5.25—the induced pitch-down moment is varying with the angle of attack and has to be trimmed.

The effect of the turbojet does not change stability, but a tendency equivalent to an aerodynamic zero-lift moment is found. Correction is possible by a preset of trailing-edge flaps or the horizontal tail—again independent of the angle of attack. For Mach numbers past the switch to ramjet propulsion the contribution of the propulsion system to the longitudinal stability is distinct with an increase of stability. That has to be corrected by a negative setting of the trailing-edge flaps or the horizontal tail.

The propulsion-system induced pitching moments found for the SÄNGER lower stage configuration are compared to the pitching moment induced by the airframe, Fig. 5.35.

The sum of both—the total system—has to be balanced by the trim surfaces of the flight vehicle. Note that for the point of design—$M_\infty = 6.8$—which was an acceleration pinch point before, no drag-producing trim-surface deflection is necessary. The configuration satisfies the design task with zero pitching moment at the design point.

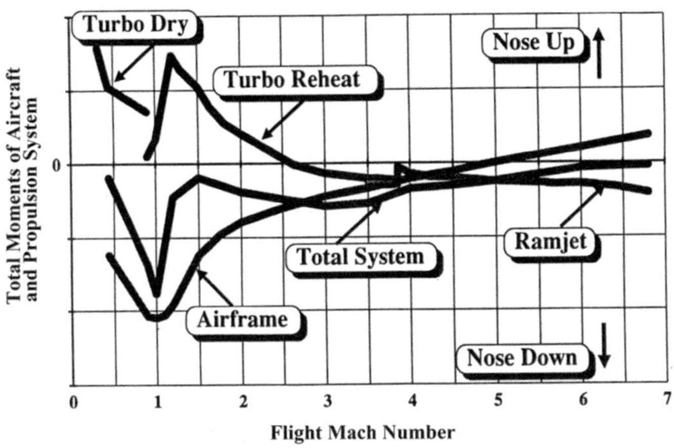

Fig. 5.35 Budget of the pitching moment for the lower stage of the TSTO-system SÄNGER [28]

5.9.2 The Trimming Balance

The joint answer given by the two departments ‘design aerodynamics’ and ‘propulsion’, as well as the practical and numerical preliminary design is summarized in Fig. 5.36. The figure shows the in the different phases of the design process found necessary flap deflection for trimming of the lower stage of SÄNGER. The three acceleration pinch points, i.e, points with minimum excessive thrust (SEP), are indicated: transonic regime, switch from turbojet to ramjet propulsion and maximum flight Mach number at upper-stage launch.

The broken drag line with the squares presents the basic design—still without engines—at the very beginning of the iterative design process. The contribution to trimming via the thrust vector was balanced by assumptions for the expected tendencies.

In a next step the meanwhile known—calculated—influences of the thrust vector on the stability and the zero-lift moment were implemented. Results from wind-tunnel tests were also used and the line with the black triangle was the outcome. The result was not yet satisfactory with regard to performance, trim drag and trim-surface deflections.

A new iteration from the side of aerodynamics was focussed on the optimization of the camber of the wing in order to find a better c_{m_0} distribution. The optimum result for the complete design is given with the line with open circles.

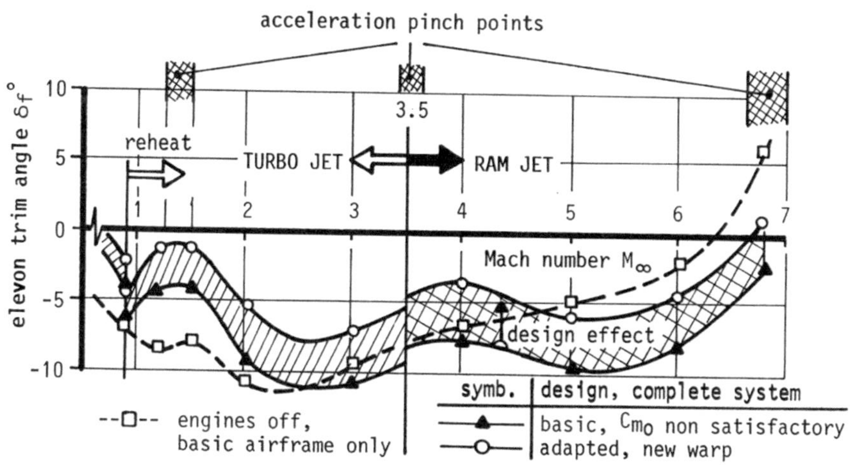

Fig. 5.36 SÄNGER lower stage: trim deflection of trailing-edge flaps along the ascent trajectory (flap surface = 6% of A_{ref}) [26]. The three acceleration pinch points are indicated

5.10 Concluding Remarks

In the sections and sub-sections of this chapter issues of propulsion and airframe-propulsion integration were presented. Tacitly it was assumed that the airframe is ideally rigid, engine failure was not considered. Concluding we have a short look at the implications of aeroelasticity and aerothermoelasticity and of propulsion failure.

Aeroelasticity and aerothermoelasticity of the airframe, i.e., airframe elasticity, are of importance for the whole flight vehicle. That is discussed in Sect. 6.5. Here only some implications regarding the propulsion systems of the four API-types are mentioned.

API-type 1 flight vehicles have the engines at wing locations, Fig. 1.2. Here the elastic behavior of the wing is the deciding factor. In particular the under-wing arrangement is of concern. Regarding the propulsion system, bending and torsion of the wing and of the pylon structure must be taken into account. Both are influencing not so much the action of the engines, but the direction of their thrust vectors, which in turn has implications for the structural layout of the wing and pylons.

With API-type 2 flight vehicles the implications of the fuselage elasticity are not clear. The working of the inward turning inlet at the upper side of the fuselage appears to be unaffected by the elastic fuselage behavior, because no pre-compression happens as with API-type 5 flight vehicles. Nevertheless, the elastic deformation of the fuselage under aerodynamic load needs attention.

Cruise missiles as API-type 3 flight vehicles are of compact design. Fore-body deformation in flight probably will be of minor influence regarding the pre-compression of the inlet onset flow.

Regarding API-type 4 and 5 flight vehicles everything depends on the size of the airframe. With the compact experimental vehicle X-43A no noteworthy deformation of the forebody is to be expected. That is different with the 55 m long forebody of the lower stage of the TSTO-system SÄNGER. In view of the high net-thrust sensitivity on changes of the effective angle of attack $\overline{\alpha}$ of the lower side of the forebody, Sect. 5.5.3, the elasticity of the forebody is of high importance. The tools to handle that in the flight vehicle design is a major topic of Chap. 7.

Propulsion failure is shortly addressed, but without regard to the different flight phases take-off, subsonic, supersonic and hypersonic climb, hypersonic cruise, upper stage release, return flight, approach and landing. Consequences in view of flyability and controllability or even mission abort have not been treated in detail.[6]

To begin with we note that a complete engine failure of course means the end of a mission. If the flight vehicle remains controllable, emergency landing or splashdown over sea should be possible. This holds for any type of flight vehicle, except for API-type 3 flight vehicles, i.e., cruise missiles, which with engine failure anyway are ending their mission.

[6] For the STRATOFLY MR3 flight vehicle, being of API-type 2, a detailed investigation of out-of-nominal scenarios can be found in [29].

The question is what happens with the different API-types of flight vehicle if one or more, but not all engines fail. For type 1 vehicles it is the question, if for instance a wing-tip engine fails, whether the tail volume is sufficient to assure controllable flight. If not, the engine at the other wing tip must be shut off.

With API-type 2–5 flight vehicles failure of only one engine already implies end of the mission. However, flyability and controllability should be assured, even if only one engine is left operating.

In any case, like for every type of aircraft, also for hypersonic aircraft all possible out-of-nominal scenarios must be identified and investigated. Appropriate measures must be developed in order to cope with such scenarios.

5.11 Problems

Problem 5.1 (*API Types and Airframe-Propulsion Integration*) Compare the main benefits and disadvantages of API-Type 1 (e.g., SR-71) and API-Type 5 (e.g., SÄNGER lower stage) vehicles in terms of airframe-propulsion integration. Why is API-Type 5 considered more complex in terms of integration?

Problem 5.2 (*Forebody Pre-Compression*) Explain how the forebody of a hypersonic vehicle, such as SÄNGER, contributes to pre-compression. Flat-plate case: what is the relationship between the angle of attack α and the stream-tube height h_2/h_1? How is the configuration of the flight vehicle affected? How does pre-compression affect engine performance?

Problem 5.3 (*Lower side of the Forebody*) Why are the pre-compression surfaces, i.e., the lower sides of the forebodies of API-type 4 and API-Type 5 ahead of the inlet or inlets flat?

Problem 5.4 (*Inlet-Onset Flow*) Why two-dimensional inlet-onset flow?

Problem 5.5 (*Boundary-Layer Diverter*) Why does the lower stage of the TSTO-system SÄNGER has a boundary-layer diverter and the X-43A not?

Problem 5.6 (*Net Thrust Sensitivity*) Describe the issue of net thrust sensitivity for hypersonic vehicles like the SÄNGER lower stage. How does a small change in the angle of attack $\overline{\alpha}$ affect the net thrust, and why is this sensitivity more pronounced at higher Mach numbers?

Problem 5.7 (*Thrust-Vector Angle*) What is the significance of the thrust-vector angle in API-type 4 and 5 vehicles? How does the changing thrust-vector angle affect vehicle stability and trim?

Problem 5.8 (*Shock-on Lip*) What are the benefits and the drawbacks of the shock-on-lip situation of the inlet?

Problem 5.9 (*g-Reduction*) How large is the g-reduction at flight with $M_\infty = 8$ at $H = 30$ km? Is the passenger comfort affected?

Problem 5.10 (*Coolant Flow*) Discuss the coolant flow situation shown in Fig. 5.32.

Problem 5.11 (*Acceleration Pinch Points*) What are the causes of the three acceleration pinch points?

References

1. Peebles, C.: Road to Mach 10-Lessons Learned from the X-43A Flight Research Program. Library of Flight, AIAA, Reston, Va (2008)
2. Reubush, D.E., Nguyen, L.T., Rausch, V.L.: Review of X-43A Return to Flight Activities and Current Status. NASA-2003-RTF (2003)
3. Kliche, D.: Multidisziplinäre Analyse und Bewertung von Staustrahltriebwerken mit Überschallverbrennung (Multidisciplinary Analysis and Rating of Ramjet Engines with Supersonic Combustion). Doctoral Thesis, Universität der Bundeswehr, Neubiberg. Verlag Dr. Hut, München, Germany (2012)
4. Cockrell, C.E. Jr., Engelund, W.C., Bittner, R.D., Dilley, A.D., Jentink, T.N., Frendi, A.: Integrated Aero-Propulsive CFD Methodology for the Hyper-X Flight Experiment. AIAA-Paper 2000-4010 (2000)
5. Kuczera, H., Sacher, P.W.: Reusable Space Transportation Systems. Springer, Berlin Heidelberg (2011)
6. Sacher, P.W.: Hypersonic Technology Programme. Space Course Stuttgart (1995)
7. Hirschel, E.H., Rizzi, A., Breitsamter, C., Staudacher, W.: Separated and Vortical Flow in Aircraft Wing Aerodynamics. Springer, Heidelberg (2021)
8. N.N.: Deutsche Aerospace (DASA) Raumfahrt: Hyperschall-Technologieprogramm, Staubrennkammer und Versuchsanlagen. Flyer printed and distributed in Germany by DASA-R 3076/11.93 Kr (1993)
9. Kuczera, H.: Konzeptstudie Sänger und Fortführung. Workshop Hypersonics Technology, Spitzingsee, Germany (1992)
10. Viola, N., Fusaro, R., Saracoglu, B., Schram, C., Grewe, V., Martinez, J., Marini, M., Hernandez, S., Lammers, K., Vincent, A., Hauglustaine, D., Liebhardt, B., Linke, F., Fureby, C.: Main challenges and goals of the H2020 STRATOFLY project. Aerotec. Missili and Spazio **100**, 95–110 (2021)
11. Steelant, J.: Personal communication (2022)
12. Hirschel, E.H., Weiland, C.: Selected Aerothermodynamic Design Problems of Hypersonic Flight Vehicles. Progress in Astronautics and Aeronautics, AIAA, Reston, Va., vol. 229. Springer, Heidelberg (2009)
13. Heiser, W.H., Pratt, D.T.: Hypersonic Airbreathing Propulsion. Education Series, AIAA, Washington, D.C., 1994, online edition (2012)
14. Seddon, J., Goldsmith, E.L.: Intake Aerodynamics. Education Series, AIAA, Reston, Va (1999)
15. Babu, V.: Fundamentals of Propulsion. Springer, Cham (2022)
16. Bissinger, N.C.: Inlet and Propulsion Integration of RAM Propelled Vehicles. AGARD-R-813, Aerothermodynamics and Propulsion Integration for Hypersonic Vehicles, 8-1–8-45 (1996)
17. Berry, S., Daryabeigi, K., Wurster, K., Bittner, R.D.: Boundary Layer Transition on X-43A. AIAA-Paper 2008-3736 (2008)
18. Hirschel, E.H.: Basics of Aerothermodynamics. Springer, Berlin, and, Vol. 204, Progress in Astronautics and Aeronautics, AIAA, Reston, Va.,: Second, revised Springer, Heidelberg (2004). (2015)

19. Johnston, P.J., Whitehead, A.H., Chapman, G.T.: Fitting Aerodynamics and Propulsion into the Puzzle. Aerospace America **1987**, 32–42 (1987)
20. Hirschel, E.H.: Aerothermodynamic Phenomena and the Design of Atmospheric Hypersonic Airplanes. In: Bertin, J.J., Periaux, J., Ballmann, J. (eds.) Advances in Hypersonics, vol. 1, pp. 1–39. Defining the Hypersonic Environment. Birkhäuser, Boston (1992)
21. Zellner, B.: Auftretende Einzelkräfte am SÄNGER Antriebssystem. Arbeitsnotiz (1991)
22. Herrmann, O.: Integration of Turbo-Ramjet Engines for Hypersonic Aircraft. AGARD CP-498, Paper 32 (1992)
23. Schaber, R.: Einfluss entscheidender Triebwerksparameter auf das Leistungsverhalten eines Hyperschall-Antriebs. MTU-N94-EP-0001, MTU, München/Karlsfeld, Germany (1994)
24. Luber, W.: Eigenschwingungsanalyse am SÄNGER. DASA-LME24-Interner Bericht, Ottobrunn, Germany (1995)
25. Sacher, P.W.: Status Referenzkoncept SÄNGER. DASA-LME13, Ottobrunn, Germany (1993)
26. Staudacher, W.: Entwurfsproblematik luftatmender Raumtransportsysteme. Space Course Stuttgart (1995)
27. Herrmann, O., Rick, H.: Propulsion Aspects of Hypersonic Turbo-Ramjet-Engines with Special Emphasis on Nozzle/Afterbody Integration. ASME 91-6.T-395 (1991)
28. Heitmeier, F.: SÄNGER Studie Antrieb. Phase 1b. Abschlussbericht MTU München. B93EP-0002 (1993)
29. Bajeli, B.: STRATOFLY Project: Analysis of the Concept of Operations of a Hypersonic Civil Passenger Transport Aircraft. Master Degree Thesis, Politechnico Di Torino, Italy (2019)

Chapter 6
Structure and Materials Aspects

This chapter is devoted to a short consideration of structure and materials aspects of hypersonic airbreathers. Neither detailed design instructions are given, nor are flight operational issues treated. Overall the basic structural concepts of hypersonic airbreathers are the same as those of general aircraft: 'build the structure as light as possible, but at the same time as durable as necessary'. The big differences are due to the thermal loads, i.e., the wall temperature T_w and the wall-heat flux q_w, which are to be dealt with.

The reader should remember: (a) the wall temperature T_w—affected strongly by surface radiation cooling—governs the wall material including the surface coating, as well as the properties of the flow along the wall, (b) the wall-heat flux q_w—also affected strongly by surface radiation cooling—in the case of a cold primary structure determines the type and thickness of the thermal protection system, or with a hot primary structure, the structure and material concept.

We begin, because the reader may be more familiar with them, with a look at the loads on low and medium Earth orbit re-entry vehicles. That permits to obtain a general picture of their structure and materials issues. These, of course, are related to their particular missions, hence to their shapes and in particular to the thermal loads, which they are exposed to. The shapes are different for non-winged and winged re-entry vehicles, but not the thermal loads.

Then basic structural and material issues of hypersonic airbreathers are considered. The main sections of this chapter are dealing with structure and materials aspects of an API-type 4 and an API-type 5 flight vehicle. These are the X-43A and the lower stage of the TSTO-system SÄNGER, see the overview in Sect. 1.1. The lower stage of the TSTO-system SÄNGER essentially was a reference concept, whereas the X-43A was an experimental vehicle, which flew at $M_\infty = 6.83$ and $M_\infty = 9.68$.

After a short look at the all deciding topic of aeroelasticity and aerothermoelasticity, final remarks close the chapter, pointing to the much wider range of structure and material problems than superficially touched upon here.

E. H. Hirschel et al., *Elements of Hypersonic Airbreather Design and Development*,
https://doi.org/10.1007/978-3-031-94219-8_6

6.1 Re-entry Vehicles: Shapes, Structures, Materials

The mission not only dictates the outer shape of a flight vehicle, but also its structural topology and the materials. The classical hypersonic flight vehicle is the atmospheric re-entry vehicle, either non-winged as capsule or winged, as for instance the Space Shuttle Orbiter, see Sect. 1.2 and for details [1]. The re-entry flight basically is an aero-braking flight from low Earth orbit (LEO), or in the case of Apollo, from a Moon mission. For general low Earth orbit re-entry trajectory issues see, e.g., [1] and in particular [2].

The fundamental problem of re-entry flight are the extremely high thermal loads in terms of the surface temperature T_w and the heat flux into the wall q_w. The radiation-adiabatic temperature can be considered as conservative estimate of the wall temperature in reality, Sect. 3.7. Re-entry vehicles as a rule have a cold primary structure and a thermal protection system (TPS)—heat shield—at their windward side in order to cope with the thermal loads.[1] A heat front creeps into the heat shield and the re-entry mission must have ended before that reaches the cold primary structure. Remember that during the final approach thermal reversal happens, Sect. 3.10.4.

The capsule as non-winged re-entry vehicle generally can be a cone, with its base being the blunt bow with the TPS on it, Fig. 6.1. The capsule during re-entry flies at small angle of attack and has a lift-to-drag ratio in the range $0.3 \lessapprox L/D \lessapprox 0.4$ in order to allow trim and also trajectory changes [1].

Other shapes have been proposed and also flown [1]. Normally capsules do not have control surfaces, but attitude control thrusters. The structural topology of capsules is rather simple. The conical shape with a structure from light metal, parts of it maybe from steel, houses the payload, the blunt bow is the heat shield, i.e., the thermal protection system. That can be an ablative or a non-ablative system.

Winged re-entry vehicles, Fig. 6.2, see also Fig. 3.3, with the lifting body a special case, on purpose fly a considerable part of their re-entry trajectory at high angle of

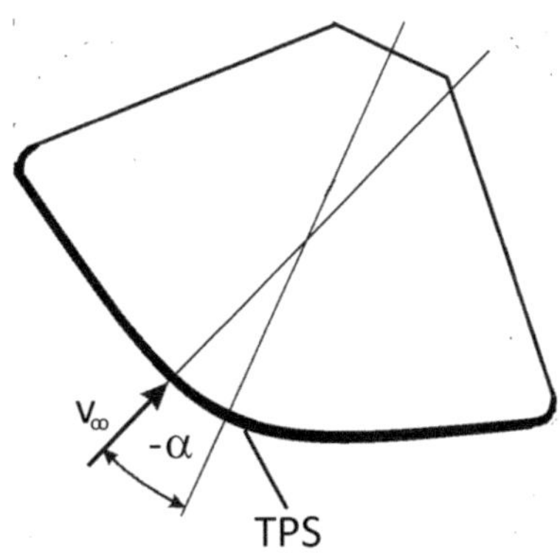

Fig. 6.1 Principle shape of a non-winged re-entry vehicle, i.e., a capsule with the heat shield (TPS) in front

[1] Certain structural elements, parts of wings, trim and control surfaces may have a hot primary structure.

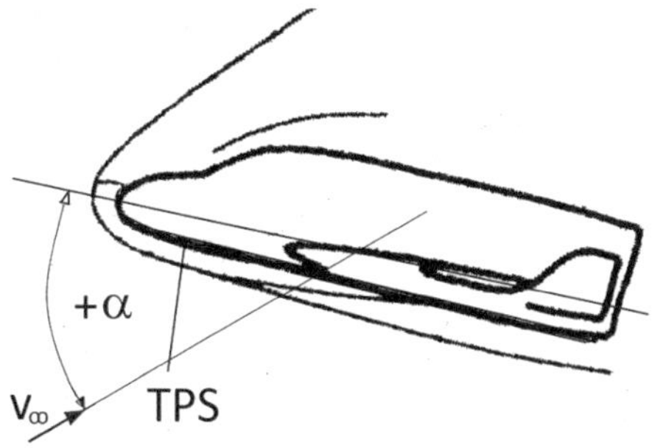

Fig. 6.2 Principle shape of a winged re-entry vehicle with the TPS at the bow and at the lower side and at all lift and control surface edges

attack in order to create the drag needed for the deceleration of the vehicle. The lift-to-drag ratio then is $L/D \approx 1$. In the supersonic flight regime the lift-to-drag ratio may be up to ≈ 2 and in the subsonic flight regime up to ≈ 5 [1].

The great operational example of a winged re-entry vehicle is the former Space Shuttle Orbiter, which became more or less worldwide the blueprint for all following studies and projects of winged re-entry vehicles.

The orbiter's cold primary fuselage structure basically was a fuselage frame-stringer-skin one, like transport aircraft have. The material primarily was an aluminum alloy. The lower side including the wings, as well as the nose and all leading edges had a thermal protection system consisting of reusable surface insulation tiles.

Special problems: avoidance of buckling of the skin of the cold primary structure beneath the TPS under mechanical and thermal loads in view of the attached tiles, the large payload bay doors, payload attachment, the floating crew compartment, acoustic loads and related fatigue during take-off, pogo oscillation prevention, etc.

Detailed "Lessons Learned" about the shuttle's orbiter performance are given in [3]. Subject areas regarding structural topics are the "Aerothermal Environment" and "Thermal Protection". Noteworthy in view of the particular development problems touched on in Sect. 1.6 is the remark that the orbiter's center of gravity at the beginning of the low-speed flight tests was not known as accurately as desired. This remark also concerns the longitudinal stability problem experienced during the first orbiter re-entry, but does not explain that problem. Regarding that problem we point to [1].

In the Orbiter Experiments (OEX) Program [4], structural topics included catalytic surface effects, gap heating, wing leading edge heating and others. A compact overview of the structural design of the Space Shuttle Orbiter can be found in [5].

Figure 6.3 shows as an example of re-entry trajectory data of winged re-entry vehicles those of the STS-5 mission of the Space Shuttle Orbiter. After the initial descent down to about 80 km altitude, the trajectory part with peak heating began at $t \approx 400$ s and lasted up to $t \approx 1{,}200$ s. The mission ended at $t = 1{,}821$ s. Note the high angle of attack $\alpha = 40°$ for more than half of the flight time.

Figure 6.4 finally illustrates thermal loads in terms of computed surface temperatures at a point one meter downstream of the stagnation point at the lower side of the HERMES configuration. The data were obtained by F. Monnoyer with the

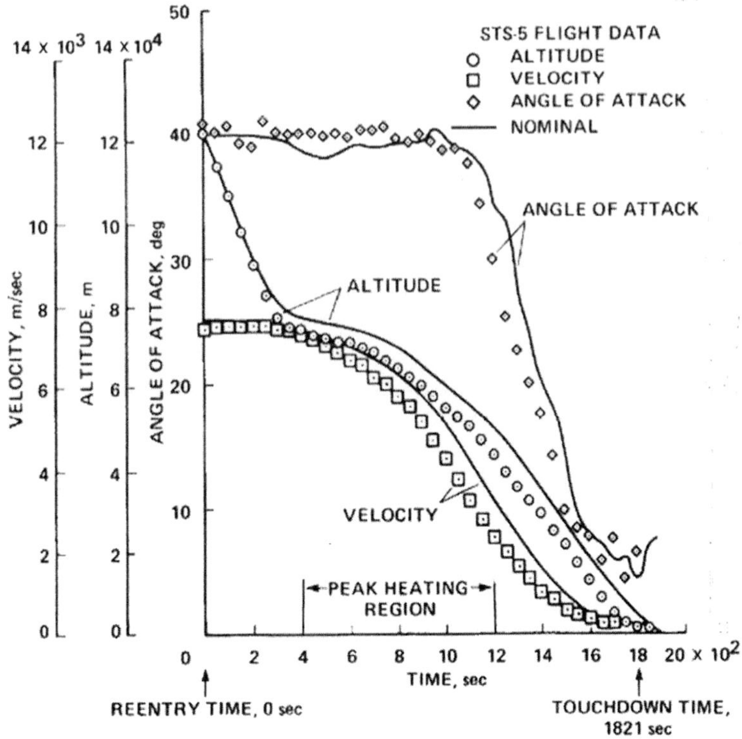

Fig. 6.3 Re-entry trajectory data of the STS-5 mission of the space shuttle orbiter [6]

at that time most advanced second-order three-dimensional boundary-layer method coupled to an Euler solution with a full thermo-chemical model, see also [7].

The upper black circles show the recovery temperature T_r on the re-entry trajectory at selected altitudes with the corresponding Mach numbers. The highest value is found at 60 km altitude with about 5,800 K. Surface radiation cooling with the assumed emissivity coefficient $\epsilon = 0.85$ brings the surface temperature at that altitude down to the radiation-adiabatic temperature $T_{ra} \approx 1{,}900$ K, black square.

This figure demonstrates the high efficiency of surface radiation cooling. At the outer edge of the viscous layer in the external inviscid flow field of course the very high temperatures still are present.

6.2 Airbreathing Hypersonic Vehicles: Basic Issues

Other than re-entry vehicles, cruise and acceleration vehicles (CAVs), and ascent and re-entry vehicles (ARVs)—regardless of their airframe-propulsion integration (API-) type, Sect. 1.5—have slender shapes and fly at low angle of attack in order

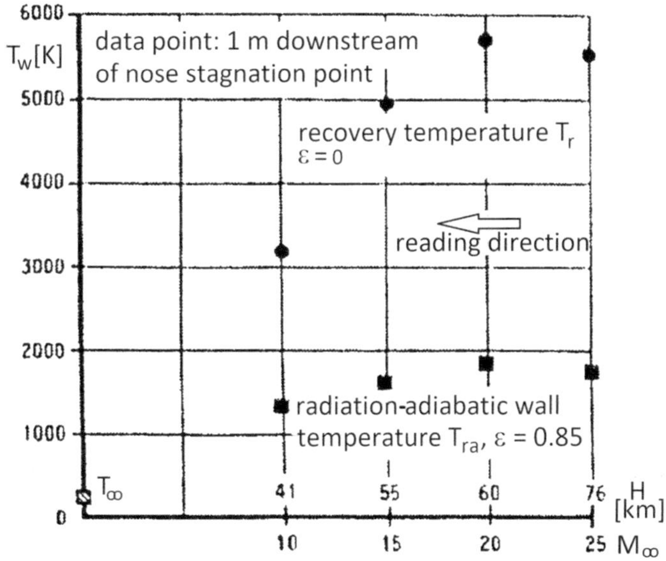

Fig. 6.4 Effect of surface radiation cooling close to the stagnation point of the projected re-entry vehicle HERMES. Angle of attack $\alpha = 40° =$ const., laminar flow, different trajectory points H, M_∞

to keep the aerodynamic drag as low as possible. Also these flight vehicles face the principal problem of thermal loads in terms of the surface temperature T_w and the heat flux into the wall q_w.

Three facts are to be kept in mind:

1. The radii of the fuselage nose and the leading edges of the wing and stabilization surface should be as small as possible in view of the wave drag and as large as possible in view of radiation cooling, Sect. 3.5.
2. Radiation cooling in the case of turbulent flow ought to take into account that the vehicle surface should be as hot as possible in order to reduce the skin-friction drag, Sect. 3.10.1.
3. It is still not possible to predict to the needed accuracy location, shape and extent of the laminar-turbulent transition zone, Sect. 3.11.1.

These facts have consequences regarding the thermal loads problem, hence the question of hot or cold primary structure and choice of the surface/skin material. However, there are other demands.

It is necessary to consider the following effects:

– It is typical for API-type 4 and 5 flight vehicles, that the radiation-adiabatic temperature at the upper side of the configuration is distinctly lower than that at the lower side. Figure 3.31 in Sect. 3.8 shows that at $M_\infty = 6.8$ this amounts to about

200 K for the turbulent case. An explanation of the effect can be found below that figure together with some considerations regarding possible consequences.
- At compression ramps like the external inlet ramps and at trim and control surfaces a stepwise or nearly stepwise increase of the wall pressure, the unit Reynolds number and the thermal loads happens, even also of the wall shear stress. Figure 3.37 in Sect. 3.10.3 shows the increase of the radiation-adiabatic temperature over a three-ramp inlet, for general ramp flow, over, e.g., flaps, the situation is similar.
- At a single expansion ramp nozzle (SERN) surface-radiation cooling is possible. The question is, whether the hot jet gas does curtail that to a degree.
- At the roots of wings, stabilizers and so on, non-convex effects occur, even if the root region is well rounded. Figure 3.33 in Sect. 3.9 demonstrates that the effectiveness of surface radiation cooling is severely reduced. The situation can even be worse, if longitudinal corner flow is present, as for instance at the external inlet ramps.

The above sketched effects in general go along with an influence on the mechanical loads, e.g., the wall pressure p_w and on the wall shear stress τ_w.

The **wall-pressure field** on the one hand governs the aerodynamic properties of the configuration. On the other hand it represents the main part of the structural loads.

The **wall shear-stress field**, along with the form drag, also governs most of the aerodynamic properties. Important is, that it can erode surface properties, in particular surface coating, Sect. 3.2.3, and even can strip off surface material.

Other flow-related causes of damage, partly already mentioned, are:

- shock/boundary-layer interaction,
- vortex impingement,
- jet impingement,
- flutter,
- panel flutter,
- buckling,
- noise damage.

Impingement of dust and dirt particles during take-off and landing as well as high speed droplet impingement in adverse weather conditions are also potential causes of damage.

This short discussion shows that the overall consideration has to be supplemented by very detailed studies, in order to obtain the full picture. In the SÄNGER case, for instance, it immediately became clear that the external inlet ramps also need ceramic material in order to cope with the thermal loads. Similarly non-convex situations made necessary locally a turning-away from the titanium skin approach.

With hypersonic airbreathers it is no more adequate to identify critical structural load cases—which include payload weight, taxying loads, maneuver loads, gust loads, landing shock, etc.—with the classical estimation methods. The modern discrete numerical solution methods in the fields of fluid mechanics, aerothermodynamics, all when necessary coupled with adequate structural methods, allow for thorough investigations. Of course, the joint problem, the problem of laminar-turbulent transition as well that of turbulent flow separation remains.

In the following two sections we have a brief look at the structure and materials concepts of two flight vehicles: the lower stage of the TSTO-system SÄNGER and the X-43A.

6.3 SÄNGER Lower Stage

6.3.1 Thermal Loads and Material Choice: Overall Consideration

The lower stage of the TSTO-system SÄNGER, Sect. 2.8, is a large flight vehicle of API-type 5. The take-off mass was assumed to be 410 to, the fuel mass 134 to, and the mass of the orbiter 115 to, Sect. 2.8. The structural design of the lower stage is sketched for instance in [8]. Here we give a short overview regarding the global structure and materials concept.

Consider Fig. 6.5, where along the lower-side symmetry line for two flight Mach numbers the radiation-adiabatic temperature distribution is given. Curve (a) refers to the cruise-out part of the trajectory with $M_\infty = 4.5$, Fig. 2.13, curve (b) to the maximum Mach number $M_\infty = 6.8$ at the launch of the upper stage. The surface emissivity coefficient was assumed to be $\epsilon = 0.85$.

Figure 6.5 presents a highly simplified picture of the situation at the lower side of the lower stage. At the upper side of the lower stage the orbiter—the upper stage—is located during the cruise-out part of the trajectory. For the related aerothermodynamic issues of upper-stage integration and release see Sect. 3.12. During cruise-back trough-flow issues are of importance.

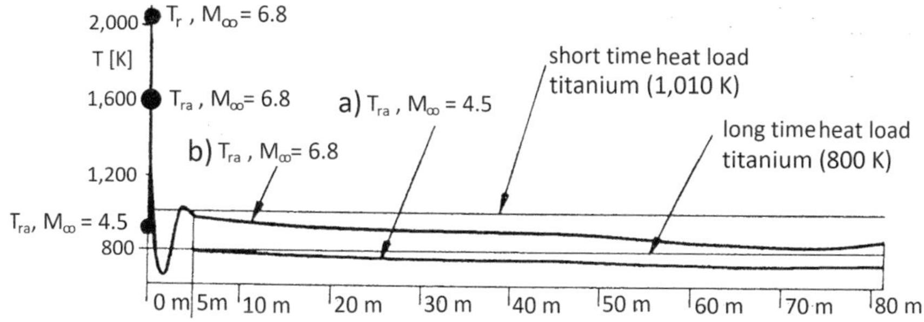

Fig. 6.5 SÄNGER lower stage: radiation-adiabatic temperatures along the symmetry line of the lower side [8]

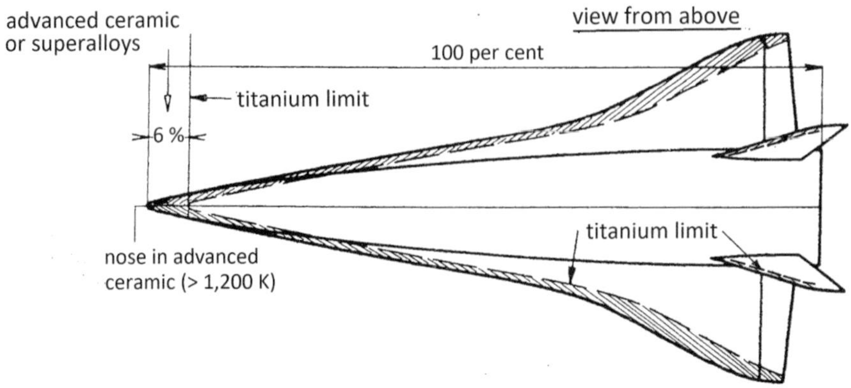

Fig. 6.6 SÄNGER lower stage upper side without trough: possible choice of the materials concept [8]

For both cases in Fig. 6.5 the location of laminar-turbulent transition was assumed at about 2.5 m downstream of the nose, which was supposed to have a radius of 0.2 m. This transition location is, regarding the thermal loads, a conservative assumption. For the generally ensuing problems in view of laminar-turbulent transition see Sect. 3.11.1.

Consider case (a). The radiation-adiabatic temperature at the nose is $T_{ra} \approx 900\,\text{K}$. Along the lower side of the configuration we have $T_{ra} < 800\,\text{K}$. This means that the wall temperature is below the long-time heat load limit of titanium.

For case (b) much higher temperatures are resulting. The recovery temperature is $T_r \approx 2{,}100\,\text{K}$. With radiation cooling the temperature at the nose is $T \approx 1{,}600\,\text{K}$. Downstream of the transition location the radiation-adiabatic temperature is well below 1,000 K, compare with Fig. 3.31. This means that the temperature is below the short-time heat load limit of titanium with $T \approx 1{,}010\,\text{K}$.

Figure 6.6 shows the final result regarding the materials concept.

Except for the nose—about six per cent of the total length of the vehicle—and the leading edges of the wing and the vertical stabilizers titanium can be employed. For the nose and the leading edges advanced ceramics were seen to be necessary [8]. We note that expansion joints are present between the ceramic and the titanium parts. For the possible implications see Sect. 3.11.3.

6.3.2 Structural Approach of the Lower Stage of SÄNGER

The flight time of the Space Shuttle Orbiter re-entry was about 1,800 s and the peak-heating time period was about 800 s, Fig. 6.3. The flight time of a typical mission of a TSTO system is much longer. Figure 6.7 gives an example from an investigation of a SÄNGER-like TSTO space transportation system [9], see also [1].

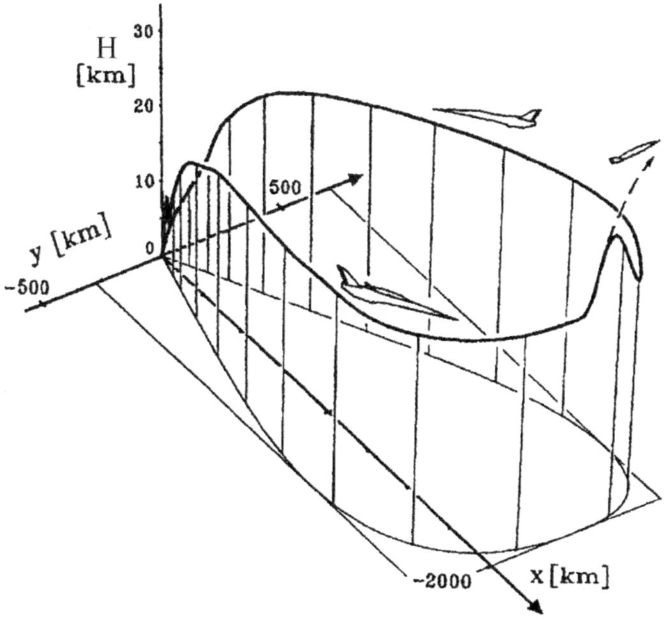

Fig. 6.7 TSTO space transportation system [9]: the flight altitude on the flight path. Upper stage separation is initiated at $t = 1{,}850\,\text{s}$

In this example the long-time thermal load with $M_\infty \approx 3.5$ to 5.5, would be about 1,500 s, and the short time load with $M_\infty \approx 6.8$ about 300 s, Fig. 6.8.

This assessment admittedly is superficial. It nevertheless shows that the lower stage of the TSTO space transportation system has to cope with rather high thermal loads over quite a long flight-time period.

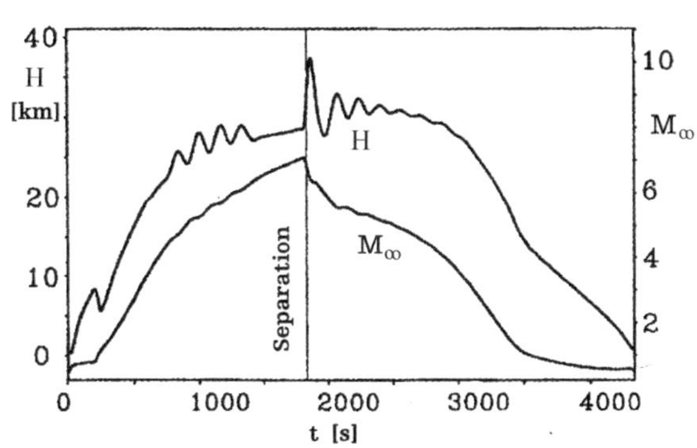

Fig. 6.8 TSTO space transportation system [9]: flight altitude H and flight Mach number M_∞ as function of the flight time t

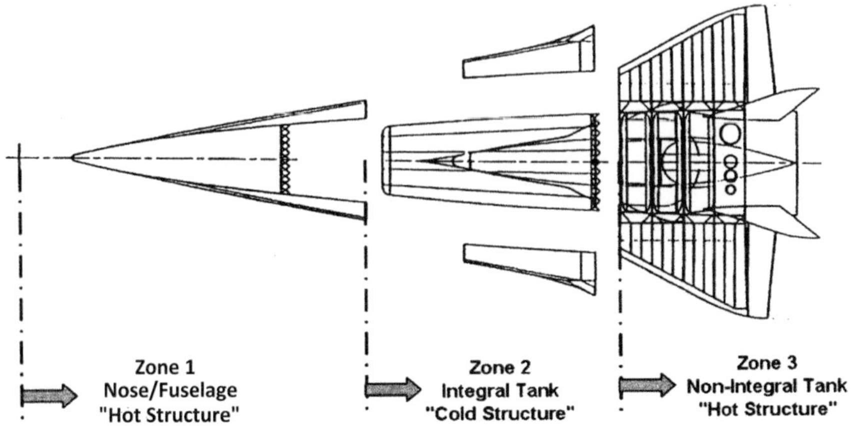

Fig. 6.9 The three major structural zones of the lower stage [10]

Taking into account the estimated mechanical and thermal loads and the demands of the mission, for the lower stage a mixture of hot and cold structure zones was chosen, Fig. 6.9.

Zone 1 essentially encompasses as hot structure the nose cone and a fuselage portion. Zone 2 is a cold structure, which houses a liquid hydrogen—cryogenic—integral tank, Zone 3 finally again is a hot structure with a liquid hydrogen non-integral tank.

The Zone 2 concept, Fig. 6.10, has the following features [8]: (a) an undisturbed structure, (b) equal structural dimension of tank and outer fuselage, (c) minor load carrying in terms of bending moment and shear load, (d) load introduction due to the nose undercarriage and the upper stage is considered as minor, (e) heat-protected integral tank, (f) maximum weight and volume effectiveness at undisturbed structure.

Fuel tanks for cryogenic hydrogen face two major problems: (1) the content must be insulated against any heat source, the ambient air included, (2) an evaporation barrier is necessary in order to avoid evaporation—boil-off—losses of the content.

A big challenge is the insulation against the outer hot wall. That is, a barrier is needed to insulate against the heat flux q_w, which comes from the vehicle's surface.

Fig. 6.10 Zone 2: cold structure with integral cryogenic tank [11]

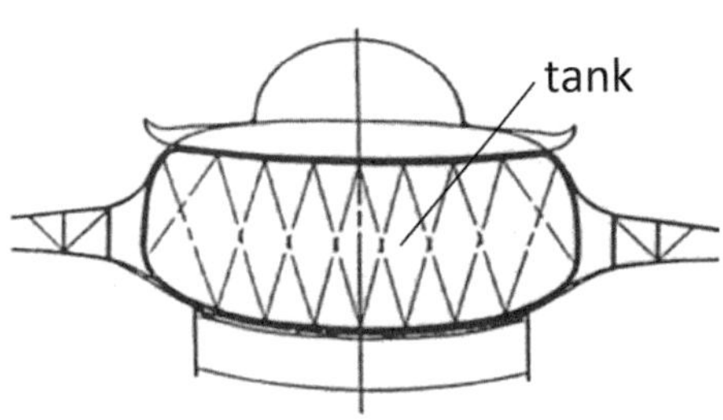

Two concepts were proposed [11]:

1. An integral concept with an evacuated sandwich tank wall with carbon fiber-reinforced polymer (CFRP) as upper and lower face sheets. All this also to serve as H_2 evaporation barrier. CFRP stiffeners are to be positioned inside the sandwich core. Above this, located between the upper face sheet and the fuselage skin, a high-temperature insulation, not further specified at that time, is located, also with stand-off elements.
2. The other concept was a cryoinsulation CFRP tank wall with internal stiffeners. The wall at the inner side had a vapor barrier. Above the tank wall also a high-temperature insulation was foreseen.

Zone 3 was to have a hot primary structure with non-integral cryogenic bubble tanks, Fig. 6.11. The features of Zone 3 are [8]: (a) intersection of fuselage, wing and vertical stabilizers, (b) integration of inlet, engine, nozzle and upper stage, (c) major load transformation in terms of bending and torsion moments of the wing, and bending moments of the fuselage, (d) major load introduction due to engine and nozzle, main undercarriage and upper stage, (e) non-integral bubble tanks, (f) optimum weight and volume effectiveness of a complex structure.

After the dimensioning load cases, as well as bending and torsion moments were established, the structure and materials concept emerged. Not yet established were the aerothermoelastic properties in particular those of the long forebody. They are of high importance in view of the net-thrust sensitivity, Sect. 5.5.3. The follow-up studies, where the concept was to be refined, were not realized.

The structure concept is depicted in Fig. 6.12. The meanings of the acronyms of the proposed materials are: SiC = silicon carbide, Fr-aluminum = flame retardant aluminum, SPFDB-SW titanium = superplastic-forming diffusion bonding titanium, CFRP = carbon-fiber-reinforced polymer.

A particular problem is the move of the center of gravity during the release of the upper stage, and that of the center of pressure as function of the flight Mach number. A fuel management, encompassing the forward integral tank in Zone 1 and the five independent non-integral bubble tanks in Zone 3, was foreseen to cope with that problem.

Regarding the aerothermodynamic design of the flight vehicle, especially the nose region with a ceramic nose cone followed by a titanium skin is a critical issue

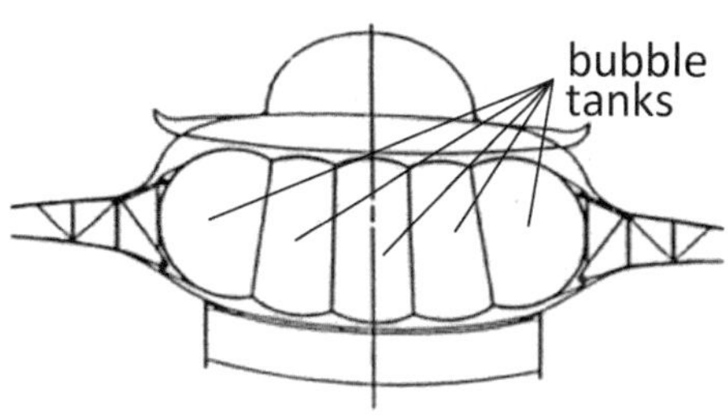

Fig. 6.11 Zone 3: hot structure with non-integral cryogenic bubble tanks [11]

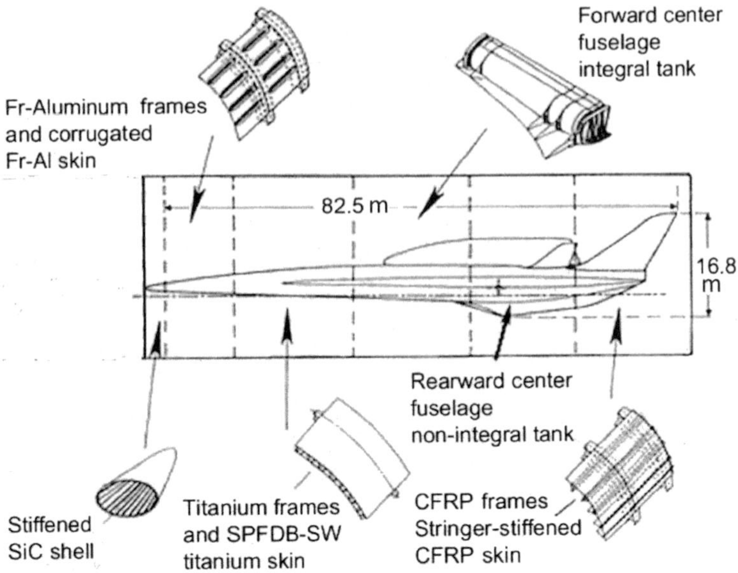

Fig. 6.12 The structure concept of the lower stage with the choice of materials [12]

because of the expansion joint between them. That is a surface imperfection which in particular can play a role in the laminar-turbulent transition process.

6.3.3 A Look at Two API-Type 5 SSTO-Systems

It may be of interest to sketch besides the structural concept of the API-type 5 TSTO-system SÄNGER, Fig. 6.9, also that of two API-type 5 SSTO-systems, the NASP/X-30, configuration 201, 1990 [13–15] and the Scram 5 [16].

The NASP/X-30, scramjet-propelled, was to reach low Earth orbit with a Mach number $M_\infty = 25$, Fig. 6.13. Its take-off mass was about 140 to, the fuel mass plus oxydator about 80 to.

The wing of the about 43 m long flight vehicle was a double delta one, reflecting the need for adequate low-speed flight properties. The side view in Fig. 6.14, upper part, reveals the large liquid H_2 tanks, and the side by side four engine arrangement. Indicated are the turbojet engines (red) and below the ram/scram engines. The view from above is given in the lower part of the figure.

Again the tank arrangement is shown, also that of the oxygen tanks (green circles). Not indicated are the reaction control devices for the low Earth orbit flight regime, nothing is mentioned about the return flight from the low Earth orbit. An inverted re-entry flight, as in the case of the Scram 5, can be ruled out because of the canopy at the upper side and the vertical tails that need to maintain their integration.

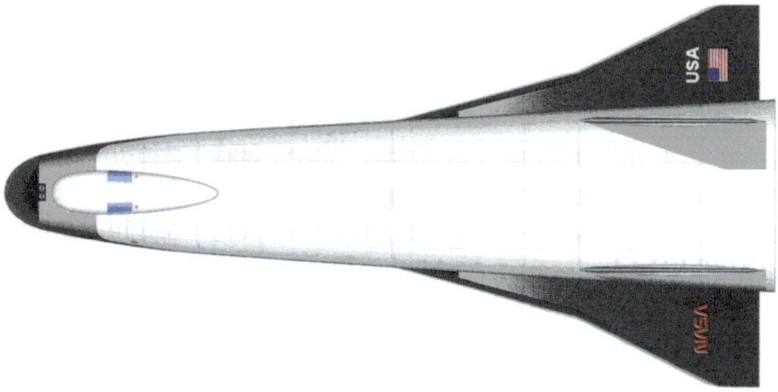

Fig. 6.13 The NASP/X-30, configuration 201, view from above [15]

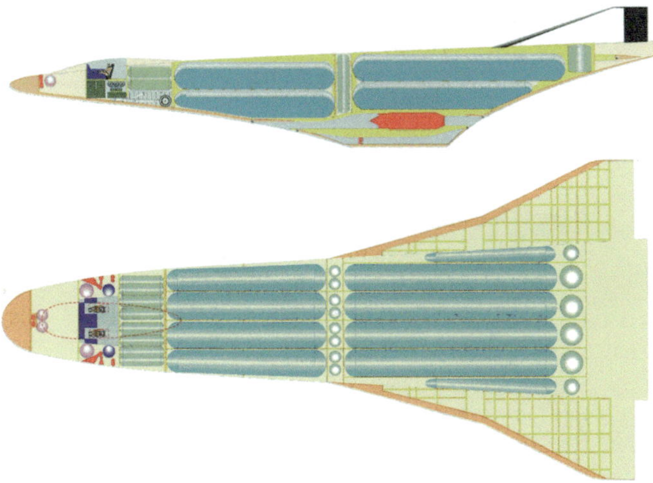

Fig. 6.14 The NASP/X-30, configuration 201 [15]. Upper part: longitudinal section—side view, lower part: longitudinal section—view from above

The Scram 5, Fig. 6.15, with a length of 44.9 m was also to reach orbit, but as an airbreather up to $M_\infty \approx 12$ at about 35 km altitude, and then as rocket propelled vehicle [16], see also Sect. 2.1.2. It was to take-off from a lower stage at $M_\infty = 4$ and at about 22 km altitude. The mass at separation from the lower stage was about 178 to, with 46 to LH_2 and 72 to LO_2.

The wing has a downward anhedral of the outboard part. That reduces the danger of pitch-up, enlarges longitudinal stability and improves the flight characteristics in general [17]. A major point is that the unobstructed upper surface—with a proper thermal protection system—allows an inverted re-entry flight. No part of the configuration and in particular not the scramjet propulsion system needs a special thermal protection. During the critical inverted re-entry phase all lies at the upper side—then the lee-side—in the hypersonic shadow.

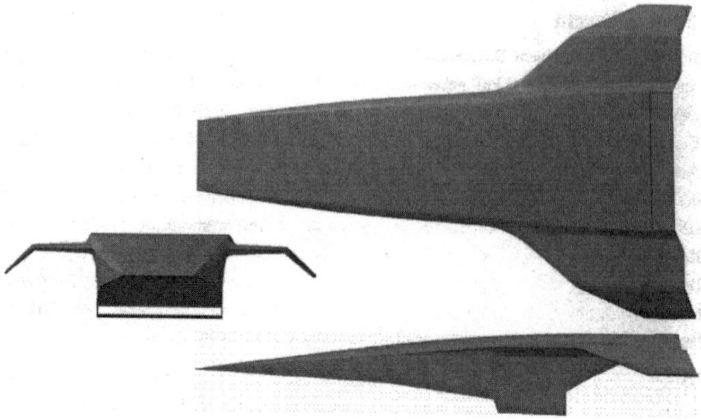

Fig. 6.15 The Scram 5 configuration, three-side view [16]

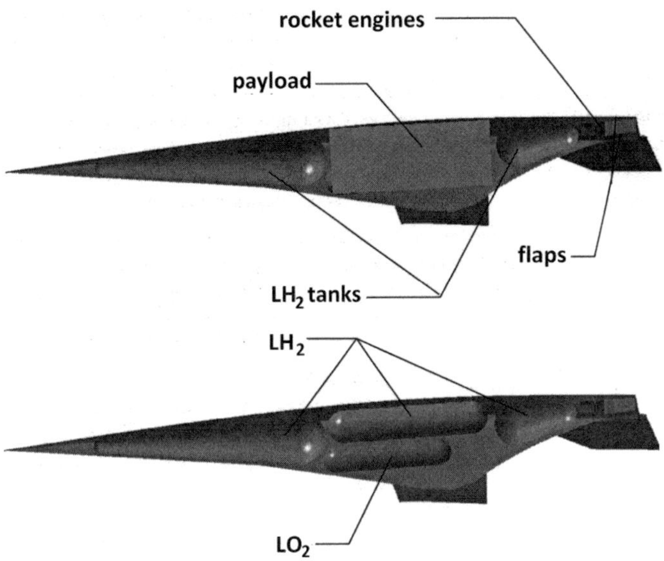

Fig. 6.16 The Scram 5 [16]. Upper part: internal layout, lower part: layout of the additional tanks

The internal layout of Scram 5 is shown in Fig. 6.16. As API-type 5 flight vehicle the scramjet package is at the lower side of the configuration, upper part of the figure. Five rocket engines are located at the upper end of the configuration. Reaction control devices are not indicated.

The payload is located above the scram engine. The LH_2 tanks are ahead and behind the scram engine. At the two sides of the payload bay additional LH_2 tanks are located and also the LO_2 tanks, lower part of Fig. 6.16.

6.4 X-43A

6.4.1 *Thermal Loads and Material Choice: Overall Consideration*

In the HYPER-X program the two successful flights of the 3.658 m long API-type 4 vehicle X-43A, flight 2, January 26, 2004, with $M_{\infty,max} = 6.83$, and flight 3, November 16, 2004, with $M_{\infty,max} = 9.68$ usually are considered as short single stage missions, which had a duration of about 10 s and 11 s.

Actually, however, the missions were three-stage missions, see, e.g.. [18]. The first stage was a B-52B aircraft, the second one the Hyper-X launch vehicle (HXLV) [19], and the third one finally the X-43A. The rocket propelled HXLV, Fig. 6.17, a one-stage variant of the Pegasus rocket, had a length of about 10.5 m.

Hence, when considering the thermal loads on the X-43A one has to take into account the whole launch sequence. The mission profile of the nominal $M_\infty = 10$ flight, Fig. 6.18, [21], shows that it takes about 100 s until the test altitude at 33,223 m (109,000 ft) is reached after launch from the B-52B. Stage separation happens at 95 s.

Figure 6.19 indicates that nominally after 35 s the flight Mach number $M_\infty = 3.2$ is reached. It takes then about 54 s until separation happens of the X-43A from the HXLV at $M_\infty = 9.63$.

In view of the structure concept of the X-43A—cold primary structure and TPS all around, see below—one can assume that a surface temperature close to the radiation-adiabatic one is reached at separation. This does not mean that strong interaction phenomena—e.g., bow shock interference with wing, like sketched in Fig. 3.22—are not causing problems. The inlet cowl is closed in order to protect the engine from high thermal loads during the boost period. When the cowl opens at about $t = 100$ s, the internal propulsion flow path, as well as the SERN, at that time probably are not at operating temperature.

An investigation in [22]—with a simplified representation of the external flow path of the X-43A—regarding the thermal loads and the skin-friction yielded the picture shown in Fig. 6.20. The Mach number is $M_\infty = 10$ at $H = 29$ km, the nose radius is 1.27 mm, that of the $M_\infty = 9.68$ flight of the X-43A.

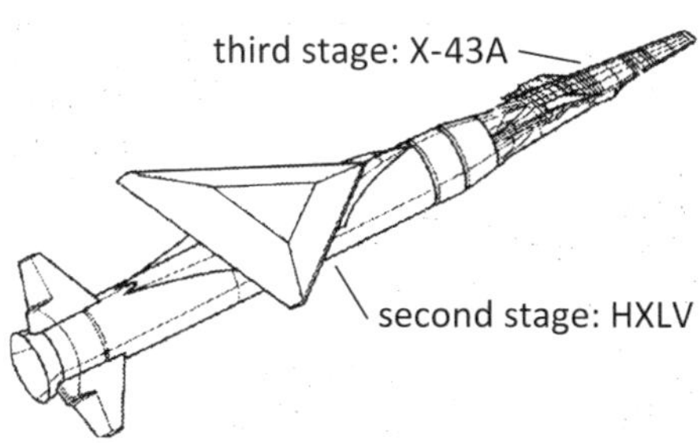

Fig. 6.17 Sketch of the hyper-X launch vehicle [20]

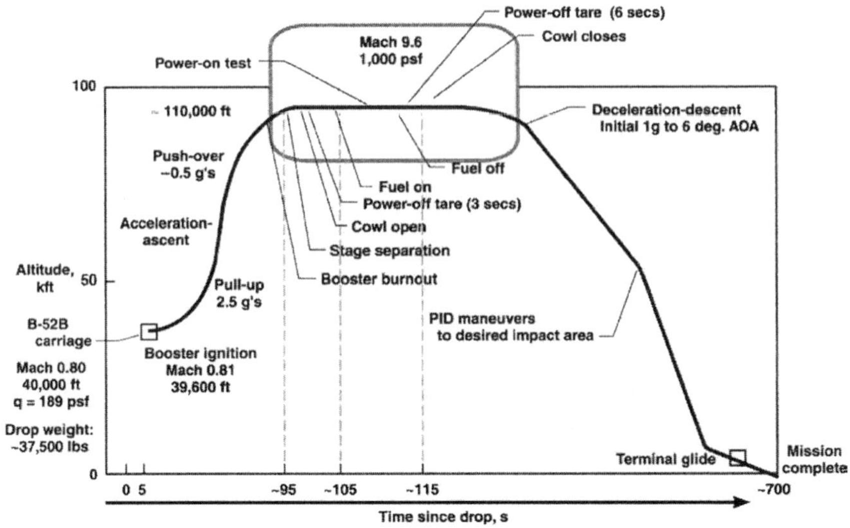

Fig. 6.18 The nominal Mach 10 flight mission profile [21]

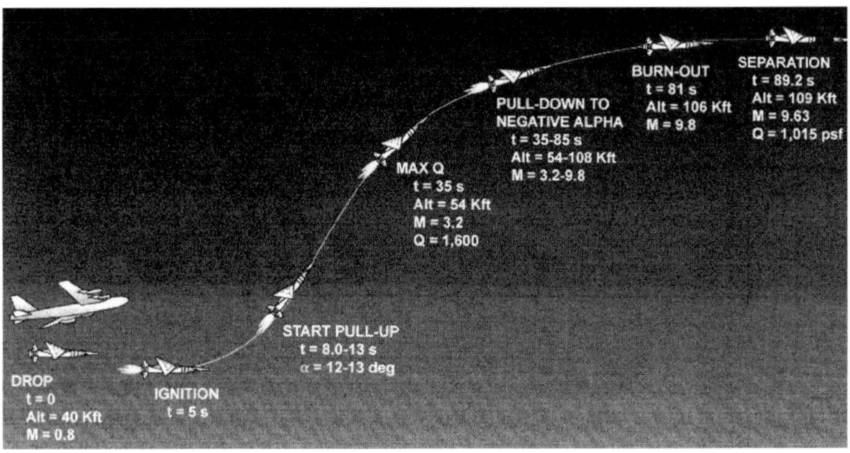

Fig. 6.19 The Mach 10 mission ascent profile with Mach number indications [19]

The loads shown in the figure are the radiation-adiabatic temperature T_{ra} and the skin-friction coefficient c_f as function of the location x.

At the nose, $x = 0$, the radiation-adiabatic temperature is $T_{ra} \approx 2{,}500$ K. The temperature then drops fast to a level of $T_{ra} \approx 1{,}500$ K at both the lower and the upper side. At the upper side—not being a free-stream surface—laminar flow is assumed and the temperature slowly drops to under 1,000 K.

At the lower side laminar-turbulent transition is enforced by boundary-layer tripping at $x = 0.46$ m, like in the case of the X-43A. The temperature increases to

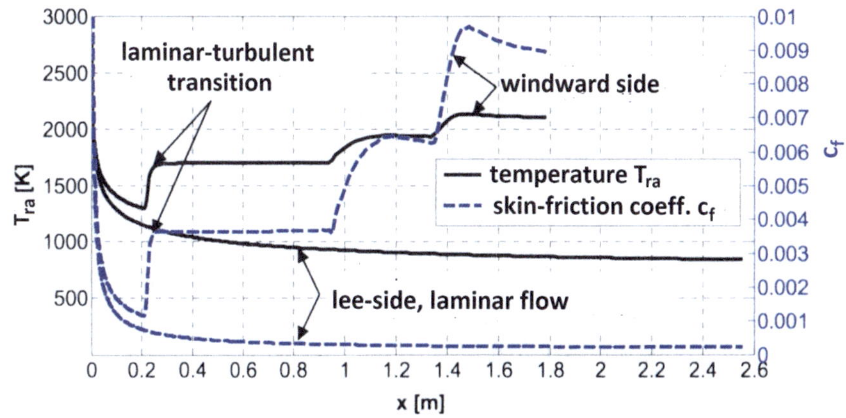

Fig. 6.20 X-43A like two-dimensional forebody and external inlet-ramp configuration: radiation-adiabatic surface temperature ($\epsilon = 0.8$) and skin-friction coefficient along the lower and the upper surface at $M_\infty = 10$ and $H = 29$ km [22]

$T_{ra} \approx 1,700$ K. At the two ramps the ramp effect eventually increases the temperature up to more than 2,000 K.

These results are not fully representative for the X-43A. But in view of the fact that its fuselage is fully packed with engine, fuel tank, measurement and control gear, it is obvious that a hot primary structure is not feasible.

6.4.2 Structural Approach of the X-43A

The X-43A basically has a cold primary structure, Fig. 6.21, together with a thermal protection system. The cold primary structure with longerons divided by fuselage frames and bulkheads mostly is from aluminum [23].

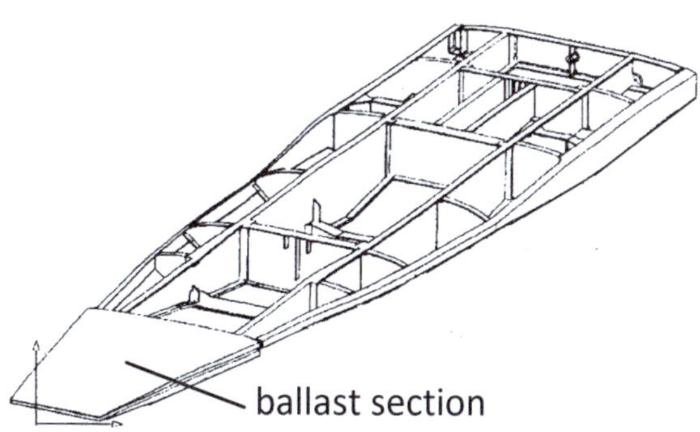

Fig. 6.21 The cold primary structure of the X-43A [23]

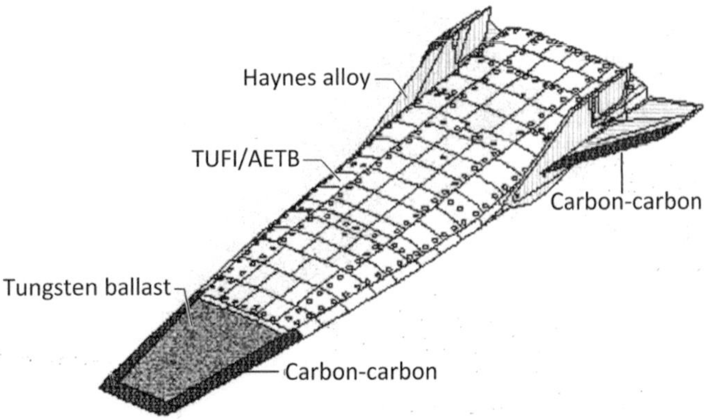

Fig. 6.22 The elements of the thermal protection system of the $M_\infty = 9.68$ X-43A [20]

The flight vehicle had a severe center of gravity problem. To solve that a ballast section was introduced at the nose [18]. That was a piece of tungsten with a mass of about 390 kg, very large compared to the total mass of about 1,350 kg.

A detailed overview of vehicle design and manufacture is given in [20]. Figure 6.22 shows the elements of the entirely passive thermal protection system.

To minimize drag at the high Mach numbers to be flown, the unswept leading edge and the swept leading edges of the horizontal and vertical stabilizers had to be very thin. Consequently, the thermal loads are high, Sect. 3.5. Figure 6.22 shows that around the nose and at the leading edges of the horizontal stabilizers a carbon-carbon material was employed. The stabilizers themselves, as well as the leading edges of the vertical stabilizers were made from Haynes alloy—a cobalt-nickel-chromium-tungsten alloy.

The fuselage itself all around was protected by TUFI/AETB tiles (TUFI: Toughened Uni-Piece Fibrous Insulation, AETB: aluminum-enhanced thermal barrier). The tiles were installed on the airframe and after that machined into the final outer mold line (OML) [20]. The surface coating is applied after that machining. The inlet cowl and the sidewalls, made from a copper alloy, are water cooled.

6.5 About Aerothermoelasticity

Aeroelasticity, respectively aerothermoelasticity is a topic which regards any flight vehicle. This topic is of utmost importance too for each of the here considered hypersonic flight vehicles, regardless of their API-type.

In Sect. 2.10.6 the "Banana-effect" was an issue, i.e., the aerothermoelastic deformation of the forebody, and in Sect. 5.5.3 the very high sensitivity of the net thrust on the effective angle of attack of the lower side of the forebody.

Regarding the X-43A a special problem was the stack stiffness [20].The stack is the HXLV booster plus the adapter plus the X-43A. The adapter and the vehicle had to be sufficiently stiff in view of the booster control system. The structures of the X-43A and the adapter had to be reinforced by a "considerable amount of steel" [20], and the vehicle needed a larger skin thickness. This led to a change of the originally scheduled trajectory. On the other hand the potential flexing of the skin could be reduced and the otherwise needed gaps between the TUFI/AETB tiles be eliminated.

We have a short look at the issues of classical aeroelasticity. The forces acting on an aircraft, Fig. 6.23—see also Fig. 2.14—are

- Aerodynamic forces **A**.
- Elastic forces **E**.
- Inertial forces (mass forces) **I**.

They are among each other in equilibrium and the aeroelastic equilibrium equation holds, see, e.g., [24]:

$$E(q) + A(q) + I(q) = Q. \tag{6.1}$$

with Q being system-independent external disturbance forces. E, A and I are functions of the elastic deflections (q).

The triangle of aeroelastic forces shows the aeroelastic forces and problems divided into static and dynamic problems, Fig. 6.23.

The static aeroelastic problems between **A** and **E** are:

- Static flight stability **SS**: the effect of an elastic deformation of structure elements on the static flight stability and the control behavior.

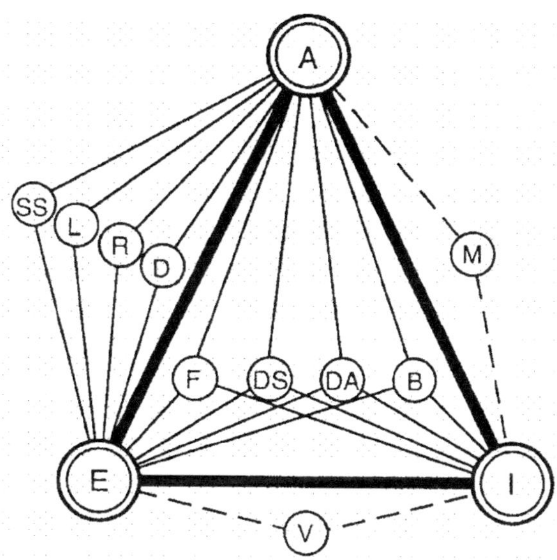

Fig. 6.23 The triangle of aeroelastic forces [24]

- Distribution of lift **L**: effect of an elastic wing deformation on the aerodynamic surface-pressure distribution and hence the lift.
- Rudder effectiveness **R**: effect of an elastic deformation on the effectiveness of control surfaces.
- Divergence **D**: static aeroelastic stability problem where a lifting system—after reaching a critical velocity—abruptly is deformed up to fracture.

Between **A**, **E** and **I** the dynamic aeroelastic problems are indicated:

- Flutter **F**: dynamic aeroelastic instability. After reaching a critical flight velocity, abruptly self-exited flutter of a lifting surface or the whole flight vehicle happens. The amplitude grows rapidly with generally catastrophic consequences.
- Dynamic flight stability **DS**: the effect of a structural deformation on the dynamic flight stability.
- Dynamic Response **DA**: externally excited flutter of individual flight vehicle components due to outer dynamic loads. These can be gusts, landing shocks, abruptly activated control surfaces and so on.
- Buffeting **B**: flutter of the flight vehicle or individual components of it due to flow separation (buffet) in the transonic flight regime or due to impingement of wakes or vortices on for instance the empennage or elements of it.

The dotted lines between **A** and **I** as well as between **E** and **I** point to two items closely related to aeroelasticity:

- Aero-servo elasticity **M**: in particular for large-scale flight vehicles with active flight control and stability systems a coupling of classical flight mechanics and aero-servo elastic effects exists.
- Natural flutter **V**: the knowledge of the natural frequencies of the fully assembled flight vehicle or its components (fuselage, wing, empennage) at rest (**A** = 0) is the starting point of the analytical treatment of all dynamic aeroelastic problems of the complete aircraft [25].

Aero-servo elasticity **M** needs a closer look, see, e.g., [26]. Automatic flight control and stabilization systems potentially can have large effects on the aeroelastic behavior of an aircraft. Figure 6.24 schematically shows the interactions of the dynamics of the flight control system with the unsteady aerodynamics and the structure dynamics, all resulting in the topic of aero-servo elasticity.

The reader interested in the topic of aircraft aeroelasticity should consult also [27–29].

When due to high flight speed thermal loads and accordingly structural interactions come in, one speaks of aerothermoelasticity. It plays a role of course, whether a hypersonic flight vehicle has a cold or a hot primary structure. A mixture of both can be present. A vehicle with a cold primary structure may have wings and stabilization-, trim- and control-surfaces as hot primary structures. In the case of the lower stage of SÄNGER, the forebody partly has a hot and partly a cold primary structure.

Thermal loads, both as wall temperature and as heat flux into the wall, influence the structure—including the properties of the structure's material—and lead together

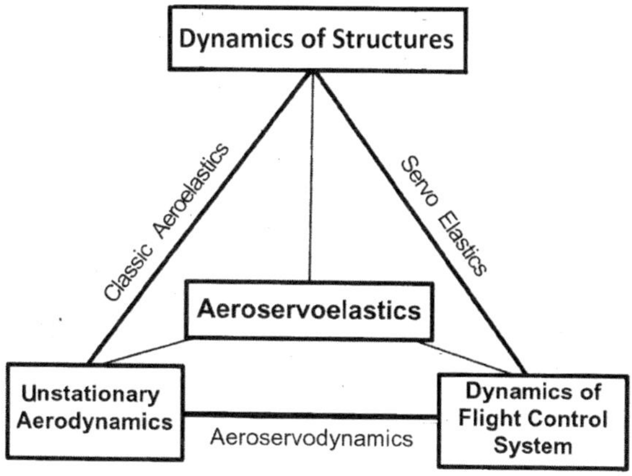

Fig. 6.24 Aero-servo elasticity triangle [26]

with the mechanical loads to deformations: all that must be in the tolerable limit and partly is a matter of time exposure.

The hypersonic topic is treated in [28, 29] in the section "Fluid/ Structure/Thermal/Dynamics Interaction (FSTDI) in Hypersonic Flow" by E.H. Dowell, pp. 717–749. The emphasis, however, mostly is on flutter phenomena of flight vehicle components. The problem of testing a fully assembled airbreathing hypersonic flight vehicle, as is common practice for transonic transport aircraft, see, e.g., [25], is mentioned, but usually not treated in depth.

Anyway, depending on the API-type of the hypersonic airbreather under consideration with a hot or partly hot primary structure the transport-aircraft approach, i.e., testing of the fully assembled aircraft, is not feasible. In particular for API-type 4 and 5 flight vehicles the high coupling of lift and propulsion makes necessary an approach, which from the beginning of the design has the topic of aerothermoelasticity as a central design challenge.

However, hot testing of separable sub-systems, like stabilization-, trim- and control-surfaces, as well as lift enhancement systems should be feasible.

6.6 Final Remarks

To enter unchartered territory has its risks. Hypersonic flight vehicles, regardless of their API-type, are such a territory. Reports of the respective endeavors are not abundant. Notable exceptions are the "Lessons Learned" of the Space Shuttle Orbiter, and that of the X-43A with Peebles' book [18], with the host of related papers attached.

The present monograph makes use of that material, but also of the material which is available to the authors from the German Hypersonics Technology Programme with its reference concept, the lower stage of the two-stage-to orbit space transportation system SÄNGER. Regarding the structural aspects of the lower stage of the SÄNGER system, relevant material is contained in the report "The Technology Development and Verification Concept of the German Hypersonics Technology Programme" [30], see Chap. 8.

6.7 Problems

Problem 6.1 (*Thermal Protection Systems in Re-Entry Vehicles*) Compare the structural and thermal protection approaches of non-winged re-entry capsules (such as the Apollo capsule) and winged re-entry vehicles (such as the Space Shuttle Orbiter). Why are their thermal protection needs different, and how are the structures designed to meet those needs?

Problem 6.2 (*Materials Choice for Hypersonic Vehicles*) What materials are suitable for the structure of hypersonic airbreathing vehicles such as the SÄNGER lower stage? Discuss the considerations for choosing materials for both the hot and the cold primary structure in terms of thermal loads and flight conditions.

Problem 6.3 (*Aerothermoelasticity in Hypersonic Flight Vehicles*) Explain the concept of aerothermoelasticity and its importance in hypersonic airbreathing vehicles. How does the deformation of the structure due to thermal loads affect the vehicle's flight stability and performance?

Problem 6.4 (*Hypersonic Thermal Loads and Transition Zones*) How do the location, shape, and extent of the laminar-turbulent transition zone affect the thermal loads on a hypersonic vehicle such as the SÄNGER lower stage? Why is it challenging to predict this zone accurately?

Problem 6.5 (*Structural Zones in Hypersonic Vehicles*) Describe the three structural zones of the SÄNGER lower stage. What are the key challenges faced by each zone in terms of structural and thermal loads?

Problem 6.6 (*Ballast*) Why has the X-43A the large ballast in the nose? Discuss the reasons and the implications.

Problem 6.7 (*Comparison*) The X-51A, of API-type 3, is about 4.3 m long (without the booster). The flat lower side of the forebody serves as pre-compression surface for the inlet. Its length is about one fourth of the missile length. The $M_\infty = 5$ flight lasted about 240 s. The booster phase was about 30 s. Compare the aeroelastic/aerothermoelastic issues with those of the SÄNGER lower stage.

References

1. Hirschel, E.H., Weiland, C.: Selected aerothermodynamic design problems of hypersonic flight vehicles. In: Progress in Astronautics and Aeronautics, AIAA, Reston, VA, vol. 229. Springer, Heidelberg (2009)
2. Weiland, C.: Computational Space Flight Mechanics. Springer, Berlin, Heidelberg (2010)
3. Arrington, J.P., Jones, J.J. (Eds.): Shuttle Performance: Lessons Learned. NASA CP-2283 (1983)
4. Throckmorton, D.A. (Ed.): Orbiter Experiments (OEX) Aerothermodynamics Symposium. NASA CP-3248 (1995)
5. N.N.: Structural Design. Engineering Innovations, pp. 270–285. https://www.nasa.gov/centers/johnson/pdf
6. Ko, W.L., Quinn, R.D., Gong, L.: Finite-Element Reentry Heat-Transfer Analysis of Space Shuttle Orbiter. NASA TN 2657 (1986)
7. Hirschel, E.H.: Basics of Aerothermodynamics. Progress in Astronautics and Aeronautics, AIAA, Reston, VA (2004), 2nd revised edn, vol. 204. Springer, Berlin, Heidelberg (2015)
8. Kuczera, H., Sacher, P.W.: Reusable Space Transportation Systems. Springer, Berlin, Heidelberg (2009)
9. Bayer, R.: Optimierung von Flugbahnen der Unterstufe eines luftatmenden Raumtransporters (Optimization of the Flight Trajectories of the Lower Stage of an Airbreathing Space Transportion Vehicle). Doctoral thesis, Technical University München, Germany (1993)
10. Kuczera, H., Hauck, H., Krammer, P., Sacher, P.W.: The German Hypersonics Technology Programme, Status 1993 and Perspectives. IAF-93-V.4.629 (1993)
11. Kuczera, H., (Ed.): Sänger Space Transportation System, Concept Study and System Definition. Final Report (Phase Ia, January 1989 until August 1990. BMFT, Referat 513, Reference No. HK 8900-8. Dezember 1991, Bonn, Germany (1991)
12. Kuczera, H., Hauck, H.: The German Hypersonics Technology Programme, Status Report 1992. IAF-92-0867 (1992)
13. Schweikart, L.: The Quest for the Orbital Jet: The National Aero-Space Plane Program (1983–1995). Air Force History and Museums Program, Bolling AFB, DC 20332-1111 (1998)
14. Miller, J.: The X-Planes: X-1 to X-45. Midland Pub. Ltd., 3rd edn (2001)
15. Gorn, M.H., De Chiara, G.: X-Planes from the X-1 to the X-60: An Illustrated History. Springer Praxis Books (2021)
16. Hornung, M.: Entwurf einer luftatmenden Oberstufe und Gesamtoptimierung eines transatmosphärischen Raumtransportsystems (Design of an Airbreathing Upper Stage and Global Optimization of a Transatmospheric Space Transportation System). Doctoral thesis, Universität der Bundeswehr, Neubiberg/München, Germany (2003)
17. Hirschel, E.H., Rizzi, A., Breitsamter, C., Staudacher, W.: Separated and Vortical Flow in Aircraft Wing Aerodynamics. Springer, Heidelberg (2021)
18. Peebles, C.: Road to Mach 10-Lessons Learned from the X-43A Flight Research Program. Library of Flight, AIAA, Reston, VA (2008)
19. Joyce, P.J., Pomroy, J.B., Grindle, L.: The Hyper-X Launch Vehicle; Challenges and Design Considerations for Hypersonic Flight Testing. AIAA-Paper 2005-3333 (2005)
20. Harsha, P.T., Keel, L.C., Castrogiovanni, A., Sherrill, R.T.: X-43A Vehicle Design and Manufacture. AIAA-Paper 2005-3334 (2005)
21. Marshall, L.A., Bahm, C., Corpening, G.P., Sherrill, R.T.: Overview with Results and Lessons Learned of the X-43A Mach 10 Flight. AIAA-Paper 2005-3336 (2005)
22. Kliche, D.: Multidisziplinäre Analyse und Bewertung von Staustrahltriebwerken mit Überschallverbrennung (Multidisciplinary Analysis and Rating of Ramjet Engines with Supersonic Combustion). Doctoral Thesis, Universität der Bundeswehr, Neubiberg. Verlag Dr. Hut, München, Germany (2012)
23. Freeman, D.C. Jr., Reubush, D.E., McClinton, C.R., Rausch, V.L., Crawford, J.L.: The NASA Hyper-X Program. IAF-97-V.4.07 (1997)

24. Försching, H.: Aeroelasticity-the interaction of the elastic aircraft structure with aerodynamic forces. In: Hirschel, E.H., Prem, H., Madelung, G. (eds.) Aeronautical Research in Germany-From Lilienthal Until Today, pp. 278–293. Springer, Berlin, Heidelberg, New York (2004)
25. Göge, D., Böswald, M., Füllekrug, U., Lubrina, P.: Ground vibration testing of large aircraft–state-of-the-art and future perspectives. In: Proceedings of the IMAC XXV, Orlando (FL) (2007)
26. Försching, H.: Aeroelasticity-new challenges and problems. In: Hirschel, E.H., Prem, H., Madelung, G. (eds.) Aeronautical Research in Germany-From Lilienthal Until Today, pp. 556–572. Springer, Berlin, Heidelberg, New York (2004)
27. Wright, J.R., Cooper, J.E.: Introduction to Aircraft Aeroelasticity and Loads, 2nd edn. Wiley (2014)
28. McNamara, J.J., Friedmann, P.P.: Aeroelastic and aerothermoelastic analysis in hypersonic flow: past, present and future. AIAA J. **49**(6), 1089–1122 (2011)
29. Dowell, E.H. (ed.): A Modern Course in Aeroelasticity, 6th edn. Springer, Cham (2022)
30. Hirschel, E.H.: The Technology Development and Verification Concept of the German Hypersonics Technology Programme. DASA-LME12-HYPAC-STY-0017-A (1995)

Chapter 7
Issues of Technology Development and Verification

To create the concept of an airbreathing hypersonic aircraft or even that of a globe-spanning airbreathing hypersonic cruiser is one thing, to design, develop and finally to produce it, is another one.

This chapter is dedicated to a sketch of relevant flight vehicle design, development and engineering issues. That to a degree is incomplete, even lacking important details. Significant differences exist compared to the design issues of winged and non-winged re-entry vehicles.

The introductory remarks give a short overview of the design and development of the basic elements of fixed-wing aircraft with emphasis on the mathematization of aerodynamics, structural mechanics and propulsion.

The challenges of flight-vehicle design and development are sketched next. They concern any flight vehicle, in particular also airbreathing hypersonic flight vehicles. Important is to understand how the different phases of 'product definition with the sub-phases predesign and design' and 'product development' are organized in principle and that the final steps of the development are made in the flight-envelope opening.[1]

The capabilities and problems of the design technologies and of some of the development technologies of airbreathing hypersonic flight vehicles are then discussed in some detail. The design technologies encompass aerodynamics and aerothermodynamics, propulsion, structures and materials, the joint problem in the field of structure-physics modeling—a matter more of the development phase—and subsystems. The joint problem is a basic issue like the prediction of laminar-turbulent transition and turbulence, the latter in particular regarding separated flow.

[1] The terms "definition", "design" and "development" are not universal.

E. H. Hirschel et al., *Elements of Hypersonic Airbreather Design and Development*,
https://doi.org/10.1007/978-3-031-94219-8_7

After a look at issucs of experimental flight vehicles the holistic computational approach to air-vehicle design is sketched. The enormous development of computer power, information technologies and solution algorithms in all fields are leading to formerly unimaginable possibilities in flight vehicle development. Thesc developments are then enabling elements to design hypersonic airbreathers.

The chapter closes with a look at the original concept of the Virtual Product.

7.1 Introductory Remarks

The development of the shape of the modern aircraft probably began with Sir George Cayley in the first half of the 1880s [1]. Otto Lilienthal in the 1890s advanced the art of heavier than air flying and the Wright brothers in the year 1903 finally made the first motorized flight.

About 13 years later with the Junkers J 3 and its cantilever wing the shape of the modern aircraft was beginning to arise. At the same time the modern wind tunnel emerged and experimental and theoretical aerodynamics evolved. Modern theoretical aerodynamics began with the formulation of the Kutta-Joukowsky condition in 1902, the discovery of the boundary layer by Ludwig Prandtl in 1904 and the development of the theory of the lifting finite-span wing by F.W. Lanchester and Ludwig Prandtl at the end of the 1920s, see, e.g., [2, 3].

The emerging turbojet propulsion in the 1930s, see, e.g., [4], led to efforts to shift the transonic drag divergence to higher subsonic flight Mach numbers. In Germany the swept wing, the supercritical airfoil—at that time termed Schnellflugprofil or fast-flight airfoil—and the area rule emerged [2, 5]. All this led to the airplane concepts as we see them today.

With Prandtl's work a first wave of mathematization of fluid mechanics and aerodynamics was to begin. Some writers consider the work of Leonard Euler 1707–1783 as the beginning of the first wave of mathematization of fluid mechanics [6, 7].

The digital computer era, which was beginning with Konrad Zuse [8], led to the second wave of mathematization.[2] In practice the second wave began in the late 1960s, when enough computer power became available and algorithm development had advanced sufficiently. In the field of fluid mechanics and aerodynamics already Prandtl did make first steps in that direction much earlier [9].

The appearance and rapid advancement of the digital computer led to the development of discrete numerical methods for the solution of the governing equations of fluid mechanics. That happened, like in the case of structural mechanics, first for the linear domain.

With increasing computer power—at the end of the 1970s the vector computer emerged, later the massively parallel computer—and with large advances in solution algorithms also the non-linear domain of fluid mechanics and aerodynamics was addressed. A host of different methods up to scale-resolving methods for massively

[2] Konrad Zuse presented in the year 1941 the first programmable computer, which he called Z3.

separated turbulent flow, see, e.g., [10], is now in use. The bottleneck still is the modeling of laminar-turbulent transition and turbulent separation. The latter seems to be overcome with the scale-resolving methods.

The wind tunnel of course still has its role in design verification and aerodynamic data-set generation. The cryogenic tunnel offers very accurate and repeatable aerodynamic simulations in realistic and independently representative Mach number, Reynolds number and dynamic pressure regimes. It has to be emphasized that experimental fluid mechanics and aerodynamics also got large benefits from the computer developments. Optoelectronic methods, for instance, now offer formerly unimagined possibilities of flow observation and interpretation.

Digital computing did revolutionize the design and development also in all other relevant technical fields of aircraft evolution. To be named are the main fields of structural mechanics and propulsion systems, as well as sub-systems, guidance and control, in summary finally the whole aircraft in all its operation modes.

7.2 The Challenges of Flight Vehicle Design and Development

The design and development of airbreathing hypersonic flight vehicles basically follows the methodology approach of subsonic, transonic and low supersonic flight vehicle design.

Figure 7.1 shows—strongly simplified for a large transonic passenger transport aircraft—that first the product definition with its design phases happens, which is followed by the development phase including the envelope opening. After that is completed successfully, full product manufacturing takes place and once the aircraft is operational, the support phase begins for the lifetime of the product.

In Fig. 7.2, see also [12], typically the definition and the development phase of a product are shown. The abscissa to a degree represents the time axis with the advancing phases and sub-phases. Most of these phases are passed through in an iterative manner.

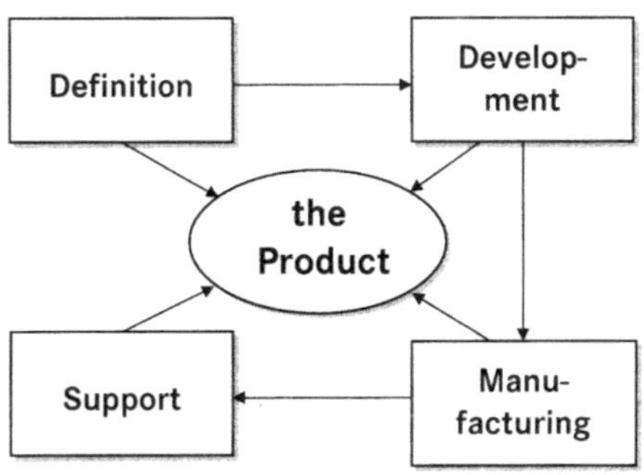

Fig. 7.1 Schematic of aircraft product phases [11]

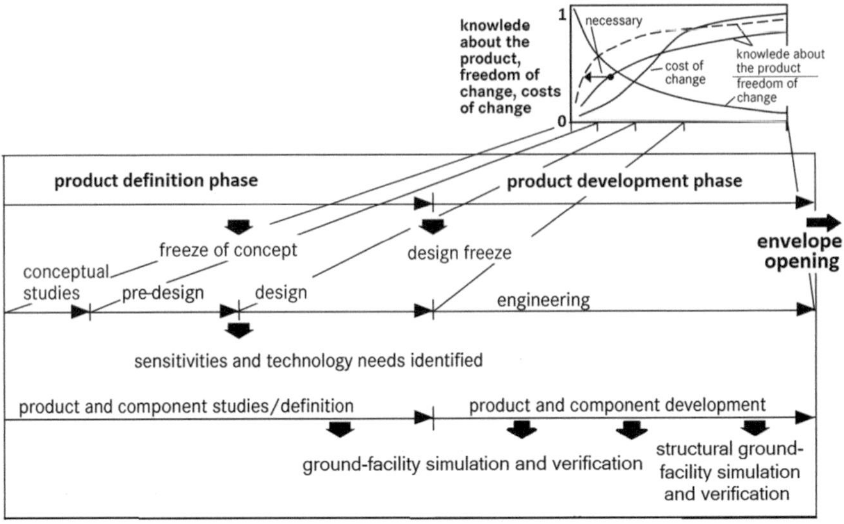

Fig. 7.2 Ideal typical schematic of product definition and development phases

The ordinate in the upper right inserted graph represents a dimensionless scale from zero at the bottom to one at the top. In that graph schematically the evolution of the knowledge about the product, the freedom to make design changes and the costs of design changes are sketched. The broken curve indicates the need to increase the knowledge about the product earlier in the two main phases. Important is that the full knowledge about the product is not reached at the end of the development phase. Uncertainties of the aerodynamic data, of the structural design (the problem to describe the influence of joints —non-linearities and damping), of the performance of control devices, the propulsion system etc. make the so-called system identification necessary.

That is performed with the flight-envelope opening, i.e., a systematic investigation of the operational properties and limits of one or more prototypes in view of the intended mission/missions of the aircraft and its final certification. The flight testing may lead to design corrections before the full expansion of the flight envelope has happened.

The product definition phase in Fig. 7.2 usually consists of the sub-phases conceptual studies, pre-design and design. The freeze of concept happens at the end of the pre-design phase. At that time the design sensitivities of the targeted product are identified. Initial data sets are available, system simulations have been performed, design sensitivities are known and technology needs are identified. The design tools are computation and optimization methods of different capabilities, also different in the involved technical departments.

The design phase, a sub-phase of the product definition phase, Fig. 7.2, is dedicated to work on the details. It may have three or even more iteration loops through all technical departments until all is settled. Ground-facility experimental simulation and presently increasing, high-performance computational simulation is performed in order to obtain the required data.

At the end of the design phase the aerodynamic design, the structural design and so on have been settled and verified and the data sets are established. The aerodynamic data set or model is an "elastified" one. That means that the structural deformation of the wing, of stabilization, trim, control and lift-enhancement surfaces under aerodynamic load has been taken into account. In earlier times this was made with linear aerodynamic methods, now non-linear methods are in use, see, e.g., [13].

Load mechanics also regards dynamic occurrences like maneuver loads, gusts, air turbulence and the landing event. The approach and results in [14] for a high aspect ratio flexible wing with gust loads show the current potential.

The knowledge about the physical properties and functions of the product rises in the definition phase and further in the development phase. Counted as belonging to the topics of the development phase is the issue of technical equipment. That encompasses for instance electric and electronic systems, computer systems, sensors, display technologies, hydraulic systems, fuel tanks, cooling systems, defroster systems, flight control systems, drive systems and so on. A high level is reached only late in the product development phase. Mistakes in these two phases must be avoided. The full knowledge will be achieved with the system identification in the flight-envelope opening process.

Evidently the freedom of changes decreases during the two major phases. The later a major change is made, the higher are the costs of the change. Changes of the airframe, in the worst case "repair solutions", if its properties do not meet the requirements, can be very costly and also may increase the mass of the vehicle, see also Sect. 2.10. This holds the more, the later the need for a change is identified in the development phase or even during the envelope opening.

The structural ground-facility experimental simulation finally verifies the design and establishes the real aeroelastic properties of the airframe. Once the design is settled one or more prototypes are built and the envelope opening happens. Usually design corrections become necessary until the aircraft is ready for certification and production. For the critical topics "requirements creep", "obsolescence", and "mass increase" see Sect. 1.6.

Functions, sub-systems, components and so on are studied in all phases and subphases. They are simulated on the computer and tested in ground-simulation facilities. The real static and dynamic aeroelastic properties of the airframe are found late in the development phase, once it has been fully assembled.

In the product definition phase the shape, the performance, the flying qualities and so on of the aircraft are fixed to a large degree. Importantly, in addition also fixed implicitly and to a large degree are the final production, operation and maintenance costs.

These considerations hold for any product, any aircraft and here in particular also for airbreathing hypersonic flight vehicles. In the following section the design technologies for airbreathing hypersonic flight vehicles are considered.

7.3 Capabilities and Problems of the Design Technologies of Airbreathing Hypersonic Flight Vehicles

Under design technologies we understand the fields of aerodynamics and aerothermodynamics, propulsion, structure and materials, guidance and control, and subsystems. This section is devoted to a short consideration of the capabilities and problems of these technologies in view of the development of airbreathing hypersonic flight vehicles. Special topics like book-keeping of aerodynamic and propulsion forces and so on are not treated.

The demands on the individual technologies of course strongly depend on the API-type of the vehicle under consideration, as well as on its overall size. The highest demands certainly are given with the large API-type 5 vehicles.

7.3.1 The Size of the Flight Vehicles and the Flight Environment

The size in terms of the length of the here considered flight vehicles, Table 7.1, varies between 3.658 and 94 m. Therefore, the size of the flight vehicle under consideration is one of the major issues besides the operational flight Mach number. It governs ground-facility size in particular regarding structural tests, but also regarding aerothermodynamic tests. Other than in low-speed flight, up to low supersonic Mach numbers, it is not the wing span, which is the determining length. Here, because slender configurations are the rule, the length of the whole flight vehicle—from tip to tail, overall—matters, Table 7.1.

Table 7.1 Comparison of flight vehicle length and propulsion system length of different hypersonic flight vehicles

No.	Flight vehicle	API-type	$M_{\infty,design}$	$L_{overall}$	$L_{prop.system}$
1	X-43A	4	6.83/9.68	3.658 m	$\approx$2.7 m
2	Missile	3	5–7	5–10 m	
3	SR-72	4	6	$\gtrsim$30 m	
4	X-30	5	25	48.768 m	
5	SÄNGER lower stage	5	6.8	82.5 m	35 m
6	STRATOFLY MR3	3	8	94 m	$\approx$40 m?

The airbreathing hypersonic flight vehicles fly at Mach number up to $M_\infty \approx 12$ at altitude H up to about 50 km. The flight vehicles are viscous-effects dominated.[3] The surface temperature levels to be expected are up to about 6,000 K, Fig. 2.8. Since the surfaces of the external flow path of these vehicles are radiation cooled, the resulting actual temperature levels are much lower and manageable. A number of thermal surface effects exists, Sect. 3.10.

The atmosphere with local—depending on the operational location over the surface of Earth (latitude and longitude)—and instantaneous altitude-depending properties,[4] and gusts and free-air turbulence as well as wakes of other aircraft of course is in thermodynamic and chemical equilibrium.

However, in the flow over the flight vehicle high-temperature real-gas effects in terms of vibration excitation and O_2-dissociation are present, Fig. 2.9, as well as non-equilibrium and surface catalycity effects. At the flight altitudes of hypersonic airbreathers which are below approximately $H \approx 50$ km laminar-turbulent transition occurs and, hence, turbulent boundary-layer flow exists, Fig. 2.8.

7.3.2 Aerodynamics and Aerothermodynamics

The tools and tasks of aerodynamic design are sketched in Fig. 7.3. The tools are handbook methods, numerical methods and the wind tunnel. They address aerodynamic shape definition, verification of design and performance, data-set generation, problem diagnosis and certification. The tools and tasks are the same for the aerothermodynamic design.

Important is that with increasing computer power and advancing solution algorithms the capabilities of the methods of numerical disciplinary and multidisciplinary aerodynamics are changing the picture in favor of the computational design. The wind tunnel, besides its role in research, will remain to be a very important verification tool, but it will be less used in the design and development process.

7.3.2.1 The Role of Aerothermodynamics

Aerothermodynamics defines the outer (airframe) contour of the flow path, including the resulting mechanical and thermal loads on the aerothermoelastic structure. Regarding the airframe the flow field must almost always be considered in its entirety. To collect the demands for every configuration/component item separately in general makes no sense. Exceptions are the forebody and the rear fuselage, which are in a sense separate components due to their direct coupling to the propulsion system.

---

[3] This is in contrast to winged or non-winged re-entry vehicles, which are compressibility-effects dominated, see Table 1.2.

[4] In Appendix A only the standard atmosphere is given.

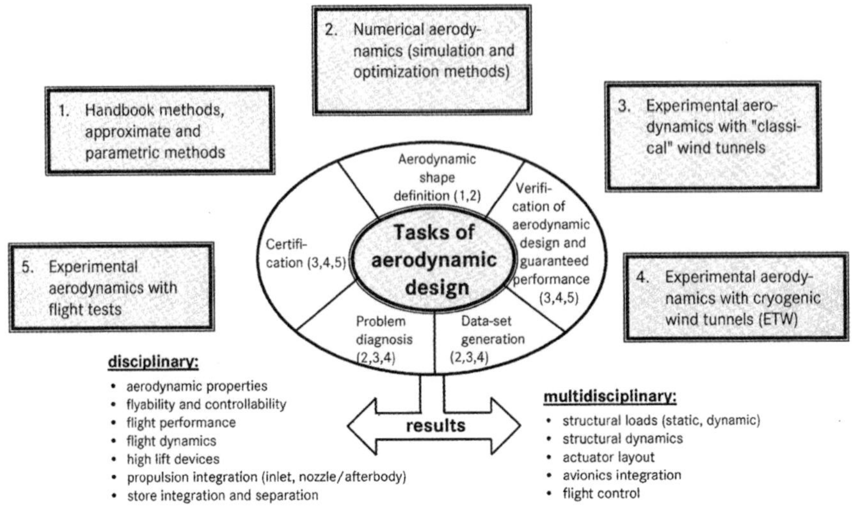

Fig. 7.3 The tasks and tools of classical aerodynamic design [9]

The lower side of the airframe of API-type 3 (partly), 4 and 5 flight vehicles to a large degree has the tasks of lift generation and the accommodation of the propulsion system. The forebody also must accommodate the sensors of the flight control and guidance system.

The mechanical and the thermal loads on the elastic airframe including the control surfaces of all kinds can be considered as the "hard" data. The integral data and their derivatives of the aerodynamic forces and moments characterize the performance of the configuration and its components. For hypersonic flight vehicles also surface properties (see Sect. 3.2.3) in terms of permissible roughness, waviness, step heights, gap dimensions, necessary surface heat emissivity coefficient, permissible surface catalycity, must be counted to the hard data, because they govern the surface material/coating and the manufacturing tolerances.

"Soft" data are the flow topology, flow-field data etc., which the designer must know in order to understand and "shape" the functionality of the airframe, its components, and most importantly, the airframe-propulsion integration.

7.3.2.2 Capabilities and Shortcomings of Computational Design and Verification

1. Computational aerothermodynamics, also coupled to structural methods has made and still makes large progress. Elastic configuration elements as well as moving trim and control surfaces etc. can be incorporated.

2. The model-building of high-temperature real-gas effects is well advanced. Surface radiation cooling also with non-convex effects can be simulated. Surface roughness is treatable to a degree.
3. The prediction of location, shape and extent of the laminar-turbulent transition zone, however is a severe bottleneck.
4. Simulation of massive separation of turbulent boundary-layer flow is possible with scale-resolving methods, see, e.g., [10].

7.3.2.3 Capabilities and Shortcomings of Ground-Facility Simulation

Besides the classical subsonic and transonic wind tunnels, facilities as high-enthalpy shock tunnels, Ludwig-tubes with high flow quality for the study of laminar-turbulent transition phenomena, and so on are in use. In addition to the force and moment measurement devices highly accurate and versatile opto-electronic measurement devices are available.

Several basic problems still exist, see also [15]:

1. Reynolds number and other important similarities are hard to achieve because of tunnel and hence model size limitation. In shock tunnels inertial forces may influence force and moment measurements. The higher the flight Mach number and the longer the flight vehicle is, Table 7.1, the more model-size limitations become an issue.
2. Hypersonic flight vehicles have radiation-cooled surfaces with varying temperature from the tip to the tail. A hot model technique in that regard does not exist, and appears not to be feasible to the needed degree. Viscous thermal surface effects have been and are studied, see, e.g., [16, 17].
3. Laminar-turbulent transition can well be studied in Ludwig tubes, but tripping in high-enthalpy shock tunnels appears not to be feasible. It would anyway require prior prediction of the location, shape and extent of the transition zone, which is not yet possible. For the general complex of problems see Sect. 3.11.
4. In the airbreathing flight domain up to $M_\infty \approx 12$ stagnation enthalpy and thermodynamic and chemical rate effects can play a role. Even if the stagnation enthalpy can be duplicated in the tunnel, freezing and non-equilibrium effects of vibration and dissociation remain in the test section.
5. Flow-quality problems exist in free-piston tunnels due to driver-gas contamination and tunnel-wall boundary-layer noise. Uncertainties in relevant similarity parameters due to uncertainties of the tunnel-flow parameters are a problem.
6. Test duration is a problem in particular in high-enthalpy tunnels, aggravated by small model dimensions. Flow steadiness and hence measurements-response times can become difficult to be met.

7.3.3 Propulsion

The propulsion means of hypersonic airbreathers are turbojets, ramjets and scramjets, Fig. 4.1. When considering the capabilities and hence the design problems of the propulsion system, one has to be aware of the different tasks of each propulsion system, in particular also in view of the airframe-propulsion integration (API-) type of the flight vehicle and the combination of propulsion-system types under consideration. The design challenges are:

- Definition of the internal flow path with heat addition by means of combustion in order to produce the required thrust (inlet, boundary-layer duct, turbo-engine, ram-burner, scram-burner—with a number of possible combinations of these three—and nozzle).
- Definition and development of the fixed and movable mechanical components, which sustain the flow path as partly actively cooled light-weight structures.
- Design of the control of the propulsion system as a dynamic system in all flight phases; avoidance and handling of emergency situations, etc.

The technology demands of the propulsion system—from the inlet to the nozzle—are hard data, which are characteristic for the propulsion system. The onset-flow conditions are those immediately ahead of the inlet, the integral data are the outer forces and moments, which are related to the propulsion system.

Mechanical and thermal loads, as well as the dynamic behavior, which encompasses all dynamic phenomena from combustion instabilities to control-system induced oscillations and unstart—in any case to be avoided—, are data, which are needed and can only be obtained integrated with the airframe.

The integration of the propulsion system with the airframe to form the hypersonic aircraft leads to technological demands due to the mutual interactions. These demands are different for the different airframe-propulsion integration types, suitable book-keeping concepts of aero and propulsion forces are a problem.

With API-type 5 vehicles the flow over the long elastic forebody with static and dynamic aerothermoelastic deformations couples most strongly into the propulsion system, Chap. 5. This is in terms of the onset-flow conditions, the integral data, and the dynamic behavior of the integrated airframe-propulsion system.

Other integration effects are interactions, especially in off-design situations, if more than one propulsion system unit is present. This concerns especially the inlets, the nozzles and the rear fuselage.

Onset flow conditions are the flow properties just ahead of the inlet. In a wider sense they encompass the momentum vector of the flow entering the inlet and in the turbojet mode the boundary-layer duct.

The inlet flow path has to assure the necessary inlet performance in any (quasi-) stationary and transient/dynamic mode. The fixed and variable structure components of the inlet, which is a lightweight, partially cooled item, define this flow path, with extremely small tolerances on all segments of the flight trajectory.

Position and static and dynamic aerothermoelastic properties of all fixed and variable structure items, which are a function of the mechanical and thermal loads, belong therefore to major technological problems of the inlet.

The mechanical and the thermal loads themselves partly are governed heavily by strong shock/boundary layer interaction and separation phenomena, which on the other hand influence strongly also the inlet flow and hence its overall performance.

Any variable structure element, here the inlet ramps, per definition has sealing needs. Connected to that is the leakage problem. Mass-flow losses due to leakage, like those due to bleed must be minimal, because they influence adversely (and strongly) the thrust performance of the propulsion system.

The dynamic behavior regards the structural elements, but also the control and actuation system. The most severe dynamic event possible in an inlet, the inlet unstart, must be avoided by any means, especially at high Mach numbers because of its potentially catastrophic consequences.

The boundary-layer diversion duct has not the severe performance demands, which the inlet has. However, it must function, very probably only in the turbo mode, with minimum structural weight (cooling demands at possible higher flight Mach numbers?), and must lead the diverted boundary-layer flow (and bleed flow) with a minimum of momentum loss towards the afterbody, where it is added to the nozzle exhaust flow.

Engine center modules have their own challenges. Onset-flow conditions are local entry conditions, dynamic behavior includes also vibrations. Stiffness relates to the deformation (elastic) behavior of mechanical components under static and dynamic mechanical loads in presence of thermal loads.

Entry conditions include possible distortions from upstream locations (forebody boundary layer, inlet, etc.). The ramjet or scramjet burner may be located in parallel or behind the turbo-jet, which, however, is closed during their operation.

Dynamic behavior refers to acoustic phenomena, combustion instabilities, etc. coupled to the characteristic length of the burner and its elastic structure.

The nozzle must achieve maximum net thrust performance in all the different propulsion modes. Hence the nozzle flow path has very small tolerances and mechanical and thermal loads play a large role, too. The varying thrust direction of a single expansion ramp nozzle (SERN) reigns strongly into the force and moment household of the flight vehicle.

The position and rate control of nozzle movables, their stiffness and dynamic behavior are important items, which must be controlled. A special issue is the addition of the boundary-layer diverter flow in the turbojet propulsion mode.

System items are summarized: integrated control system, H_2-cooling system, fuel and oil system, air-cooling system, and so on.

7.3.3.1 Capabilities and Shortcomings of Computational Design and Verification

1. The theoretical/numerical model building and treatment is sufficiently advanced. Moving and deforming structure elements can be handled.
2. Thermodynamic and combustion modeling is satisfactory.
3. Critical areas still are laminar-turbulent transition over the forebody and inlet, turbulent mixing in the combustion chamber, possible re-laminarisation in the nozzle.
4. Once these can be resolved satisfactorily, with increasing computer power a treatment of the whole external (inlet) and internal flow path including the processes in the combustion chamber and the nozzle should become possible. Coupling to the structural items is questionable, at least today.

7.3.3.2 Capabilities and Shortcomings of Ground-Facility Simulation

1. Ground-facility testing of the individual constituting elements of a propulsion system is possible, size may play a role.
2. In all cases the simulation of the real-flight thermal situation of at least the surfaces of the outer and the inner flow path is a critical issue.
3. Regarding the testing of the whole propulsion system including the inlet and the nozzle, the matter of size becomes the central problem. For the design and verification of the 3.658 m long X-43A including its propulsion system see the attachments to [18].
4. For large propulsion systems ground-facility tests of the complete system appear to be impossible. The length of the propulsion system of the lower stage of the TSTO-system SÄNGER, No. 5 in Table 7.1, was projected to be approximately 35 m, with a diameter of the propulsion core of about two meter. For No. 6 about 40 m length is estimated. For large propulsion systems a transfer-model approach, as sketched in Chap. 8, appears to be the only way out.

7.3.4 Structure and Materials

Structures and materials define the airframe (and the propulsion system) as a physical entity. The airframe sustains the outer flow with its shape, while carrying the static and dynamic mechanical and thermal loads. It further generates the aerodynamic lift and has to provide flyability and controllability of the vehicle.

Very important, too, is that the airframe must provide the onset-flow of the propulsion system (forebody), and the expansion of the propulsion jet (rear-fuselage and expansion ramp), in order to permit the airbreathing propulsion system to generate thrust. And finally the airframe must carry the payload, the fuel, the sub-systems, the propulsion system, and last but not least, its own gravity and inertial loads.

On the whole flight trajectory besides the mechanical loads also the thermal loads and their influence on the structural properties are of deciding importance. The thermal loads initially build up, reach their maximum, decrease, but due to thermal reversal, Sect. 3.10.4, do not disappear in the final parts of the trajectory down to approach and landing.

The characteristic properties of the airframe are strength, stiffness and fatigue life of the configuration and its components. Important are its static and dynamic elastic properties with regard to the external flow including the parts of the propulsion system and the sub-systems, including the notch filters. The terms "stiffness" and "elastic properties" are not used as opposites. "Stiffness" is used in the classical sense of structural engineering, "elastic properties" with regard to the problem of flow/structure interactions (aerothermodynamics-structure coupling) and systems (notch filters, etc.).

The flow/structure interactions are important because the—necessary—lightweight construction of the airframe and its components must not lead to static and dynamic elastic deformations, which have adverse influence on both the aerodynamic (outer) and the propulsion (outer and inner) flow path.

To determine the properties of the airframe, the mechanical and the thermal loads must be known, which are generated by the flight through the atmosphere including gusts and wakes of other aircraft, and also by the inertial forces and any other mechanical and heat loads, which are generated by the vehicle itself.

Surface properties, see Sect. 3.2.3, of the external and the internal flow path are very important. They permit via surface radiation the passive cooling of the airframe. On the other hand they must minimize thermal loads due to catalytic surface recombination of dissociated propulsion-system exhaust and air constituents. If necessary anti-oxidation coating must be applied.

Very important is that the surface properties influence laminar-turbulent transition and turbulent flow characteristics via roughness, waviness etc. The surface properties must not be optimized independently, but in such a manner, that the vehicle's performance, operation costs, life cycle costs, ecological compatibility etc. are optimized.

Important is the global structural approach:

(a) cold primary structure with a thermal protection system,
(b) partly hot and partly cold primary structure,
(c) fully hot primary structure.

in all cases with hot sub-structures, for instance stabilization surfaces, flaps, lift-enhancement systems.

The focus now lies on the design approach of fully hot or partly hot primary structures of hypersonic airbreathers. Structural design happens with the classical tools and approaches. This is the picture which emerged at the time of the German Hypersonics Technology Programme.

7.3.4.1 Capabilities and Shortcomings of Computational Design and Verification

1. Numerical simulation with thc tools of structural design is the usual approach with numerical models adjusted with data obtained in static and dynamic tests of the fully assembled aircraft, as long as such tests are possible. Changes of material properties under thermal and mechanical loads must be taken into account: strength and fatigue resistance.
2. Purely numerical simulation—without being based on data from dynamic tests—with the classical tools of structural design does not give the aerothermoelastic properties of the structure, whether cold or hot, to the needed accuracy and that also in view of the final certification.[5]
3. The reason for that is beside others—idealization errors, discretization errors, erroneous assumptions of model parameters—in particular the structural joint problem. That regards rivets, screws, welding and bonding joints. It holds for all metallic and non-metallic structural elements. These joints bring in non-linearities and damping, Sect. 7.3.5.

7.3.4.2 Capabilities and Shortcomings of Ground-Facility Simulation

1. Static tests of around 1,000 K or more hot structures seem to appear as to be possible.
2. Dynamic tests in view of the aerothermoelasticity of an around 1,000 K or more hot structure of a fully assembled flight vehicle (lengths of the items 3 to 6 in Table 7.1) appear not to be possible, see Chap. 8. For sub-systems like trim and control surfaces, lift-enhancement systems etc. due to their rather small sizes such tests in principle should be possible.

7.3.5 *Structure-Physics Modeling and the Joint Problem*

Structure-physics modeling poses very significant challenges. The main problem with the prediction of the aeroelastic and aerothermoelastic properties of an airframe is that today it can be considered only as being *perfect-elastic* with, however, a number of partly problematic assumptions and simplifications. The *real elastic* properties due to joints of all kinds cannot be modeled adequately. That holds for both metallic and non-metallic airframes, or parts of them.

[5] An overview of the numerous sources of modeling errors can be found in [19].

The joints (rivets, screws, bonded surfaces, welding seams, bearings, guidances) lead to non-linearities and damping, in general to a non-linear behavior.[6] Also post-buckling behavior of the structure appears to be an unresolved topic.

The present approach to obtain the real-elastic properties of an airframe is to assemble it and to test it in a ground facility toward the end of the product development phase.[7] Only then its static and dynamic structural data sets are obtained with the needed accuracy level. That has to be overcome in particular when dealing with hypersonic airbreathers. The tests are too late, the information is needed much earlier. Depending on the API-type, the aerothermoelasticity of the flight vehicle is a critical topic from the very beginning of the design work.

The present approach has grown over many decades. It is necessary, because the computational structure mechanics (product) models and methods in use today do not permit in full generality to describe joints of all kinds at the numbers they are incorporated in the whole structure.

Hence, early in the design process needed data cannot be obtained with the necessary accuracy. An approach with non-linear aerodynamics and non-linear structural mechanics is necessary. Here the joint modeling comes into the picture. The challenges to accomplish this are very large. Joint modeling with the large scale differences between for instance a rivet and the whole airframe, poses a problem similar to the turbulence problem, where also very different characteristic lengths scales are present.

Since a long time the situation appears to have not changed very much, in particular also because composite structures come into use. The subsequent assessment gives the prospects of joint modeling [20], see also [21–23].

"With the increased computer power it is possible to analyze models of higher order. If this is transferred to the modeling of solid bodies—solid mechanics—, respectively structures, a more refined resolution is possible. This, however, still is restricted to the linear elastic properties (stiffness) and the inertial characteristics (mass distribution).

The damping currently cannot be modeled and not "designed". The damping results from the kind of design of the structure and the quality of its manufacturing. In fact non-linear effects of interface mechanics at structural joints (joint problem) are often the reason for the damping.

Despite non-linear effects can and should be taken into account, often linear modeling is preferred. Equivalent linear models for spot-welded joints (CWELD) were introduced, as well as for fastener connections (CFAST), but their properties usually have to be fitted to experimental data. Therefore they do not present a true prediction.

A possibility to get an assessment is to use the ratio of 'mass of primary structure' to 'operational mass empty, OME'. It appears that if this is large, for instance

---

[6] Indicatively, at the airframe of an all-metal transonic airliner more than 100,000 joints are present.

[7] Tests are needed anyway due to the manufacturing specifications. There is no other way. If the simulation model is erroneous, it would be a design risk to use it further. A test with model validation hence is a means to reduce the design risk.

in the case of transport aircraft, structure models are more exact than those for aircraft with small ratio, like helicopters. Regarding the quality of the manufacturing process of a structure, not only the tolerances—they can be modeled as statistical parameters—are of importance. Thus, the mathematical tools for the respective analysis already exist. However, the quantification of the scatter of different structural parameters (such as joint stiffness) still is an unresolved problem. The scatter of structural parameters today can be taken into account with Monte-Carlo simulations or with interval arithmetic.

A special problem is with composite structures, where the actual material properties are resulting from the manufacturing process. In aircraft production the improvement of the manufacturing process happened with prefabricated fiber layers and with machine fiber deposition. This has led to even lighter structures, but unintendedly the increased fibre-volume-ratio has led to a reduction of structural damping, which must be considered as a drawback for dynamic aeroelastic stability."

This all holds for aircraft, operating up to Mach numbers where thermal loads do not pose a problem. The predictions, even with the uncertainties due to the model building, are supplemented by the static and the dynamic ground tests. Moreover, during the flight envelope opening the aeroelastic properties of the airframe finally are settled.

The joint-modeling problem of course holds for hypersonic airbreathers, too. Depending on the API-type of the flight vehicle a low or high coupling between lift and propulsion is present. Besides the mechanical loads, the thermal loads must be taken into account as well. Static and dynamic ground-facility testing, depending on the vehicle's length, may be restricted partly or fully.

In the following Chap. 8 the Hypersonic Technology Development and Verification Concept—HYTEDAV—derived for the SÄNGER TSTO system is sketched. For the lower stage a transfer-model approach was derived, making use also of dedicated experimental hypersonic vehicles, Sect. 7.4. Although the concept reaches back into the mid 1990s, it still has its validity and can and should serve as a guideline for the development of airbreathing hypersonic flight vehicles.

7.3.6 Flight Vehicle Sub-systems

Aircraft sub-systems or equipment include the actuators of trim- and control-surfaces, as well as the lift enhancement system and all hydraulic, electric and cooling systems. The flight control system, the air-data system, the navigation system, communication systems, cockpit equipment, payload components, air conditioning etc. and also the landing gear belong to this category. Regarding cooling and heating needs of hypersonic airbreathers (internal flow path, engines, tanks, cockpit, payload compartments etc.) a thermal management system of high complexity is the rule.

In any case, the types and mission of the flight vehicle under consideration govern the kind and complexity of the sub-systems. Partly the systems work in the internal environment of the flight vehicle, partly they are exposed to the harsh and hot flight environment. Some of them even run along large parts for instance of the wing.

A number of sub-systems is contributed by specialized equipment companies. All systems need to be tested and qualified. With long fligth vehicle development periods the problems sketched in Sect. 1.6, namely obsolescence, mass increase and requirements creep, are an important topic.

Some sub-systems, for instance the trim- and control surfaces, including the actuators and the electric gear, are "hot" sub-systems. They need to be tested in the simulated true flight environment. The so-called iron birds, i.e., an arrangement of them as it will be on the aircraft, can have a large extension.

Guidance, navigation and control (GNC) needs sensors, which partly work in a hot environment. The air-data system (ADS) gives the input into the flight control system. It provides the real-time atmospheric properties, flight speed vector, altitude, and Mach number. Navigation needs data of different kind, for instance position, speed over ground etc. In any case sensors are placed at positions of the flight vehicle, where they are exposed to the local environment. Ground testing in such cases should be possible.

Which sensors and their locations are suited best for hypersonic flight is an open question, also because of the still strongly advancing GNC technologies. The problem of the coupling of the oscillations of the elastic airframe (due for instance to control-surface forces or to gusts) into the flight-control system remains, with the extra coupling due to the propulsion system.

The notch filters of the flight-control system are the means to block the frequencies, which are not relevant for the flight control system. Their layout of course demands the full knowledge of the aerothermoelastic properties of the flight vehicle and its constituents.

For all sub-systems of hypersonic airbreathers the ground testing is mandatory, too. For many of the systems ground testing in a hot environment should be possible. If the structural dynamics of the whole hot flight vehicle come into play, the problem is the same as that sketched in Sect. 7.3.4.

7.3.7 The "Ilities"

The "ilities" of a hypersonic flight vehicle, like those of any other flight vehicle, are of high importance, too. The first two were already addressed: flyability and controllability. Here we mention the "reliability", "maintainability", and "testability" of the flight vehicle and its hardware components through its entire lifetime. It is evident that the "ilities" especially concern the flight vehicle operation, in particular the preparation activities before and the post-flight inspections after a flight mission. Because the "ilities" are potential operation cost drivers, they must be considered throughout the technology/development work from the beginning on.

In the context of the "ilities" the certification of the flight vehicle is to be mentioned and in a wider context the whole life cycle up to the final decommissioning. The flight vehicle may find its end in the exhibition of a museum or on the wrecking yard.

7.4 Experimental Flight Vehicles

We have a short look at the issue of hypersonic experimental vehicles. Two basic types, respectively missions can be distinguished:

1. general scientific mission,
2. part of a flight-vehicle development program.

The great examples of **general scientific missions** are the X-Planes of the USA [24, 25]. At the beginning the X-Planes were to pave the way toward supersonic flight. Up to date a number of different flight vehicles was built and flown, several of them being hypersonic flight vehicles, notably the X-43A and the X-51A. It appears, however, that in general there was no direct connection to a large flight vehicle development effort, in the end even regarding the X-30.

Known to be part of a **flight-vehicle development program** probably were three experimental vehicles or even families of them. Two of them were re-entry vehicles, the third one a hypersonic airbreather. Other ones (but not considered here) are for instance the Japanese flight test vehicles in the HOPE/HYMES programme.

First to name are the BOR vehicles. Five of them were flown to gain aerodynamic, thermal, stability and acoustic insight and data for the development of the Soviet Union's winged re-entry vehicle BURAN, see, e.g., [26].

As second to see is the experimental vehicle MAJA, which was proposed for the European winged re-entry vehicle HERMES, see, e.g., [27]. HERMES was to be launched atop an ARIANE 5, which led to a very slender delta-like configuration. There was no design experience available from other projects. The US Space Shuttle Orbiter as well as BURAN had configurations not comparable.

There was further the today almost forgotten pitching-moment problem during the first re-entry of the Space Shuttle Orbiter, see, e.g., [28]. Because no explanation of it was known at the time of the HERMES design work, this caused great concern. Possibly it was a problem of the wind tunnel experiments. At that time the discrete numerical methods (CFD) of aerothermodynamics were advancing fast, but it was felt that flight data were needed in order to validate them.

That led to the concept of the experimental vehicle MAJA, being a downsized HERMES, see, e.g., [15]. MAJA finally was discarded in favor of an unmanned HERMES, until in 1992 the entire program was abolished.

Thirdly, the airbreathing experimental vehicles proposed in the Hypersonics Technology Development and Verification Concept (HYTEDAV) are to be mentioned. The HYTEDAV concept, Chap. 8, was developed during the German Hypersonics Technology Programme, 1987–1995, see, e.g., [29].

At that time it was assumed that a space transportation concept like the two-stage concept SÄNGER would be the successor of HERMES. In view of the (as not being feasible considered) ground-simulation of the hot, 82 m long lower stage, Sect. 7.3.4, a technology development and verification strategy was developed. That was to make use of a number of hypersonic airbreathing experimental flight vehicles HYTEX 1 to 3 [30].

For details see the next chapter. Here we only point out, that the approach led to a doubly recursive structure of the technology development program, Fig. 8.2 in Sect. 8.1.

Both the reference concept of the technology programme (the lower stage of the TSTO-system SÄNGER) and the experimental vehicles required technology development. The demands on the key technologies, the potentials and the deficits of ground-facility simulation, computational simulation and in-flight simulation basically were the same for the reference concept and the experimental vehicles. Of course they were very high for the reference concept, lower, but increasing for the three smaller experimental vehicles.

For any type of airbreathing hypersonic experimental flight vehicles basic demands and aspects—with all the issues discussed in the preceding chapters in the background—are:

- representative configuration and its movables, structural concept and material with realistic aerothermoelastic properties under relevant thermal loads in order to have the critical phenomena and issues present,
- a simple down-scaling of a target vehicle in general is not possible, because the thermal loads at for instance the vehicle nose and the leading edges may become untenable,
- representative propulsion system, including inlet and nozzle and also fuel-tank system,
- representative switch-over from turbo- to ram/scram propulsion and vice-versa,
- the flight vehicle must be tolerant to inaccurate predictions, it should, for instance, be possible to correct a wrongly estimated pitching moment by means of a movable ballast,
- the flight trajectory—guidance logic—of the experimental vehicle must be chosen such that the critical phenomena and issues are present to investigate,
- a highly accurate air-data system is mandatory, as well as a reference atmosphere model and a reference gravitation model,
- customized data measurement system for the flight attitude, the verification of system reliability, the acquisition of relevant aerothermodynamic coefficients and their derivatives, of propulsion, structural etc. data,
- on-board storage of all data and transmission to ground stations,
- crash recorder, if a catastrophic flight-vehicle loss. occurs.

7.5 New Ansatz: The Holistic Computational Approach

"The Power of Digital" was the headline of Aerospace America, the journal of the American Institute of Aeronautics and Astronautics, June 2022. Two topics were presented. The first one was certification by analysis [31], with the prospect of reducing the many expensive and 'harrowing' test flights needed for aircraft certification. The second one, [32], was asking for the increase of virtual testing.

The power of digital is a topic not only of the fields of aeronautics and astronautics. It applies to almost all technical, physical, chemical, medical, even public, and other branches, the potentials and options of artificial intelligence and related fields not to be forgotten. The power of digital computers still is rising, and the quantum computer is looming.

Since several decades Computational Fluid Dynamics (CFD) and Computational Structure Mechanics (CSM) more and more are becoming enabling tools in the design and development of aircraft for all speed regimes. Important is that they increasingly permit a holistic approach to the description of the aerodynamic and structural properties of the aircraft under development.

However, their application value partly is reduced due to so far not adequate model building: laminar-turbulent transition, turbulent separation, gas radiation regarding infrared signature modeling, and combustion in the case of CFD, and modeling of joints in the case of CSM. They have, despite these shortcomings, a very large potential and find wide application in the field of low as well as high speed aeronautics.

Our topic of course is airbreathing hypersonic flight vehicles, but first we have a look at the developments in the field of lower speed flight vehicles.

Regarding CFD the first systematic approaches to assess the performance and accuracy of codes of numerical fluid mechanics probably were the GAMM Workshops in the years 1977 to 1991 [9]. Well known and published are also the AIAA Workshops on Drag Prediction, High-Lift Prediction, Aeroelastic Prediction, Multifidelity Modeling in Support of Design and Uncertainty Quantification etc., beginning in the year 2001 with the first Drag Prediction Workshop.

A holistic approach to set up 'Computational Research and Acquisition Tools and Environments (CREATE)' of the US Department of Defense is the 'High Performance Computing Modernization Program (DOD HPCMP)'. The air-vehicle part of it is CREATE-AV. The effort aims at the improvement of flight vehicle acquisition programs in terms of time line, cost, and performance [33, 34]. Besides the simulation of the mission phases of fixed wing aircraft in the subsonic, the transonic and the supersonic flight regimes also hypersonic flight vehicle analysis is addressed.

From virtual flight testing toward simulation-based certification is a topic in the frame of the guiding concept "Virtual Product" of the German Aerospace Center DLR [35], see also [36]. The aim of the Virtual Product is to virtualize the design, the development and the manufacturing processes of aircraft all the way to major

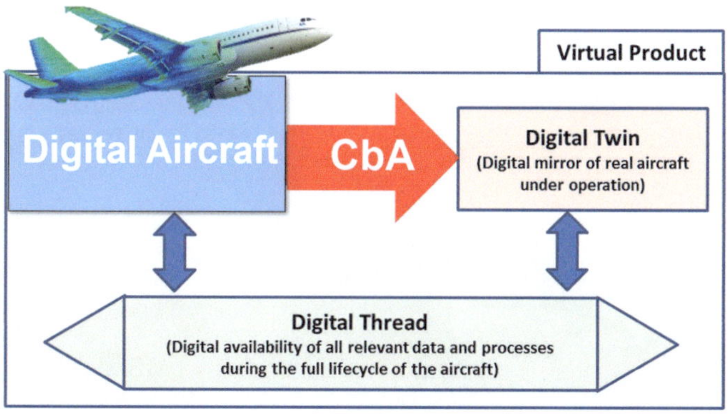

Fig. 7.4 Schematic of the DLR guiding concept Virtual Product with the certification by analysis (CbA) at the center [37]

parts of the certification process. Correspondingly an equivalent validation strategy is aimed for. High-fidelity computation tools for disciplinary and multidisciplinary simulations are the precondition.

The schematic of the simulation-based certification is shown in Fig. 7.4. The digital twin of the aircraft represents it with the relevant physical properties including the operational behavior.

The digital thread, on the other hand, includes all relevant data and processes through the life cycle, which define the aircraft as well as its digital twin. The overview of the building blocks of the digital simulation is given in Fig. 7.5.

The software architecture consists of three layers, Fig. 7.6: the control layer, the plug-in layer and the data layer.

The properties of the atmosphere, the influence of gusts and of wakes of other aircraft, the control-surface modeling, also in view of gust attenuation, inputs of the flight control system (FCS) etc. are modeled and merged into the computational presentation of the aircraft. The airframe is considered to be perfect elastic. The digital tools in terms of high-performance computation (HPC) are computational fluid dynamics (CFD), computational structure mechanics (CSM) and rigid-body dynamics (RBD). The flight-control system is denoted with FCS.

The control layer as Python script operates the simulation process. In the plug-in layer the relevant software components are assembled. They are the CFD code (TAU at that time, nowadays to be replaced by CODA), the CSM code, the RBD code, etc. The simulation of, for instance the encounter of gusts or wake vortices of other aircraft, may begin with a trimmed steady flight state.

A space-flight related simulation process is the *CONFFAS* simulator [38], which couples numerical methods of fluid mechanics, flight mechanics, structure mechanics etc. in the sense of the Virtual Product, Fig. 7.7.

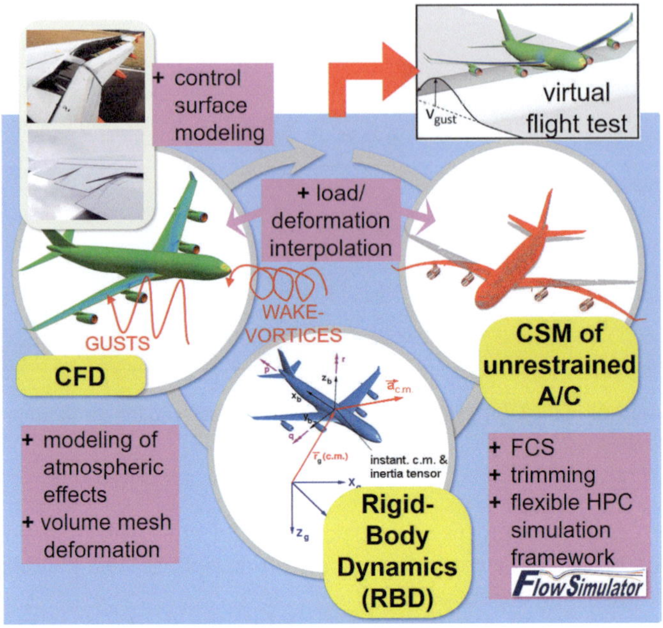

Fig. 7.5 Building blocks of the flight-testing simulation tool [37]

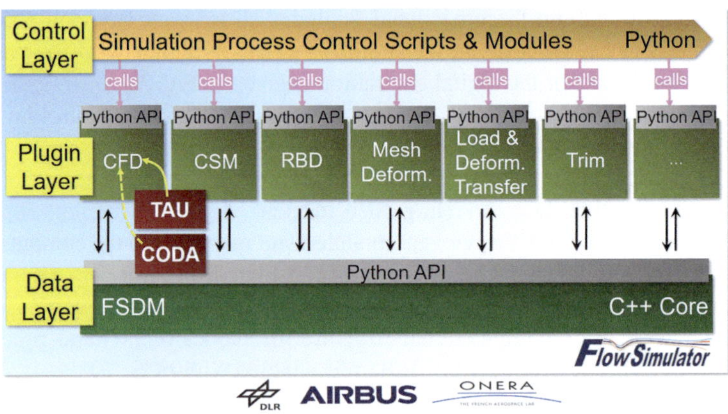

Fig. 7.6 The FlowSimulator suite [37]

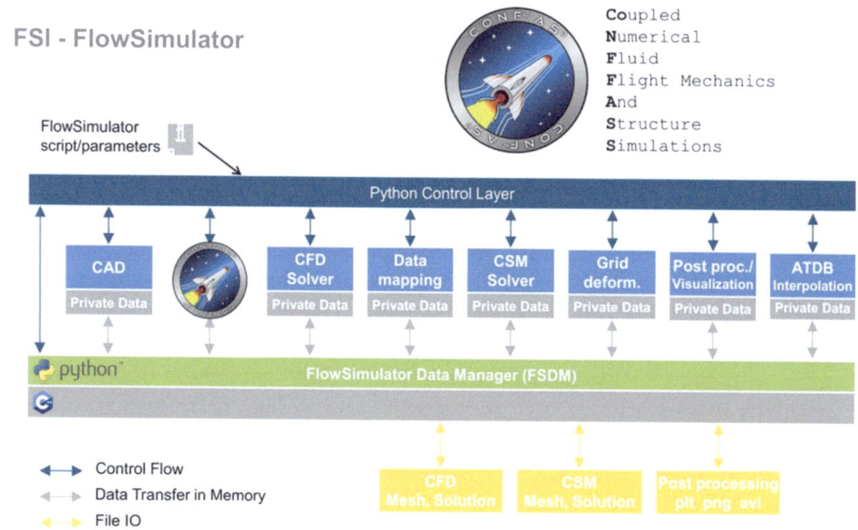

Fig. 7.7 The FlowSimulator suite *CONFFASS* [38]

The object of the simulation are flights of national and international experimental rocket systems. The flow solver (CFD solver) is the TAU solver with the relevant high-temperature real-gas model. The structural mechanics solver (CSM solver) includes thermal coupling, deformation under aerodynamic loads, both for steady and unsteady situations.

Arbitrary atmosphere models can be chosen, as well as arbitrary gravitation models. Local wind conditions can be incorporated. The flight control system is a complete system, but simpler versions are available. The airframe, mainly the fins of the rocket systems and the control surfaces, are modeled as perfect-elastic ones.

7.6 Sketch of the Original Concept of the Virtual Product

The concept of the Virtual Product (VP) evolved in the second half of the 1990s, and found its shape in, for instance, [11, 39]. It was stimulated by the simulation challenges identified for the lower stage of the TSTO-system SÄNGER, Chap. 8, and by design and development problems of other aircraft projects.

The Virtual Product (VP) is a high-fidelity mathematical/numerical representation of the physical properties and the functions of a product. The enabling elements are

- high-performance computation with local and distributed systems,
- information technologies, data-bank systems, visualization tools, etc.,
- high-performance communication,

- mathematical/numerical product models, flow-physics models, structure-physics models, etc.,
- disciplinary and multidisciplinary numerical simulation methods,
- disciplinary and multidisciplinary numerical optimization methods.

Depending on its application phase, in particular in the product definition phase, Fig. 7.2, high-fidelity computation is not always necessary.

- **Benefits to be expected of the VP**: – high-fidelity description and optimization of the functions and the physical properties of the aircraft, – holistic treatment of the aircraft in formalized computer-aided product-definition and development processes, – more effective pursuit of sensitivities, including growth factors etc., – early high fidelity product simulations, – early recognition of problems, – sensitivity-driven design tools (from simple tools to high-performance computing), – improved feasibility statements, – improved control of life-cycle costs, – improved product quality (performance guaranties, system efficiency, and operational risks), – reduction of development risks and need for repair solutions later in the development process.
- **Further effects**: – reduction of the number of design cycles, – improved definition of sub-systems, – more effective integration of suppliers, – reduction of design margins, – more reliable programme decision criteria, – effective demonstrations for the customer, – reduction of flight testing, – support of the certification process, – reduction of cost and time to market.

Behind this are the needs of the design and development of aircraft in general and hypersonic airbreathers in particular. Regarding the latter, the intended mission, the operational demands, the flight envelope and the API-type play a role as well as the **size of the to be developed airframe and that of the propulsion system**.

Necessary first of all is a serious appraisal of the capabilities of the mathematical approaches and models, as well as of the experimental ground-simulation facilities. If necessary, much research and development must be conducted in order to pave the way toward the goal. **The matter of real-elastic versus perfect-airframe plays a very significant role.**

- **Implementation and acceptance**: – Implementation and acceptance of the Virtual Product approach both in research and in industry need efforts as large as the development of the VP itself. – Industry and research today are still organized and work discipline-oriented, in accordance especially with the second aspect of Cayley's design paradigm, Sect. 2.2. – The often discussed and heralded Concurrent Engineering is indeed exactly what the Virtual Product demands. In general it faces, at least in industry, as experience in many fields shows, acceptance problems on several organization levels. – The Virtual Product entails also systems engineering and process changes—indeed, it aims for them—, and hence organization changes. – Finally, and this must not be forgotten, it will introduce risks, as any new approach does, and the cost for its implementation must be recovered in due time.

- **Who is to be addressed**: – to be addressed is the industry, which has to adopt fully the concept of the Virtual Product. – Further the research establishments and universities, – and finally also the certification agencies, which must accept and approve the new approach. – All organization levels, from the management level down to the staff level are concerned. Psychological factors must be regarded. Incentives must be given. – In industry they are cost and risk reduction in product design for management, and job security for the staff. – In research they are the big scientific challenges of the Virtual Product in engineering sciences, in applied mathematics and in computer science, which will lead to very interesting results, and also will have positive effects in funding acquisition.
- **Staff qualification**: – The Virtual Product as a mathematical/numerical representation of the physical properties and the functions of a product is a tool. – Tools are only as good as the people, who wield them. – Innovative product design first of all is a matter of imagination, knowledge and creativity of persons. However, it can be dramatically enhanced with superior tools. – Given the very high complexity and the high order of integration of the Virtual Product and the (new) processes and organizations, in which it is to be applied, it is one of the greatest challenges for industry to "build" staff with adequate qualifications. – Without that the Virtual Product will not unfold its potential, it even can become counterproductive. However, this holds for every process innovation, and therefore it is not a new problem. – On the one hand more persons of the staff will have to know more about the product and its sub-systems than now. – At the same time the use of high-fidelity multidisciplinary simulation and optimization methods demands very broad knowledge of phenomena, mathematical/numerical models, and the methods themselves.
- **Education**: – An extended education is necessary of industry staff, and at the universities on process technologies: – disciplinary and multidisciplinary model building, discrete numerical methods, information and communication technologies, product technologies, system technologies. – A problem will be with university education to fit in the new technology requirements, because the basic disciplines cannot be neglected. Hence it is the question how to shape university education without unnecessary and unwanted prolongation of it.
- **Conclusion and Outlook**: – The potential of the Virtual Product in aircraft design is large and needs to be exploited. – Regarding hypersonic airbreathers it possibly will be the only enabling approach. – The two driving factors are the overall business climate, and the weakening of Cayley's design paradigm. – The central element of the Virtual Product is high-fidelity multidisciplinary numerical simulation and optimization, which is becoming feasible because of the two enabling factors: the advances in high-performance computation and information technologies in general including artificial intelligence, and the advances in numerical simulation and optimization methods. – Substantial research and development work is necessary for its realization. – Implementation and acceptance issues in industry and research must not be underestimated. For industry the Virtual Product is one of the few approaches with a large potential to reduce design risks, and time and cost

to market. – In addition it is indispensable to overcome the design challenges of even more integrated future aircraft.

- **Last but not Least**: – The challenges to encounter the climate change, to arrive at circular economy, and in connection to the topic of this book, to reduce the ecological footprint of aircraft, to generally make air traffic climate-friendly certainly can be beneficially addressed with the Virtual Product approach, partly or fully.

7.7 Problems

Problem 7.1 (*Phases of Aircraft Development*) Describe the typical phases of aircraft product development from concept to production. What are the key milestones, and how do the knowledge of the product, freedom for changes, and cost of changes evolve throughout the development process?

Problem 7.2 (*Challenges of Developing Airbreathing Hypersonic Vehicles*) Identify and explain at least three specific challenges faced in the design and development of airbreathing hypersonic flight vehicles. How do these challenges differ from those in the development of subsonic and transonic aircraft?

Problem 7.3 (*Tool Limitations in Hypersonic Vehicle Design*) What are the limitations of computational methods and ground-facility testing when designing hypersonic airbreathing vehicles? Provide examples for both computational and experimental challenges.

Problem 7.4 (*Role of Virtual Product in Aircraft Design*) Explain the concept of the Virtual Product (VP) in aircraft design. What are the benefits and challenges of implementing this approach in the development of hypersonic airbreathers?

Problem 7.5 (*Role of Experimental Hypersonic Flight Vehicles*) Discuss the role of experimental hypersonic flight vehicles in the development of future airbreathing hypersonic vehicles. What are the essential characteristics that experimental vehicles should possess?

Problem 7.6 (*Re-Entry Vehicles and Hypersonic Airbreathers*) What are the basic differences between a re-entry vehicle and a hypersonic airbreather in terms of the aerodynamic properties and the thermal and the mechanical loads on the airframe?

References

1. Anderson Jr., J.D.: The Airplane-A History of its Technology. AIAA, Reston, VA (2002)
2. Hirschel, E.H., Prem, H., Madelung, G. (eds.): Aeronautical Research in Germany-From Lilienthal Until Today. Springer, Berlin, Heidelberg, New York (2004)
3. Bloor, D.: The Enigma of the Aerofoil-Rival Theories in Aerodynamics, 1909–1930. The University of Chicago Press, Chicago and London (2011)

4. Eckardt, D.: Jet Web-Connections in the Development History of Turbojet Engines 1920–1959. Springer Nature, Wiesbaden (2022)
5. Meier, H.U. (ed.): German Development of the Swept Wing-1935–1945. Library of Flight, AIAA, Reston, VA (2010)
6. Rizzi, A., Hirschel, E.H.: General developments of numerical fluid mechanics until the middle of the 20th century. In: Hirschel, E.H., Krause, E. (eds.) 100 Volumes of Notes on Numerical Fluid Mechanics—40 Years of Numerical Fluid Mechanics and Aerodynamics in Retrospect, pp. 61–76. Springer, Berlin Heidelberg (2009)
7. Weiland, C.: The Hierarchy of Fluid Dynamic Equations-Foundations of the Second Mathematization Wave of Fluid Dynamics. Springer Nature, Switzerland (2025)
8. Zuse, K.: https://de.wikipedia.org/wiki/Konrad-Zuse
9. Hirschel, E.H., Krause, E. (eds.): 100 Volumes of Notes on Numerical Fluid Mechanics—40 Years of Numerical Fluid Mechanics and Aerodynamics in Retrospect. Springer, Berlin, Heidelberg (2009)
10. Hirschel, E.H., Rizzi, A., Breitsamter, C., Staudacher, W.: Separated and Vortical Flow in Aircraft Wing Aerodynamics. Springer, Heidelberg (2021)
11. Hirschel, E.H.: Towards the virtual product in aircraft design? In: Periaux, J., Champion, M., Gagnepain, J.-J., Pironneau, O., Stoufflet, B., Thomas, P. (eds.) Fluid Dynamics and Aeronautics New Challenges. CIMNE Handbooks on Theory and Engineering Applications of Computational Methods, pp. 453–464. Barcelona, Spain (2003). www.ehhirschel.com (2024)
12. Hirschel, E.H., Weiland, C.: Design of hypersonic flight vehicles: some lessons from the past and future challenges. CEAS Space J. **1**, 3–22 (2011)
13. Barriety, B., Boin, J.-P., Chandre-Vila, O., Mauermann, T.: Fast fluid-structure computational method taking into account non-linear aerodynamics. In: IFASD 2019, Savannah, Georgia, USA (2019)
14. Chandre-Vila, O., Nivet, Y., Marquier, S., Morlier, J., Gourdain, N.: Study of fast aeroelastic solvers for gust load computation. In: IFASD 2022, Madrid, Spain (2022)
15. Hirschel, E.H., Grallert, H., Lafon, J., Rapuc, M.: Acquisition of an aerothermodynamic data base by means of a winged experimental vehicle. ZFW **16**(1), 15–27 (1992)
16. Simeonides, G.: Extrapolation-to-flight of aerodynamic heating measurements and determination of in-flight radiation-equilibrium surface temperature in hypersonic/high enthalpy flow conditions. Shock Waves **16**, 25–34 (2006)
17. Hirschel, E.H.: Basics of Aerothermodynamics, vol. 204, Springer, Berlin. Progress in Astronautics and Aeronautics, AIAA, Reston, VA., Second, revised. Springer, Heidelberg (2004, 2015)
18. Peebles, C.: Road to Mach 10-Lessons Learned from the X-43A Flight Research Program. Library of Flight, AIAA, Reston, VA (2008)
19. Mottershead, J.E., Link, M., Friswell, M.I.: The sensitivity method in finite element model updating: a tutorial. Mech. Syst. Signal Process. **25**, 2275–2296 (2011)
20. Böswald, M.: Personal communication (2023)
21. Worden, K., Tomlinson, G.: Nonlinearity in Structural Dynamics. CRC Press (2019)
22. Kerschen, G., Worden, K., Vakakisc, A., Golinval, J.-C.: Past, present and future of nonlinear system identification in structural dynamics. MSSP **20**(3), 505–592 (2006)
23. Popov, V.L.: Contact Mechanics and Friction, 2nd edn. Springer, Berlin, Heidelberg (2017)
24. Miller, J.: The X-Planes: X-1 to X-45, 3rd edn. Midland Publication Ltd (2001)
25. Gorn, M.H., De Chiara, G.: X-Planes from the X-1 to the X-60: An Illustrated History. Springer Praxis Books (2021)
26. Newkirk, D.: Soviet space planes. J. Spaceflight **32** (1990)
27. van den Abeelen, L.: Spaceplane HERMES—Europe's Dream of Independent Manned Spaceflight. Springer Praxis Books (2017)
28. Hirschel, E.H., Weiland, C.: Selected Aerothermodynamic Design Problems of Hypersonic Flight Vehicles. Progress in Astronautics and Aeronautics, AIAA, Reston, VA., vol. 229, Springer, Heidelberg (2009)

29. Kuczera, H., Sacher, P.W.: Reusable Space Transportation Systems. Springer, Berlin, Heidelberg (2009)
30. Hirschel, E.H.: The Technology development and verification concept of the German hypersonics technology programme. DASA-LME12-HYPAC-STY-0017-A (1995)
31. Button, K.: Flying Digitally. Aerospace America, AIAA, Reston, VA (2022)
32. Waas, A.: The case for more digitally testing. Aerospace America, AIAA, Reston, VA (2022)
33. McDaniel, D.R., Morton, S.A.: HPCMP CREATE-AV Krestel architecture, capabilities, and future directions. In: AIAA SciTech Forum, 8–12 Jan 2018, Kissimmee, Florida (2018)
34. McDaniel, D.R., Morton, S.A.: HPCMP CREATE-AV Krestel new and emerging capabilities. In: AIAA SciTech Forum, 6–10 Jan 2020, Orlando, Florida (2020)
35. Ritter, M., Handojo, V., Krüger, W.R., Roeser, M.S., Mönnich, W., Kirmse, T., Reimer, L., Heinrich, R., Geisbauer, S., Görtz, S.: Virtual aircraft technology integration platform: from virtual flight testing towards simulation-based certification. In: AIAA-Paper 2021–1203 (2021)
36. Heinrich, R. (ed.): Advanced Aircraft Understanding via the Virtual Aircraft Model. In: Notes on Numerical Fluid Mechanics and Multidiscipinary Design (NNFM), vol. 155. Springer, Cham (2025)
37. Reimer, L., Heinrich, R., Ritter, M., Krumbein, A., Geisbauer, S., Görtz, S., Leicht, T.: Virtual aircraft technology integration platform: ingredients for multidisciplinary simulation and virtual flight testing. In: AIAA-Paper 2021–1202 (2021)
38. Eggers, T., Franze, M.: Shefex I + II post-flight analysis: starting point for a virtual flight test environment. In: 68th International Astronautical Congress. Adelaide, Australia (2017)
39. Hirschel, E.H.: On the way to the virtual product. In: Hirschel, E.H., Prem, H., Madelung, G. (eds.) Aeronautical Research in Germany-From Lilienthal Until Today, pp. 643–648. Springer, Berlin, Heidelberg, New York (2004)

Chapter 8
The SÄNGER Hypersonics Technology Development and Verification Concept

This chapter gives an outline of the SÄNGER Hypersonics Technology Development and Verification (HYTEDAV) Concept, which was devised toward the end of the German Hypersonics Technology Programme. The reader may ask why look at the results of that study, which was performed nearly 30 years ago. The answer is, that the findings of the work in the four main areas of the technology programme are still of importance today. That also holds for the work in other large programs of that time, for instance the NASP program, the HOTOL program and others. And, of course, it also holds for the more recent studies and projects, which were mentioned in Sect. 1.1. Much to be found is in the lessons learned from the Space Shuttle Orbiter [1, 2] and in the lessons learned from the X-43A program, [3] with the many attached technical reports.

Important is that only in large and profound programs with long and concentrated development work definite statements about the feasibility of hypersonic flight-vehicle systems can be made. The reasons are the heavy demands put on such systems (reliable, affordable, little ground infrastructure, small environmental impact etc.), and the very high technology hurdles. Military systems of course partly have other profound demands.

Another point is, that the technology basis, which is acquired, or is being concretized during the work, is influencing the flight-vehicle system studies themselves.

The work performed so far in technology programs on airbreathing hypersonic flight systems yielded many important results. It also did show that very critical technology items exist, which need long and concentrated development effort, before final statements about the feasibility of the system under consideration can be made.

Critical items usually are, at least partially, well perceived when setting up a technology program. However, two observations are important:

- The extent of technology problems partly emerges only after deep and prolonged work. In the German Hypersonics Technology Programme this gradually led to the insight that a technology development and verification concept is necessary, which

E. H. Hirschel et al., *Elements of Hypersonic Airbreather Design and Development*,
https://doi.org/10.1007/978-3-031-94219-8_8

charts the long-term overall strategy. This became evident after the need emerged for new or improved experimental ground-simulation facilities, the role and needs of numerical methods for aerothermodynamics, propulsion, structural design etc., and finally for flight testing with experimental vehicles. Such a concept must be flexible in order to react to new results coming from accompanying reference-concept system studies, and at the same time must ensure sufficient continuity in the technology development.
- The technology base, which is gained, or is becoming more concrete during the work in the technology programme, influences the reference-concept system studies. Initial assumptions may prove to be too optimistic, and systems sensitivities emerge, which have strong repercussions on the importance, contents and direction of the technology development efforts.

8.1 The Background

The German Hypersonics Technology Programme lasted from 1987 to 1995, see, e.g., [4]. It basically was intended to provide technology insight in view of a future TSTO space transportation system as a possible successor of the European space plane HERMES. For the latter see, e.g., [5]. The lower stage of the API-type 5 TSTO-system SÄNGER, Sect. 2.8, was the reference concept of the German program.

8.1.1 The Technological Challenges of the Reference Concept

The lower stage of SÄNGER of about 80 m length and about 300 tons mass—fuel included—was to launch the upper stage with a length of about 30 m and a mass of 115 tons into orbit. The payload of the upper stage was about 8 tons. A mission would start in Europe with a hypersonic flight with $M_\infty = 4$ to 5 in southern direction for the insertion of the upper stage into orbit, see, e.g., Fig. 2.18.

The design problems of the lower stage are related to the flow path from tip to tail. A very strong coupling of propulsion, aerothermodynamics, materials and structures and flight dynamics exists. It governs the design of the vehicle and its components.

In the following the flow path from tip to tail of the lower stage is sketched in order to point out design problems and the resulting simulation problems in the design process. Many details were presented already in the preceding chapters.

Airbreathing hypersonic vehicles of API-type 5 are characterized by a highly coupled lift and propulsion system. That leads to the difficult problem of airframe-propulsion integration. The aerodynamic lift with rising Mach number is increasingly created at the lower (pressure) side of the vehicle, which houses the propulsion system, too.

A long and slender forebody is necessary in order to keep the propulsion unit well within the bow-shock surface. This ensures a minimum of interactions that would increase wave drag. The lower side of the forebody must be flat transversally, in order to achieve the necessary pre-compression for the inlet onset flow. The also longitudinally-flat lower surface of the SÄNGER lower stage would be flown at an effective angle of attack of around ten degrees. Together with the blunt nose this results in a pre-compression, which reduces the inlet-capture area by a factor of about two, which again reduces the cross-section of the inlet package and thus the length of the forebody.

The long forebody gives rise to a thick turbulent boundary layer at the inlet location (about 50–60 m downstream of the nose tip in the case of the SÄNGER lower stage). The turbo propulsion component for the Mach numbers up to 3–4 would demand a forebody boundary-layer diverter.

With the present-day knowledge the ram or a scram propulsion component would not demand a forebody boundary-layer diverter. Whether that is true in general, has not yet been established. It might be necessary for flow-stability reasons, even if the X-43A experience—API-type 4 flight vehicle with short forebody—does not show that.

Boundary-layer bleed at locations of strong shock/boundary-layer interaction (inlet ramps) appears to be necessary. Boundary-layer diverter flow and/or bleed would demand a duct above the inlet/engine package, see Fig. 5.7. It adds the diverted air mass with its momentum to the propulsion stream, which expands at the asymmetric external nozzle ramp at the tail of the vehicle. Special problems arise in the transonic Mach number range, because of large wave drag and base drag increments: acceleration pinch point 1, Fig. 5.36.

The inlet must function over a large Mach number span. At hypersonic Mach numbers it must have a large amount of internal compression. A high total-pressure recovery and a reliable avoidance of inlet unstart is necessary. A low flow distortion is mandatory at least in the turbo mode. Efficient and stable combustion is required in the ram mode. The nozzle and the expansion ramp/base (SERN) must produce optimum thrust in the whole Mach number range, however considerable thrust-angle variations must be taken into account, Sect. 5.7.

Structural mass must be as low as possible as for any flight vehicle. The aerodynamic loads at every flight state hence must be known to sufficient accuracy in order to size the structure and its elements. Aeroelastic effects, both steady and unsteady, pose extra problems, also because aerothermoelasticity comes into play at the high flight Mach numbers.

The cardinal problem of hypersonic flight lies in the thermal loads, which increase with increasing Mach number. Airbreathing hypersonic vehicles up to $M_\infty \approx 7$ can get along with passive cooling, i.e. by surface radiation. Because all viscosity effects (viscous drag at $M_\infty \approx 7$ accounts for approximately half of the total drag) would be amplified by surface roughness, such vehicles must have subcritical surface roughness.

A classical thermal protection system, as for instance the Space Shuttle Orbiter had, is not advisable, especially at the lower side of the long forebody. A possible material and structure concept of the SÄNGER lower stage was a hot primary (titanium) structure for about 90% of the fuselage, Sect. 6.3. The rest (nose cap, leading edges) could be made from ceramics.

Heat loaded structures like the inner part of the inlet, the diverter duct, the engine, the nozzle, which cannot be radiation cooled for geometric reasons, must be cooled actively, if the critical thermal loads of the respective component material are exceeded.

In all cases trade-offs are necessary to determine optimum structure, materials and cooling concepts. The thermal loads in terms of the radiation-adiabatic wall temperatures, recovery temperatures, and heat fluxes, must be determined to the necessary degree of accuracy. The problem is the strong coupling to the vehicle surface properties, Sect. 3.2.3, which is roughness/waviness, radiation emissivity, and surface catalycity, if the airstream is dissociated.

The coupling of aerothermodynamics and flight dynamics in general is not very strong. The fact, however, that the neutral point of the configuration shifts by about ten per cent of the airframe length over the Mach-number span of a $M_\infty = 7$ vehicle, poses a very large problem in vehicle design. Large trim and control surface efficiency is needed, especially if the vehicle is flown at hypersonic speed neutrally stable or slightly unstable for trim drag reduction purposes.

For airbreathing hypersonic flight vehicles of API-type 4 and 5 the change of the thrust-vector angle of the asymmetric outer nozzle (SERN) with the flight Mach number is another major problem. Two principally different approaches are possible with regard to vehicle trim, either a completely autonomous airframe trim, or an integrated airframe-propulsion trim, both with advantages and disadvantages. In the integrated case it must be assured that neither inlet unstart nor engine flame-out occurs, because this would lead to an uncontrollable situation for the flight vehicle.

A way out of the thrust-vector angle change problem would be symmetric nozzles, like for rocket propulsion. However, with this technique the nozzle-geometry changes for large Mach-number spans cannot be achieved and radiation cooling is not possible for the external part. The basic problem, however, is the need to avoid tail striking in ground proximity, Sect. 1.5. With symmetric nozzles very high landing-gear arrangements would be needed with the associated mass penalty.

The capability of the designer to handle all these design problems is to be supported by appropriate design tools, and especially simulation means. Here lies the big problem for the shape definition of the flight vehicle and its components, the propulsion system, thermal and mechanical loads, etc. That was the situation at the time of the SÄNGER studies, which today basically remains to be the same.

The two major classes of simulation means in aerothermodynamics, propulsion integration and propulsion etc. are **computational simulation** methods, which encompass scaling methods, approximate and discrete numerical methods, and **experimental ground-facilities simulation**, which encompass wind tunnels, shock tubes, etc. with appropriate sub-scales of configurations and configuration components (inlet, nozzle/base, etc.), as well as test rigs, direct-connect and free-jet facilities, and so on.

The main problem in aircraft ground facility simulation is that sub-scale models are to be employed, which also holds for the ram and scram engines of hypersonic airbreathers. That is associated with large simulation-parameter deficiencies.

Another grave problem with ground-facility simulation is that it creates an environment which is not the same as that of flight: non-representative disturbance environment regarding laminar-turbulent transition (receptivity problem) and freezing of chemical composition and internal energy states (e.g. vibration energy) in high-enthalpy wind tunnels.

In computational simulation especially flow-physics (laminar-turbulent transition, turbulence), and thermodynamics (finite rate effects, transport properties, surface catalycity) pose large modeling challenges.

8.1.2 The Hypersonics Technology Development and Verification (HYTEDAV) Concept

Toward the end of the technology program, a Hypersonics Technology Development and Verification (HYTEDAV) concept was devised as blueprint for the possible European successor project [6, 7]. The HYTEDAV concept work was integrated in the systems and technology efforts, Fig. 8.1, hence into the results and the insights of the partners involved in the SÄNGER technology program.

The HYTEDAV concept work had the following objectives with regard to the technology needs:

- structuring of the technology development and verification process,
- identification of interdependencies between the technology fields,
- identification of necessary Transfer Models (Sect. 8.3),
- identification of simulation needs (ground-facility simulation, computational simulation, in-flight simulation),
- structuring of long-term basic research and development work,
- development of a simulation-hardware master plan (ground simulation facilities, super computers, experimental vehicles), with time schedules and cost estimations.

The system concepts were the lower stage of the TSTO-system SÄNGER as reference concept and also up to three hypersonic experimental flight vehicles (EV1, EV2, EV3), which emerged as determining part of the HYTEDAV concept. The system requirements, the analysis tools and so on were common requirements, left box in Fig. 8.1.

The systems work was determining the requirements and demands on the enabling technologies, red lower box, which in return were determining the work on the system concepts, i.e., the reference concept and the three experimental vehicles. Both the reference concept—the SÄNGER lower stage—and the experimental vehicles needed technology development.

The main technology challenges in the four primary technology fields were seen to be the following:

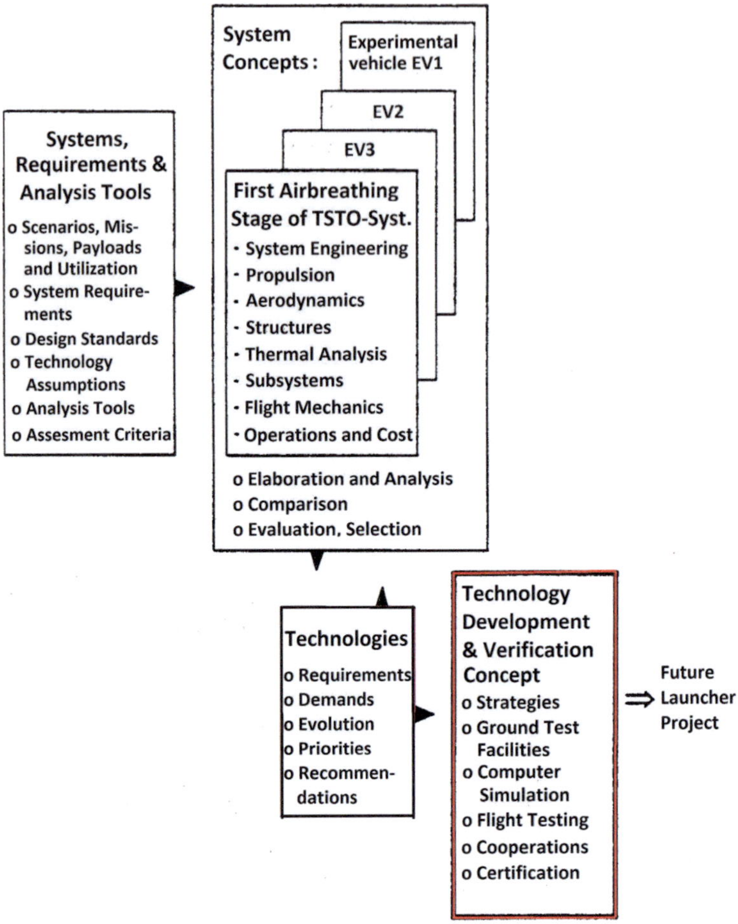

Fig. 8.1 The HYTEDAV concept and its embedding in the overall systems and technology efforts

1. Propulsion:
   - sufficient thrust for the whole flight Mach-number regime,
   - low structural mass,
   - low fuel consumption,
   - optimum integration of the whole propulsion system and the airframe with regard to total mass and performance.
2. Aerothermodynamics:
   - shape definition of the lower stage and its components (airframe, control surfaces, inlet etc.) including the airframe-propulsion integration with regard to performance, flyability and controllability,

- determination of aerodynamic and thermal loads,
- feasibility of upper-stage integration and in particular stage separation at $M_\infty \approx 6.8$,
- definition of demands on other technology areas like surface properties, surface-radiation emissivity (radiation cooling), aerothermoelasticity etc.

3. Materials and structures:

   - extremely light-weight structures with the realization of smooth and exact surface contours,
   - application of specifically light-weight materials for high-temperatures as well as for extremely low-temperatures (cryogenic fuel),
   - heat protection of the structure by radiation cooling, anti-oxidation coating with high radiation emissivity, special heat-protection systems, active cooling systems for internal flow paths.

4. Guidance and control, subsystems:

   - guidance, navigation and control,
   - air-data system for high Mach numbers,
   - notch filter layout,
   - thermal management system.

The requirements on the key technologies, the potentials and the deficits of ground-facility simulation, computational simulation and in-flight simulation basically were the same for the reference concept and the experimental vehicles. Of course they were very high for the reference concept, but lower, although increasing for the three experimental vehicles. That all led to a doubly recursive structure of the HYTEDAV concept, Fig. 8.2.

8.2 The HYTEDAV Structure

As shown with Fig. 8.2, due to the involvement of experimental flight vehicles a doubly recursive structure of the technology development and verification concept emerged. With the results and knowledge obtained in the technology program, including the reference concept studies, the requirements of the reference concept on the key technologies are identified (box "Key Technologies" at the upper left).

From these requirements **the demands** on the simulation means are derived. These technology demands on the simulation means are understood as the data, which the technology developer—and later the vehicle designer—needs to know (e.g. lift, drag, thrust, loads, strength, deformation, etc.), in order to develop the technologies, which will ensure, that the considered space transportation system meets its design and mission goals.

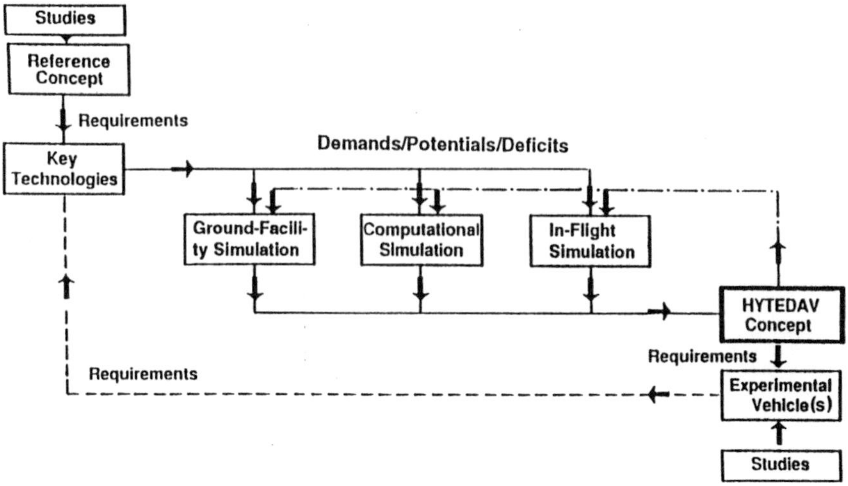

Fig. 8.2 The doubly recursive structure of the German hypersonics technology development and verification concept [6]

These data ("need to know") are to be obtained from the simulation means, i.e. ground-facility simulation, computational simulation, in-flight simulation. Hence the **potential** of these simulation means is assessed in order to clarify, whether they can meet the demands. The **deficits** of each simulation means are identified.

That results in the Hypersonics Technology Development and Verification (HYTEDAV) concept. This concept in turn leads to requirements on the simulation means—in Fig. 8.2 line above with alternating dots and dashes—, for instance new simulation facilities for basic research and development work. It also may have demands on an experimental flight vehicle—or several of such vehicles—which must allow to obtain certain data (for instance a flow-physics database from flight experiments). The experimental vehicle itself must be designed and may have its own requirements on key technologies—dashed line below—, hence the doubly recursive structure in Fig. 8.2.

The final Hypersonics Technology Development and Verification Concept puts the triad ground-facility simulation, computational simulation, and in-flight simulation into place with the goal to mature to readiness the key technologies and to develop and verify the needed Transfer Models. It identifies, justifies and directs the long-term basic research and development work.

That often is neglected by industry and sponsoring agencies, because of prevailing short-term project and funding rationales. If it has been shown, that a certain research and development topic is of decisive importance, then it must be pursued accordingly, if necessary even over long periods of time.

8.3 Sketch of the Transfer-Model Concept

At the time of the SÄNGER studies—and still today—the very large design (technical, financial) and operational (performance, personnel) risks made it necessary to find alternatives to the classical design process in the relevant technology fields. The many serious simulation problems and shortcomings encountered in hypersonic vehicle design regard the overall vehicle shape definition, the propulsion system, airframe-propulsion integration, and, if applicable, the upper-stage separation of TSTO space transportation systems.

While the objectives of the classical design process remain valid, type and role of the design tools had to be and must be redefined. In the frame of the German Hypersonics Technology Programme the Transfer-Model philosophy, [6, 7], was developed, which is valid for all involved technology areas. The main reason for that were the shortcomings of ground simulation in the verification step to cope with the parameter demands, and partly with the sheer size of the vehicles and their components.

Transfer Models at that time were numerical simulation methods, and scaling methods of such capabilities and accuracy that the demands of all design steps including the verification step are covered. They must allow to transfer knowledge and data found with ground simulation (sub-scale tests) and computational simulation to the full-scale design problem without the classical ground-facility verification step.

Transfer Models in particular were methods, which describe real life processes, which in the present context are characterized by severe flow/propulsion/structure interactions, and by dynamic phenomena due to these interactions, but also due to flow separation, acoustic phenomena, and combustion instabilities. Today they are called Virtual Product, CREATE, etc., see Sect. 7.5.

The development of such models solely from first principles was and still is utopian. Instead a pragmatic approach with much engineering judgement and a systematic use of all simulation means including experimental vehicles is necessary. Several basic requirements were seen and still exist for the successful development of Transfer Models:

- computer architectures with true teraflops and even higher performance at very low cost levels,
- flow-physics and thermodynamics models of sufficient accuracy,
- dynamic structural models with sufficient accuracy of material properties and the modeling of joints (structure-physics models),
- novel special ground-simulation facilities and test techniques for the creation of appropriate experimental databases.

Appropriate hypersonic experimental vehicles were seen necessary for the gathering of experimental data, which cannot be obtained in ground-simulation facilities, for the validation of flow-physics, thermodynamic and structure-physics models and finally for the validation of Transfer Models themselves, other design tools and the overall design strategy.

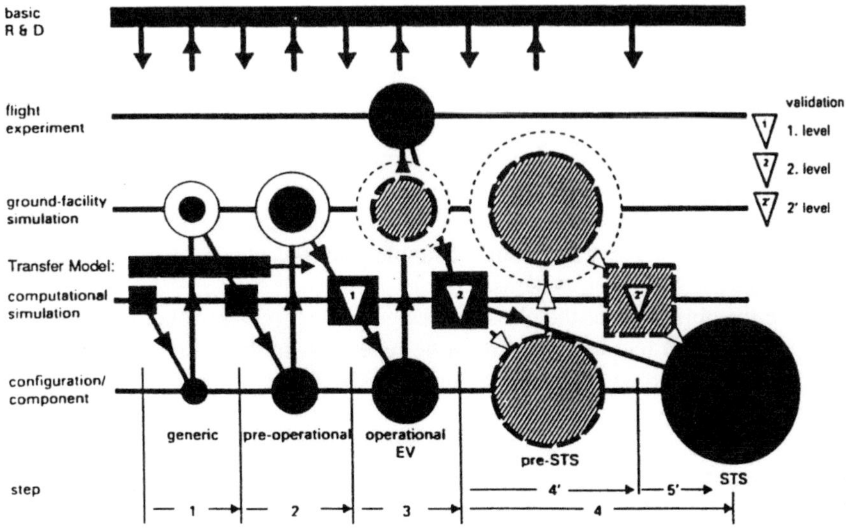

Fig. 8.3 Schematic of the development and application process of a transfer model [6]. EV: experimental vehicle, STS: space transportation system

The development and the application process of a Transfer Model in a schematic way is shown in Fig. 8.3.

The classical methodology and simulation means are employed. In Step 1 a generic configuration or component is designed. A model is tested in a suitable ground facility. With the knowledge and data gained the Transfer Model, which is a multidisciplinary computation method, or a combination of disciplinary computation methods, is improved.

With the improved Transfer Model a pre-operational design is made in Step 2, and a model is tested in a preferably larger ground facility. At the end of Step 2 the methodology, simulation tools and the Transfer Model have reached the first validation level. Important is that not only the computational tools and the Transfer Model must be validated, but also the design methodology and the ground-facility simulation.

The first decisive test happens with Step 3. An operational experimental vehicle is designed. Models of the configuration and its major components are tested, if appropriate ground facilities are available regarding size and parameter range.

The flight tests and experiments then show whether design methodology and design tools have reached a sufficient validation level (level 2). In addition the information, which cannot be acquired in experimental ground facilities (flight data as a final input into flow-physics and thermodynamics models, notably laminar-turbulent transition and turbulence data, as well as that of the aeroelastic/aerothermoelastic behavior), is obtained. In the HYTEDAV concept up to three experimental vehicles were foreseen.

Level 2-validation by definition means that the enabling technology is available, and in particular the Transfer Model, to design a space-transportation system of the reference-concept type. However, the flight tests may reveal, that level 2-validation of a particular Transfer Model cannot be reached. Then a pre-STS configuration or component must be designed, which is to be tested in a very large and costly, newly to be created, dedicated ground facility (Step $4'$).

The data and knowledge then gained should lead to the necessary level $2'$-validation of the respective Transfer Model. In the worst case the result of Step $4'$ could be, that the necessary enabling technology cannot be provided. History has seen several such situations. If all would be right, step 4 would lead to the design of the aspired space transportation system.

In any of the cases, basic research in flow-physics and thermodynamics modeling, as well as structure-physics modeling, play a major role. All research effort must be combined in continuous long-term unified approaches. The research will give input into the Transfer Models. On the other hand, it will get data from ground-facility simulation, and very important, from the flight experiments. Dedicated ground facilities, computational tools etc. are developed and employed.

All this implies, as noted above, long-term basic research and development work. A firm dedication of all involved partners, in particular also political and financial support, is the precondition.

8.4 Final Remarks

This chapter gave a look at the Hypersonics Development and Verification Concept of the former German Hypersonics Technology Programme. Although that is about three decades old, the insights gained are still valid.

The technological challenges to develop airbreathing hypersonic flight vehicles invariably are extraordinarily large. Major reasons are the very large technological steps necessary, especially in propulsion, aerothermodynamics and propulsion integration, and the significant demands on light-weight heat-resistent structures. It does not appear that in the meantime the decisive progress has been made. Information about military applications is not openly accessible.

Since ground-facility simulation is restricted in all fields, the classical approach in the aerospace field, to verify a design (vehicle and components) in ground facilities, is not possible to the needed extent. Of course, the API-type of the flight vehicle under consideration and its overall dimensions play the deciding role.

Computational simulation, in both the design and the verification step, has made large advances in the meantime. As its essence were seen Transfer Models, which are based on the triad experimental ground-facility simulation, computational simulation and flight testing. That still holds today.

8.5 Problems

Problem 8.1 (*Understanding HYTEDAV*) Explain the purpose of the HYTEDAV (Hypersonics Technology Development and Verification) concept as discussed in the chapter. What are its key objectives, and why was it developed toward the end of the German Hypersonics Technology Programme?

Problem 8.2 (*Challenges in the SÄNGER Lower Stage Development*) Identify and discuss three critical technological challenges related to the SÄNGER lower stage as outlined in the chapter. How do these challenges influence the design of the vehicle?

Problem 8.3 (*Role of Transfer Models*) Define a Transfer Model as described in the chapter. What are the key requirements for developing a successful Transfer Model, and how does it support hypersonic vehicle design?

Problem 8.4 (*Hypersonic Experimental Vehicles*) What are the primary purposes of experimental hypersonic vehicles within the HYTEDAV concept? Identify at least two specific types of data that these vehicles help gather, which cannot be obtained through ground simulation.

Problem 8.5 (*Simulation Triad in Hypersonic Vehicle Design*) Explain the triad of ground-facility simulation, computational simulation, and in-flight simulation as outlined in the HYTEDAV concept. Why is each component essential, and how do they complement each other?

References

1. Arrington, J.P., Jones, J.J. (eds.): Shuttle Performance: Lessons Learned. NASA CP-2283 (1983)
2. Throckmorton, D.A. (ed.): Orbiter Experiments (OEX) Aerothermodynamics Symposium. NASA CP-3248 (1995)
3. Peebles, C.: Road to Mach 10–Lessons Learned from the X-43A Flight Research Program. Library of Flight, AIAA, Reston, Va (2008)
4. Kuczera, H., Sacher, P.W.: Reusable Space Transportation Systems. Springer, Berlin Heidelberg (2009)
5. Abeelen, L. van den: Spaceplane HERMES - Europe's Dream of Independent Manned Spaceflight. Springer (2017)
6. Hirschel, E.H.: The Hypersonics Technology Development and Verification Strategy of the German Hypersonics Technology Programme. AIAA-Paper 1993-5072 (1993)
7. Hirschel, E.H.: The Technology Development and Verification Concept of the German Hypersonics Technology Programme. DASA-LME12-HYPAC-STY-0017-A (1995)

Chapter 9
Closing Notes

Access to space with reusable winged flight vehicles started with the Space Shuttle Orbiter in 1981. The orbiter was launched like a rocket and returned to Earth as a glider. Soon after worldwide studies were beginning which aimed at airbreathing access to space. Pure airbreathing single-stage-to-orbit (SSTO) access proved to be impossible, two-stage-to-orbit (TSTO) with an airbreathing lower stage and a rocket-propelled upper stage seemed to be the solution.

All that finally was given up, instead a focus mainly on the propulsion technology, the ramjet and in particular the scramjet, remained. The topics aerothermodynamics, airframe-propulsion integration, structure and materials more or less moved into the background. These, however, also pose problems and challenges comparable to those of the propulsion topic, needing concentrated and long-term research and development efforts. But they did not have the appeal, which the propulsion topic had and still has.

How is the situation now and what can be considered as constraints? It appears that presently airbreathing hypersonic flight everywhere is a military topic. Space transportation is a matter of rockets, reusability, however, is advancing. The matter of environmental protection, on the other hand, is strongly growing. Lost is the memory of the background of the German Hypersonics Technology Programme to have an access to space from European territory.

Civil airbreathing hypersonic flight nevertheless should be a topic. With TSTO-systems using LH2 as propellant most probably the environmental impact of launch to orbit can be reduced in comparison to pure rocket launch. The access to space from places far in the north, which was the background of the German programme, also is a viable reason to pursue airbreathing hypersonic flight. Concepts of long-haul hypersonic passenger flight, however, appear to be questionable.

The aim of this book is to guide attention to the technology topics mentioned above, however with only little regard to propulsion-specific aspects. When looking at the various concept studies of hypersonic flight vehicles, it soon became clear

E. H. Hirschel et al., *Elements of Hypersonic Airbreather Design and Development*,
https://doi.org/10.1007/978-3-031-94219-8_9

that a distinction in terms of their airframe-propulsion integration (API-) type was desirable. The six resulting API-types of hypersonic airbreathers have quite distinct design challenges and first of all several different research demands. The question of civil or military applications did not appear to be of major significance.

In view of the actual development of the different API-type hypersonic airbreathers a few very critical items are present: laminar-turbulent transition, the structural joint problem with the topic of aerothermoelasticity in the background, and the ground-facility testing of large hypersonic flight vehicles and their propulsion systems. The so called digital revolution, which happens since some decades, certainly holds large promise. But also it depends on progress in the mentioned critical items.

In general it can be observed that system studies of hypersonic airbreathing flight systems have large appeal, but the topic of actual flight-vehicle development and the needed long-term basic research are somewhat in the background. Even if for instance laminar-turbulent transition lately has found much interest in basic research, its actual significance in the vehicle design process still appears not to receive due attention considering its importance regarding concept feasibility of some of the flight vehicle types addressed in this book. And that holds for the research community, the industry, and in particular also for the national and international funding agencies. Long-term thinking is needed, but usually short-term topics draw more attention. And it is always the question of the staying power of government projects, as was noted by Jon Kelvey.

The book makes use of experience gained in the German Hypersonics Technology Programme, which appears to be rather far away from today's topics and developments. Nevertheless, the lessons learned at that time still have high significance also for today's and future airbreathing hypersonic flight vehicle development. Much depends on the API-type of a given concept, and some topics hence may not have much importance for that.

Hypersonic airbreathing flight is a fascinating topic, poses very large challenges and will remain to be a great field of research and development for all involved disciplines and their integration into multidisciplinary design and optimization.

Chapter 10
Solution Guide and Solutions of the Problems

Problems of Chapter 2

Problem 2.1

1. Speed of Sound a_∞:

$$a_\infty = \sqrt{\gamma R T_\infty} = \sqrt{1.4 \times 287.06 \times 226.5} = 301.7\,\text{m/s}.$$

2. Mach Number M_∞:

$$M_\infty = \frac{v_\infty}{a_\infty} = \frac{2500}{303.34} \approx 8.29.$$

3. Dynamic Pressure q_∞:

$$q_\infty = \frac{1}{2}\rho_\infty v_\infty^2.$$

Assuming atmospheric density at 30 km altitude $\rho_\infty \approx 0.0184\,\text{kg/m}^3$:

$$q_\infty = \frac{1}{2} \times 0.0184 \times (2500)^2 = 57.5\,\text{kPa}.$$

Problem 2.2

Scramjet Mach Range: Scramjet propulsion is typically effective in the Mach number range of $M_\infty \approx 6$ to $M_\infty \approx 24$. If one takes the thrust to weight ratio of the propulsion system into account, the upper limit for hydrogen propelled scramjets is $M_\infty \approx 12$. Above that Mach number, rocket propulsion is superior.

E. H. Hirschel et al., *Elements of Hypersonic Airbreather Design and Development*,
https://doi.org/10.1007/978-3-031-94219-8_10

Advantages of Hydrogen Fuel:

1. High Specific Impulse (Isp): Hydrogen has a higher specific impulse compared to hydrocarbon fuels due to its low molecular weight and high heat of combustion.

2. Cooling Capacity: Hydrogen provides excellent cooling properties, essential for managing thermal loads in scramjet engines at hypersonic speeds.

Problem 2.3

Cayley's Paradigm Weakening: In traditional subsonic and supersonic aircraft design, systems like wings (for lift) and propulsion (for thrust) are loosely coupled and can be optimized separately. However, in hypersonic airbreathing vehicles, such as API-type 4 and 5, the propulsion system is closely integrated with the airframe. For example, air inlet and engine are embedded in the fuselage, making it impossible to optimize one without impacting the other.

Major Challenges:

1. Thermal Integration: Managing extreme thermal loads from both aerodynamic heating and engine operation.

2. Aerodynamic-Propulsion Coupling: The shape of the airframe directly influences the performance of the propulsion system, particularly the inlet compression and exhaust flow.

3. Structural Challenges: The airframe must be lightweight but also capable of withstanding high thermal and aerodynamic loads.

Problem 2.4

Impact of Non-Propulsive LH2:

– The increased volume required for storing non-propulsive LH2 leads to a larger airframe, increased surface area, and subsequently higher drag.

– To overcome the additional drag, more thrust is required, leading to the need for a larger propulsion system and more fuel.

– Based on the growth factor analysis, the total mass would increase by approximately 2.75 times the added non-propulsive LH2 mass. Therefore, for a 5% increase in non-propulsive LH2, the overall take-off mass would increase by about 13.75% of the given non-propulsive LH2 mass.

Problem 2.5

Mach-Number Independence Principle: This principle states that for sufficiently high Mach numbers (depending on the body shape typically $M_\infty \approx 6$ to 10), aerodynamic coefficients (such as of lift and drag), the shock-wave structure and other flow properties become independent of further increases in Mach number.

Problem 2.6

For a wing with large aspect ratio the induced drag depends inversely on the aspect ratio $\Lambda = b^2/A$, with b being the wing's span width and A its surface. Hence the induced drag becomes smaller with reduced wing area at constant span.

With the shapes of the hypersonic airbreathers with the API-types 2 to 5 the classical concept of the induced drag does not hold. For the example of the API-type 1 it may hold to a degree.

Problem 2.7

The static temperatures of the atmosphere are 216.65 K at $H = 20$ km, 250.35 K at $H = 40$ km, and 247.02 K at $H = 60$ km. The speeds of sound are 295.07 m/s, 317.19 m/s and 315.08 m/s. The flight Mach numbers are accordingly $M_\infty = 6.63, 6.31, 6.35$. The changes hence are due to the temperature distribution in the atmosphere.

The two major effects are the changes of the wave drag with flight Mach number and the change of the locations of the bow-shock/wing leading-edge intersection with flight Mach number. The effects are rather small because in this case the variation of the flight Mach number is small.

Problem 2.8

The dynamic pressure $q_\infty = 0.5 \rho_\infty v_\infty^2$ governs the aerodynamic loads on the airframe and the performance of the propulsion system. The values for the three altitudes are 177.82 kPa, 7.99 kPa and 0.62 kPa. The value at 20 km altitude is much higher than the usually assumed 90 to 100 kPa, the value at 40 km altitude is at the border of acceptance, that at 60 km is much too low.

The acceptable level of q_∞ is between 10 and 90 kPa.

Problems of Chapter 3

Problem 3.1

Both the flight situation and the wind-tunnel scenario represent the same physical phenomena but in different reference frames. According to the principle of Galilean invariance, the laws of motion are the same in all inertial frames. In the flight scenario, the vehicle moves through the atmosphere, while in the wind-tunnel, the vehicle is stationary, and the air moves past it. Since the flow problem is Galilean invariant, the relative motion of the vehicle with respect to the air or the air with respect to the vehicle doesn't change the nature of the flow dynamics. Therefore, both scenarios are valid for studying aerothermodynamic phenomena.

Problem 3.2

At higher flight speeds (Mach numbers), vehicles typically fly at higher altitudes where the atmospheric density is lower. The reason for that it the demand that the dynamic pressure q_∞ stays in the range of 10 kPa $\lessapprox q_\infty \lessapprox$ 90 kPa.

The Reynolds number is given by $Re = \rho v L / \mu$, where ρ is the density, v is velocity, L is the characteristic length, and μ is the dynamic viscosity. As altitude increases, ρ decreases significantly, leading to lower Reynolds numbers despite the increase in velocity.

In hypersonic flight, lower Reynolds numbers indicate a shift towards more laminar flow, which affects boundary-layer characteristics, heat transfer, and skin friction. Consequently, this requires careful design consideration to manage thermal loads, to predict boundary-layer transition, and to optimize vehicle performance.

In the ram/sramjet domain up to flight Mach numbers $M_\infty \approx 12$ the Reynolds numbers are always high enough such that laminar-turbulent transition happens. The sufficiently accurate prediction of the location and the shape of the transition zone is the still unmastered problem.

Problem 3.3

The surface temperature of a hypersonic vehicle influences the boundary-layer flow. A hotter surface reduces the wall shear stress due to the boundary layer, leading to lower skin-friction drag. This effect is much stronger in turbulent flow than in laminar flow. Of course, the influence on the form drag, i.e., the viscous-effects induced pressure drag, must be taken into account. High surface temperatures can affect the laminar-turbulent transition, because certain stability modes can be influenced [1].

Conversely, a cooler surface increases wall shear stress, causing higher skin-friction drag. For the vehicle design, it is critical to balance thermal protection with the need to reduce drag, especially for surfaces where the boundary-layer flow is turbulent.

Problem 3.4

The Oswatitsch Mach-number independence principle states that for sufficiently high flight Mach numbers (typically $M_\infty \gtrapprox 5$), the force coefficients of a body (such as drag and pressure coefficients) become independent of M_∞ [1, 2]. That is because at high flight Mach numbers the flow properties downstream of the bow shock become independent of the Mach number. All, however, depends on the shape of the body under consideration. The shock structure, force coefficients and so on stabilize, leading to this independence.

The principle simplifies the analysis and design of hypersonic vehicles by allowing predictions of aerodynamic behavior without needing to account for changes in Mach number beyond a certain threshold. That, however, must be known with sufficient accuracy.

Problem 3.5

The Weiland mapping rule says that for the blunt shapes of re-entry vehicles the force coefficients as functions of the angle of attack for different flight Mach numbers can be collapsed into one curve. Whether and to what extend it can be applied to the slender shapes of the hypersonic airbreathers considered here, has not yet been established. For the SÄNGER configuration at least the results look promising.

The value of the mapping rule basically is to check generated data, not to abstain from generating data.

Problem 3.6

Surface radiation emissivity plays a crucial role in thermal protection by allowing a vehicle surface to radiate heat away. A high emissivity coefficient (ϵ close to 1) effectively reduces the thermal loads (wall temperature T_w and heat flux into the structure q_w) on the vehicle's structure. This allows to maintain surface temperatures during flight, which can be handled by metallic and non-metallic structures. Actually it is the means that allows hypersonic flight at all, which also holds for re-entry flight from Earth orbit [1].

Important is that non-convex effects, for instance at the root of a stabilizing surface, strongly reduce the cooling effect.

Surfaces with high emissivity may lead to higher viscous drag. This holds for turbulent boundary layers, where skin-friction drag increases with decreasing wall temperature. A surface with low emissivity will lead to a higher T_w, but possibly requiring active cooling means or more robust thermal protection material. The optimal design for a hypersonic vehicle must balance surface radiation emissivity to manage thermal loads while minimizing performance penalties such as increased viscous drag.

Problem 3.7

Bluntness of the nose and of leading edges is a critical feature in hypersonic vehicle design. A sharper nose or leading edge reduces wave drag, because it minimizes the strength of the bow shock, which is formed at supersonic speeds. However, the smaller the radius of the nose or edge is, the thinner is the boundary layer there, and in consequence, the higher are the thermal loads.

Hence a trade-off is needed between wave drag and thermal loads. In the extreme case, when wave drag minimization has highest priority, active cooling must be applied. That, however, leads to increased systems complexity. Designers for a given mission hence must master the trade-off between wave drag, thermal loads and systems complexity.

Problem 3.8

The surface of the external flow path of the flight vehicle must have several properties.

(1) The most important one is that an effective surface-radiation cooling is assured. That is given with a high radiation emissivity coefficient—$\epsilon \rightarrow 1$, actually $\epsilon \approx 0.8$ to 0.9—, which is achieved with a coating of the vehicle surface. That coating leads to an almost black appearance of the surface material.

(2) Surface roughness and surface irregularities of any kind must be sub-critical, such that in laminar flow no premature laminar-turbulent transition is triggered, and in turbulent flow no increase of the skin friction and the heat transport happens.

(3) The surface must be anti-catalytic in order to avoid extra thermal loads. It further needs, depending on the surface material, also anti-oxidation properties. Both are achieved with appropriate properties of the coating.

The properties (2) and (3) also hold for the surfaces of the internal flow path—i.e., those of the engine flow path.

Problem 3.9

Expansion joints are needed, if different surface materials are employed. An example for instance is that of the lower stage of the TSTO-system SÄNGER. The hot primary structure can be of titanium, nose and leading edges, due to their small radii, must be of an advanced ceramic. The material temperature increases with increasing Mach number and flight time. The materials expand and the expansion joints are needed in order to avoid mutual deformation and damage.

Problem 3.10

Regarding boundary-layer flow separation, two basic forms, ordinary separation and flow-off separation, can be distinguished [3]. The first one is the classical separation, the other for instance that, if the flow leaves a trailing edge. Trailing edges are usually considered as being sharp. In the reality of aircraft structures they have a thickness, which can reach up into the centimeter range.

If the boundary-layer flow is laminar, it can be computed with high accuracy with numerical solutions of the Navier-Stokes (NS) equations. That holds for both basic forms of separation. At a thick trailing edge the flow can be vortex-street like, which can be handled.

Turbulent boundary-layer flow and mild separation is treatable with the Reynolds-averaged (RANS), or the Favre-averaged (FANS) equations. The usually present unsteadiness can be treated with the unsteady forms of these equations (URANS or UFANS).

Once massive unsteady separation is given, scale-resolving methods like the large-eddy-simulation (LES), for which a number of variants exist, must be employed. Because of the high computer demands they can be used with zonal approaches—RANS for the attached flow portion, LES for the separated flow portion. However, it must be noted, that a practical universal approach is not yet reached.

Another approach is the direct numerical simulation (DNS), which, however, needs very high computer power and storage. Its fields of application therefore are only small-scale scientific simulations of transition and turbulence phenomena.

Problems of Chapter 4

Problem 4.1

The transition from ramjet to scramjet propulsion occurs at flight Mach number $M_\infty \approx 7$. This is because, at higher Mach numbers, the air entering the ramjet is decelerated to subsonic speeds, causing excessive heating. Oxygen dissociation starts occurring, which hampers combustion. The scramjet, which operates with supersonic combustion, avoids these issues by allowing the airflow through the engine to remain supersonic.

Problem 4.2

Substitute the given values into the dynamic pressure formula: $q_\infty = \frac{1}{2} \times 1.4 \times 1200$ Pa $\times 8^2 = 53.76$ kPa.

Thus, the dynamic pressure at Mach 8 is $q_\infty = 53{,}76$ kPa.

Problem 4.3

In a scramjet engine, the airflow remains supersonic throughout the engine, resulting in very short residence times for the air in the combustion chamber. Consequently, fuel injection, mixing, and combustion must occur very rapidly. The challenges include: 1. Injecting fuel into the high-speed airstream. 2. Ensuring proper mixing of fuel with air within a short distance. 3. Achieving ignition and sustained combustion in a highly compressed and fast-moving air stream.

Problem 4.4

Thermal blockage occurs when excessive heat is added to the supersonic airflow in the combustion chamber, causing the flow velocity to decrease towards Mach 1 (sonic speed). As heat is added, the flow decelerates, and if enough heat is added, the Mach number can drop below 1, leading to subsonic conditions. This can cause structural damage, changes in pressure distribution, and potentially leads to inlet unstart, where the engine stops functioning correctly.

Problem 4.5

When $\Phi > 1$ (rich mixture), more fuel is present than needed for complete combustion, leading to increased thrust. However, this also results in higher fuel consumption, reducing fuel economy. – When $\Phi < 1$ (lean mixture), less fuel is available than needed for complete combustion, leading to decreased thrust but improved fuel efficiency. The trade-off is between achieving maximum thrust for high-speed operation (at the cost of fuel economy) and conserving fuel for extended missions at the expense of thrust.

Problem 4.6

The inlet isolator in a scramjet engine has two main functions: 1. To increase static pressure ahead of the combustion chamber, ensuring efficient combustion. 2. To prevent disturbances (such as pressure fluctuations from the combustion process) from propagating upstream to the inlet, which could cause inlet unstart and disrupt the entire propulsion process. Maintaining stable supersonic flow is critical for engine operation, and the isolator plays a vital role in preserving this stability.

Problem 4.7

Shock-wave/boundary-layer interaction can lead to boundary-layer separation, causing an increase in pressure and thermal loads on the engine structure. This can reduce the efficiency of the inlet, increase drag, and potentially lead to blockage of the airflow, reducing thrust. At high speeds, such interactions become more severe, increasing the likelihood of engine unstart and causing damage due to the excessive thermal and mechanical stresses.

Problem 4.8

Relaminarization or reverse transition can happen with several mechanisms, one of them being strong acceleration of the boundary-layer flow [1].

The boundary-layer flow leaving the combustion chamber most probably is turbulent. In the nozzle, here a SERN, the flow is highly accelerated, which holds for the boundary layer, too. A turbulent boundary layer, when strongly accelerated, undergoes relaminarization. Hence the flow can be assumed to be laminar. The problem is a robust criterion at which location it happens.

Problems of Chapter 5

Problem 5.1

Benefits of API-Type 1 (SR-71):

– Reduced aerodynamic loads on the wing due to engine positioning.

– Undisturbed flow to the engine inlets, therefore no need for a boundary-layer diverter.

Disadvantages of API-Type 1:

– Pre-compression happens mainly in the inlet, no contribution from the forebody.

– Exposed to asymmetric flow conditions during changes in angle of attack or sideslip.

– Higher moments of inertia, making control more challenging.

– High yawing moments if one engine fails.

Benefits of API-Type 5 (SÄNGER):

– Synergistic pre-compression by the forebody, reducing inlet capture area and providing lift.

– More compact design with the engine integrated into the airframe.

– Smaller engine due to pre-compression from the forebody, improving performance.

Disadvantages of API-Type 5:

– Thick boundary layer at the inlet location, requiring a boundary-layer diverter for turbojet mode.

– Aerothermoelastic effects from the forebody affect the engine flow path and overall stability. API-Type 5 is more complex because it involves a higher level of interaction between the airframe and the propulsion system, including the problem of aerothermoelastic effects, which influence thrust, lift, and overall vehicle stability.

Problem 5.2

The forebody of hypersonic vehicles like SÄNGER is designed to act as a pre-compression surface, channeling and compressing the incoming airflow ahead of the inlets. The pre-compression reduces the size of the inlet capture area and enhances the overall performance by reducing drag and increasing the efficiency of the propulsion system. The ratio h_2/h_1 represents the compression of the stream tube, where h_1 is the initial height of the stream tube and h_2 is the compressed height after the airflow passes the bow shock of the inclined forebody. The relationship is:

– As the inclination angle of the lower side of the forebody $\overline{\alpha}$ increases, the ratio h_2/h_1 decreases, meaning the stream tube is compressed more, leading to higher pre-compression.

– Higher pre-compression improves engine performance by increasing mass flow through the engine, leading to higher thrust. However, it also increases drag, lift, and thermal loads on the vehicle.

Problem 5.3

The lower side of the forebody at angle of attack $\overline{\alpha}$ being flat achieves optimum pre-compression, lift and—between the two separation lines near and below the leading edges—two-dimensional inlet onset flow.

Problem 5.4

Two-dimensional inlet onset flow leads to maximum mass flow toward the external inlet ramps, hence onto the inlet. Studies have been made with a conical axisymmetric forebody and hence an circular arrangement of the inlets. That also leads to maximum mass flow toward the inlets.

Problem 5.5

Hypersonic airbreathers starting from the ground have turbojet engines, which are operational up to $M_\infty \approx 3$ to 4. The flow entering the compressor of the engine must be free of distortions, due, for instance, to the forebody boundary layer. If the engine is located in a gondola, i.e., a podded engine like the ones shown in Fig. 1.2, no such distortions are present.

A ram or scram engine is not sensitive to distortions, hence a boundary-layer diverter is not needed.

The propulsion system of the SÄNGER lower stage is a turbo/ramjet system. In the turbo mode a boundary-layer diverter is necessary, not in the ram mode. In the latter mode the diverter is closed.

The X-43A has only a scram engine, hence a boundary-layer diverter is not needed. However, in order to have unambiguous flow conditions in the combustion chamber, a boundary-layer trip was placed at some distance ahead of the inlet in order to provide for turbulent flow in the combustion chamber.

Problem 5.6

Net thrust sensitivity refers to the high sensitivity of the vehicle's net thrust to small changes in angle of attack α. For hypersonic vehicles like the SÄNGER lower stage, a small increase or decrease in α significantly affects pre-compression, which in turn alters the mass flow into the engine, thereby affecting thrust.

At higher Mach numbers (e.g., $M_\infty = 6.8$), even a 1-degree change in α can lead to a 7% reduction or 5% increase in net thrust. This is because:

– High Mach numbers amplify the effects of pre-compression, making the propulsion system more sensitive to changes in airflow.

– The forces acting on the vehicle (inlet drag, gross thrust) are large, and the net thrust is the small difference between these forces, so small changes in airflow can have a significant impact.

– The aerothermoelastic properties of the forebody play a deciding role.

Problem 5.7

In API-type 4 and 5 vehicles, such as the SÄNGER lower stage, the thrust-vector angle is not fixed due to the use of Single Expansion Ramp Nozzles (SERN). The thrust-vector angle influences:

– Performance balance: The vertical and horizontal components of the thrust vector affect overall performance.

– Longitudinal stability: If the thrust-vector angle changes with the angle of attack, it can either increase or decrease stability.

– Trim balance: Changes in thrust-vector angle can necessitate trim adjustments to maintain balance.

For example, in the turbojet mode at $M_\infty = 1.2$, the thrust-vector angle is highly negative (–32.7°), leading to a strong pitch-up moment. In contrast, in ramjet mode at higher Mach numbers, the thrust-vector angle becomes smaller (e.g., 3.7° at $M_\infty = 6.8$), leading to more balanced performance.

Problem 5.8

The shock-on-lip situation, the critical mode, reflects the design condition. The mass flow demanded by the engine is the maximum mass flow the inlet can capture. The drawback is that an Edney interaction of type IV exists at the cowl lip. The vehicle bow shock and or the ramp shocks interact with the bow shock of the cowl lip. Resulting are high pressure and thermal loads. The cowl lip on the one hand should be thin in order to reduce wave drag, on the other hand it should be thick enough to allow sufficient radiation cooling. Active cooling means demand internal space and leads to increased systems complexity.

Problem 5.9

At 30 km altitude the static temperature of the standard atmosphere is $T = 226.509$ K. Accordingly the speed of sound is $a_\infty = \sqrt{\gamma R T_\infty} = 301.7$ m/s. The flight speed hence is $v_\infty = 2413{,}6$ m/s. The g-reduction, Eq. (5.1), then is $g_{eff}/g = 0.906\ g$.

The passenger effort should be affected only mildly. Catering should be possible in the usual way. Moving around, however, should happen with some care.

Problem 5.10

Up to the flight Mach number $M_\infty \approx 4.5$ the total H2 coolant flow of the propulsion system is rather small compared to the combustion H2 flow. Then it increases strongly, because of the cooling needs in particular of the ram duct. Nozzle cooling demands remain more or less constant up to the release of the upper stage. After release of the upper stage, the combustion H2 flow goes down strongly, whereas the other cooling needs remain high until they drop strongly for the return flight with decreasing flight Mach number.

Problem 5.11

Acceleration pinch points on the flight trajectory are those where the thrust surplus is small. The first of those pinch points is the transonic flight regime around $M_\infty = 1$ due to the transonic drag divergence. The second one is around $M_\infty = 3.5$ to 4, when the switch over from turbojet to ramjet propulsion happens. The third one is around $M_\infty \approx 7$, when the upper stage is to be released and ramjet propulsion comes to its limits.

Problems of Chapter 6

Problem 6.1

Non-winged Re-entry Capsules:

– Structure: Simple, conical shape with a blunt base. The structure is made from lightweight metals, possibly steel in some parts.

– Thermal Protection: A thermal protection system (TPS) or heat shield is placed at the blunt nose (windward side) to handle high heat flux during re-entry. The rest of the structure is less exposed to high thermal loads.

– Reason for Design: These vehicles experience less complex aerodynamic loads due to their small lift-to-drag ratio (L/D $\approx$ 0.3–0.4) and do not require control surfaces for aerodynamic maneuvering during re-entry.

Winged Re-entry Vehicles:

– Structure: More complex, with fuselage frame-stringer-skin structures similar to transport aircraft. A large part of the structure can be made from aluminum alloys.

– Thermal Protection: Requires more extensive thermal protection, including reusable surface insulation tiles (e.g., on the Space Shuttle Orbiter). The lower side, leading edges, and control surfaces are particularly protected.

– Reason for Design: The lift-to-drag ratio (L/D $\approx$ 1–5 depending on speed) is higher, and the vehicle flies at a high angle of attack to generate drag and reduce speed. This increases thermal loads across larger parts of the vehicle.

The thermal protection needs are different because capsules have more centralized and concentrated thermal loads at the blunt nose, while winged vehicles require distributed protection due to their complex aerodynamic maneuvers and extended exposure during re-entry.

Problem 6.2

Hot Primary Structure:

– Materials: Advanced ceramics, superplastic-forming diffusion bonding (SPFDB) titanium, and silicon carbide (SiC) are commonly used for areas experiencing high thermal loads, such as the nose and leading edges of the wing and vertical stabilizers.

– Considerations: These areas face the highest temperatures during flight (up to 1,600 K at the nose in some cases), requiring materials with high temperature resistance and structural integrity under extreme conditions. Expansion joints may be necessary between ceramic and metal parts due to thermal expansion differences.

Cold Primary Structure:

– Materials: Titanium, flame retardant aluminum, and carbon fiber-reinforced polymer (CFRP) are used for the body of the vehicle, particularly in the fuselage and internal structures.

– Considerations: These materials are chosen for their ability to withstand moderate thermal loads (below 1,000 K) while providing the structural strength required for supporting the cryogenic fuel tanks and internal components. Weight minimization is crucial, especially since these parts experience lower thermal stress compared to the hot structure zones. The material choice reflects the need to manage both mechanical and thermal loads throughout the vehicle's structure.

Problem 6.3

Aerothermoelasticity refers to the interaction between aerodynamic forces, thermal loads, and elastic deformation of the structure in flight vehicles, especially those flying at hypersonic speeds.

– Impact of Thermal Loads: In hypersonic flight, the structure of the vehicle is subjected to extreme temperatures, which can cause material expansion and deformation. This deformation affects the aerodynamic properties, such as lift and drag, and changes the vehicle's attitude, possibly leading to control issues.

– Effects on Flight Stability: The deformation due to thermal loads, such as the "banana effect" (aerothermoelastic deformation of the forebody), can alter the effective angle of attack. With vehicles like the SÄNGER lower stage, even small changes in the angle of attack can significantly affect net thrust and stability, creating a highly sensitive system that requires careful control to maintain stability.

– Performance Impacts: If not properly accounted for, aerothermoelasticity can lead to reduced structural integrity, affecting performance parameters such as thrust, lift, and aerodynamic efficiency. Designing to account for these factors is critical to ensure safe and effective hypersonic flight.

In summary, aerothermoelasticity is essential for the design of hypersonic vehicles as it ensures that the structure can withstand thermal and mechanical loads while maintaining aerodynamic performance.

Problem 6.4

Effects on Thermal Loads:

– The transition from laminar to turbulent flow leads to an increase in surface friction and thermal loads on the vehicle's surface. Turbulent flow generates more skin friction drag and higher heat flux into the wall compared to laminar flow.

– On vehicles like the SÄNGER lower stage, the transition zone is critical, especially around areas such as external inlet ramps or leading edges. Incorrect predictions about where the transition occurs can lead to underestimating or overestimating thermal loads, which may compromise structural integrity.

Challenges in predicting location, shape and extent of the transition zone:

– The transition zone is influenced by numerous factors, basically flight conditions, local Mach number and Reynolds number, the thermal state of the surface, surface properties, vehicle geometry.

– Small disturbances in the airflow, such as shock-wave/boundary-layer interactions, can cause the transition to occur earlier or later than expected, making accurate prediction very difficult.

– Current computational models are still limited in their ability to capture the complexities of the transition process. Accurate prediction of location, shape and extent of the transition zone is critical in view of the aerothermodynamic properties of the flow, the design of thermal protection means and the structural concept, all in all in view of the flight performance of the vehicle including the performance of the propulsion system.

– Experimental simulation faces the fact that the radiation-adiabatic surface situation as conservative one cannot be simulated. The same holds for the receptivity environment. That is different from that in flight.

Problem 6.5

Zone 1 (Hot Structure):

– Location: Nose cone and fuselage section.

– Challenges: This zone experiences the highest thermal loads, particularly at the nose, where temperatures can reach up to 1,600 K. Advanced ceramics are required to handle these loads, and expansion joints must accommodate thermal expansion between different materials (ceramics and titanium).

Zone 2 (Cold Structure):

– Location: The central fuselage, housing the integral cryogenic liquid hydrogen (LH2) tank.

– Challenges: Insulating the cryogenic fuel from external heat sources is critical to prevent boil-off and ensure fuel efficiency. The structure must support the weight of the fuel while remaining insulated from the high thermal loads on the external fuselage.

Zone 3 (Hot Structure with Non-integral Tanks):

– Location: Aft fuselage, including the engine, nozzle, and non-integral bubble tanks for hydrogen.

– Challenges: This zone experiences both mechanical loads from engine thrust and thermal loads from the propulsion system. The structure must accommodate large bending and torsion moments while ensuring that the cryogenic tanks remain insulated from the surrounding hot structure.

Each zone requires careful material selection and structural design to balance the need for thermal protection, structural integrity, and weight efficiency.

Problem 6.6

Once the aerodynamic shape of the flight vehicle is fixed and with that the center of pressure ($cop(x, y, z)$), i.e., where the resultant aerodynamic force vector acts—changing somewhat with angle of attack and sideslip—the location of the center of gravity ($\mathrm{cog}(x, y, z)$) of the configuration comes in.[1] Changes of this location due, for instance, to the emptying of fuel tanks, or the release of the upper stage of a TSTO-system, need meticulous handling.

The location of the center of gravity of the flight vehicle now plays a large role. If center of pressure lies downstream of the center of gravity, the flight vehicle is considered to be statically stable, i.e., longitudinal stability (weathercock stability) is given. The so-called trimming is a means to correct an initially not ideal situation. If the *cop* lies upstream, the configuration is considered to be longitudinally unstable.

After the internal layout of the flight vehicle is finished, with that the arrangement of the masses of the subsystems etc. including that of the fuel, the situation may arise that the cog lies downstream of the cop. The resulting unstable flight can be handled with artificial stability. That would lead to a large increase of complexity and is an option only for fighter aircraft.

The way out is the application of ballast as far forward in the airframe as possible. This was the situation with the X-43A. The fact that about 390 kg of tungsten had to be placed in the nose region of the only about 3.7 m long flight vehicle indicates the very large cog problem, which arose during the internal layout of the flight vehicle.

Problem 6.7

The X-51A with $L = 7.6$ m is a rather short missile-type flight vehicle of API-type 3. The airframe appears to have a nearly rectangular cross-section. The pre-compression surface appears to have a length of one fourth of the total vehicle length. Compared to the forebody length of the lower stage of SÄNGER with about 55 m this is short and compact.

The flight time of the X-51A was about 200 s, the flight time with the booster up to start of the engine probably was of the same order. This is short compared to the flight time of SÄNGER up to upper stage release, which was about 1,800 s. Hence the time-integrates thermal loads on the airframe were comparably small.

[1] Note that the location of both the cog and the cop ideally should be a function of the location x only.

All in all this indicates that the aeroelastic/aerothermoelatic effects present with the X-51A were much smaller than those present with the lower stage of SÄNGER. The forebody pre-compression hence was less affected and hence also the total thrust of the ramjet propulsion system.

Problems of Chapter 7

Problem 7.1

The typical phases of aircraft product development include:

1. Product Definition:

– This phase includes conceptual studies, pre-design, and detailed design.

– The concept freeze happens at the end of the pre-design phase.

– Initial data sets, system simulations, and design sensitivities are identified during this phase.

– The design phase involves iterative loops across technical departments to settle the final design.

2. Development:

– Verification through ground simulation and high-performance computational simulation is carried out.

– System identification occurs through flight-envelope opening (testing prototypes to investigate operational properties and limits).

– Any design corrections are made based on test results.

3. Production:

– After flight testing and final design corrections, the aircraft is certified and mass production begins.

4. Support:

– Once operational, the aircraft enters the support phase, covering maintenance and operational lifetime support.

Key Milestones:

– Concept freeze (end of pre-design).

– Design verification (end of the design phase).

– Flight-envelope opening (during development).

– Certification and production.

Knowledge, Freedom for Changes, and Costs:

– Knowledge about the product increases gradually through the definition and development phases, reaching its highest level during flight-envelope opening.

– Freedom for changes decreases as the design solidifies, with fewer opportunities for significant changes once production begins.

– Cost of changes increases as development progresses, making early design corrections more cost-effective than late-stage modifications.

Problem 7.2

1. Thermal Loads and Material Selection:

– Hypersonic vehicles experience extremely high thermal loads (up to 6,000 K in certain areas), requiring advanced materials and coatings for radiation cooling and thermal protection.

– In contrast, subsonic and transonic aircraft do not face such extreme temperatures, making material selection simpler and less critical to structural integrity.

2. Aerothermodynamics:

– Hypersonic vehicles must deal with real-gas effects (e.g., oxygen dissociation), non-equilibrium effects, and complex boundary-layer behaviors (laminar-turbulent transition).

– Subsonic and transonic aircraft design does not need to account for such extreme high-speed aerodynamic phenomena, focusing more on simpler flow behavior.

3. Airframe-Propulsion Integration (API):

– Hypersonic airbreathers require a high degree of integration between the airframe and propulsion system, especially for API-type 4 and 5 vehicles, where forebody deformations influence propulsion performance.

– Subsonic and transonic aircraft generally have less complex propulsion integration issues, as engines are more independent of the airframe structure.

These challenges are unique to hypersonic vehicles due to their extreme flight conditions and the need for advanced computational and experimental tools to predict behavior accurately.

Problem 7.3

Computational Limitations:

1. Laminar-Turbulent Transition Prediction: Accurately predicting the location, shape and extent of the laminar-turbulent transition zone in hypersonic flow remains a major challenge. Computational models struggle to simulate this complex phenomenon with the required precision, impacting thermal load and aerodynamic predictions.

2. Combustion Modeling: Modeling the turbulent mixing and combustion processes within the scramjet engine accurately is still difficult. This limits the ability to simulate propulsion performance under varying flight conditions.

Experimental Limitations:

1. Size and Scaling Issues in Wind Tunnels: For large hypersonic vehicles like the SÄNGER lower stage (up to 82.5 m), accurately simulating flight conditions in wind tunnels is challenging due to model size limitations and the difficulty of maintaining appropriate Reynolds and Mach number similarity.

2. Thermal Surface Effects: Ground facilities struggle to replicate the true thermal conditions that hypersonic vehicles experience, particularly for radiation-cooled surfaces. Creating hot models with surface properties that simulate real-flight conditions is not yet feasible. These limitations mean that a combination of computational simulation and scaled-down or partial experimental testing is required to overcome these challenges.

Problem 7.4

Concept of the Virtual Product (VP): The Virtual Product is a high-fidelity, mathematical, and numerical representation of the physical properties and functions of an aircraft. It enables the simulation of the aircraft's design, development, and operational behavior in a virtual environment, minimizing the need for physical prototypes and testing.

Benefits: 1. Early Problem Detection: VP allows for early identification of design issues, reducing the need for costly late-stage corrections. 2. Reduced Development Time and Costs: By relying on high-fidelity simulations, the number of physical prototypes and test flights needed for certification is reduced, leading to faster development cycles and lower costs. 3. Optimization of Performance and Life-Cycle Costs: VP enables more efficient design optimization, ensuring that performance, operation, and maintenance costs are minimized throughout the product's life cycle.

Challenges: 1. Model Accuracy: Current computational models for hypersonic flight still have limitations, particularly in predicting non-linear phenomena such as laminar-turbulent transition and joint behavior in structures. 2. Integration into Industry: Implementing the Virtual Product approach requires significant changes in industry practices, from design methodologies to certification processes. This includes overcoming resistance to organizational and procedural changes.

Overall, while VP has the potential to revolutionize the development of hypersonic airbreathers, its successful implementation depends on improving simulation accuracy and fostering industry-wide acceptance.

Problem 7.5

Experimental hypersonic flight vehicles play a critical role in validating the technologies and designs proposed for future operational hypersonic vehicles. These experimental platforms allow designers to gather real-world data that complements computational simulations and ground-facility testing.
Essential Characteristics:

1. Representative Aerodynamics and Structures: The vehicle should closely mimic the aerodynamic shape and structural properties of the target operational vehicle to ensure relevant data on flow behavior, thermal loads, and aerothermoelasticity.

2. Scalability and Tolerance for Errors: Experimental vehicles must be designed to handle uncertainties in predictions, with provisions such as movable ballast to correct misjudged forces (e.g., pitching moments).

3. Accurate Data Measurement and Collection Systems: Advanced onboard sensors and systems must be in place to measure real-time flight conditions, including aerodynamic coefficients, structural behavior, and propulsion performance. Data should be stored onboard and transmitted to ground stations.

4. Simplified Propulsion and Control Systems: While retaining key features such as the ability to switch between propulsion modes (turbojet, ramjet, scramjet), the vehicle should have simplified systems to ensure reliability and ease of testing.

These characteristics enable experimental vehicles to simulate critical flight phenomena and provide valuable data for refining and validating designs for larger, operational hypersonic systems.

Problem 7.6

A re-entry vehicle flies a braking mission for instance from an Earth orbit, a hypersonic airbreather flies a cruise or an acceleration mission. The re-entry mission demands high drag, which is achieved by a blunt shape, and in the case of a winged re-entry vehicle by high angles of attack. Due to the high entry velocity thermal loads are extremely high and also the mechanical loads.

The airbreather flies at much lower speeds, but is drag critical. Thermal and mechanical loads are not as high as in the re-entry case, but are deciding in view of the structural concept. Light-weight and, depending on the API-type, sufficiently stiff structures are demanded.

Problems of Chapter 8

Problem 8.1

The HYTEDAV concept was developed toward the end of the German Hypersonics Technology Programme to provide a structured approach for developing and verifying the key technologies required for airbreathing hypersonic flight systems. Its key objectives were:

– Structuring the technology development and verification process:
It aimed to organize the process in a systematic way, ensuring that all necessary steps for technology maturation were considered.

– Identifying interdependencies: Understanding how various technology fields (e.g., propulsion, aerothermodynamics, structures) interact is critical for successful system design. – Identifying Transfer Models: Transfer Models were crucial for transferring data from sub-scale simulations to full-scale systems without requiring costly and complex ground testing at full scale.

– Addressing simulation needs: This included identifying the needs for ground-facility simulation, computational simulation, and in-flight simulation.

– Developing a simulation-hardware master plan: Creating a long-term plan for the necessary simulation tools, including wind tunnels, supercomputers, and experimental vehicles. HYTEDAV was developed because the complex challenges of hypersonic flight require a flexible, well-structured approach to ensure that critical technology gaps are addressed, while leveraging insights gained during the program.

Problem 8.2

1. Airframe/Propulsion Integration: The SÄNGER lower stage involves a highly coupled lift and propulsion system. The forebody of the vehicle must compress the airflow properly to feed the propulsion system, which leads to complex aerodynamic and structural integration issues. Any instability in the inlet could cause performance losses or even catastrophic unstart situations.
2. Thermal Loads: Hypersonic vehicles like the SÄNGER lower stage face extreme thermal loads. The vehicle must be designed to handle temperatures up to 2,000 K at the nose and leading edges. The selection of materials, passive cooling methods (radiation cooling), and active cooling systems in critical areas such as the inlet and nozzle are essential to maintain the structural integrity.
3. Structural Mass and Aeroelasticity: The need to minimize structural mass while ensuring structural integrity under both aerodynamic and thermal loads poses a significant challenge. Aeroelastic effects, especially at high Mach numbers, must be considered to prevent deformation that could affect the vehicle's performance. Aerothermoelasticity, where thermal and structural deformations are coupled, adds complexity to the design.

These challenges influence the vehicle's design by requiring advanced materials, precise aerodynamic shaping, and detailed computational and experimental simulations to predict and mitigate the effects of high-speed flight.

Problem 8.3

A Transfer Model is a multidisciplinary simulation tool or method designed to bridge the gap between sub-scale ground simulations and full-scale vehicle design, especially in cases where classical ground-facility verification is not feasible. The key role of a Transfer Model is to allow the accurate prediction of real-life vehicle behavior using data from simulations, reducing the reliance on full-scale testing.

Key requirements for developing a successful Transfer Model:

1. Advanced computational power: True teraflop-scale computing is necessary to handle the complex simulations involved in hypersonic vehicle design.

2. Accurate flow-physics and thermodynamics models: The Transfer Model must account for high-temperature gas dynamics, turbulence, and other flow phenomena.

3. Accurate structural models: Material properties, dynamic structural behavior, and joint modeling must be included to simulate the real behavior of the vehicle under combined mechanical and thermal loads.

4. Experimental validation: Experimental data from ground-facility tests and experimental vehicles are essential for validating the Transfer Model.

Transfer Models support hypersonic vehicle design by enabling designers to extrapolate results from smaller-scale tests to full-scale designs, ensuring the vehicle meets performance requirements without needing prohibitively expensive full-scale tests.

Problem 8.4

The primary purpose of experimental hypersonic vehicles within the HYTEDAV concept is to collect real-flight data that cannot be obtained from ground-based facilities. These vehicles serve as platforms for validating computational models, Transfer Models, and design methodologies in actual hypersonic flight conditions.

Two specific types of data that experimental vehicles help gather:

1. Laminar-Turbulent Transition Data: The behavior of boundary layers and the location, shape and extent of laminar-turbulent transition in high-speed, high-enthalpy environments can only be observed accurately in flight. This data is critical for refining flow-physics models and improving surface material and thermal protection system designs.

2. Aerothermoelastic Behavior: Experimental vehicles provide insights into how structural components behave under the combined influence of aerodynamic forces and thermal loads (aerothermoelastic effects). Ground facilities cannot replicate the exact coupling of these forces over long durations, high temperature levels, and at the required scales.

Experimental vehicles are vital for validating key assumptions and improving the accuracy of simulation tools.

Problem 8.5

The triad of ground-facility simulation, computational simulation, and in-flight simulation forms the foundation for the design and verification of hypersonic vehicles:

1. Ground-Facility Simulation: This includes wind tunnels, shock tubes, and other physical testing setups that allow sub-scale models to be tested under controlled conditions. Ground facilities are essential for understanding basic aerodynamic, structural, and propulsion characteristics of components, but they are limited by size and the inability to replicate all flight conditions, such as high-enthalpy environments.

2. Computational Simulation: Computational tools (CFD, CSM) allow for detailed analysis of aerodynamics, structural mechanics, and thermodynamics. These simulations can explore a wide range of conditions beyond what is feasible in ground facilities, but they rely on accurate modeling and validation against experimental data.

3. In-Flight Simulation: In-flight testing with experimental vehicles provides real-world data that is impossible to replicate fully in ground-facility simulations or computational models. This includes laminar-turbulent transition, turbulence, and full aerothermoelastic interactions.

Complementarity: Each component addresses the limitations of the others. Ground simulations provide the experimental foundation for validating computational models. Computational simulations fill in gaps and extend testing beyond the capabilities of physical experiments. In-flight testing validates both ground and computational methods by providing data under real operating conditions. Together, they ensure a comprehensive and validated design process

References

1. Hirschel, E.H.: Basics of Aerothermodynamics: Second, Revised. Springer, Berlin (2015). Progress in Astronautics and Aeronautics, AIAA, Reston, Va, vol. 204. Springer, Heidelberg (2004)
2. Hirschel, E.H., Weiland, C.: Selected Aerothermodynamic Design Problems of Hypersonic Flight Vehicles. Progress in Astronautics and Aeronautics, AIAA, Reston, Va, vol. 229. Springer, Heidelberg (2009)
3. Hirschel, E.H., Rizzi, A., Breitsamter, C., Staudacher, W.: Separated and Vortical Flow in Aircraft Wing Aerodynamics. Springer, Heidelberg (2021)

Appendix A
The Earth Atmosphere

Airbreathing hypersonic flight with the currently envisaged flight Mach numbers is to happen in the Earth atmosphere with altitudes up to about 50 to 60 km. In this appendix some basic features and properties of the Earth atmosphere are provided. The reader must be aware of the fact that these numbers are average numbers, which partly depend strongly on the geographical latitude, and that they are changing with time (seasons, atmospheric tides, sun-spot activities). A large number of reference and standard atmosphere models is discussed in [1], where also model uncertainties and limitations are noted.

The Earth atmosphere consists of several layers, Fig. A.1. The weather phenomena occur mainly in the troposphere, where introduced contaminants are mixed and dispersed by fluctuations, which are only weakly present at higher altitudes.

The temperature first decreases with altitude, but reaches a temperature plateau of around 220 K to 230 K in the stratosphere. In the mesosphere again it becomes colder, in the thermosphere the temperature rises fast with altitude, Table A.1. Between 18 and 25 km altitude the ecologically important vulnerable ozone layer is found.

The pressure p and also the density ρ decrease fast with increasing altitude. Because the temperature T does not change much, the curves of p and ρ look similar. The dynamic viscosity μ is a function of the temperature alone and can be determined with the Sutherland formula. That formula becomes inaccurate at high altitudes. Therefore the data are restricted to altitudes $H \lessapprox 86$ km. An empirical relation comparable to Sutherland's formula is used for the calculation of the thermal conductivity k. Again the data are restricted to altitudes $H \lessapprox 86$ km. For details and approximations see [2].

E. H. Hirschel et al., *Elements of Hypersonic Airbreather Design and Development*,
https://doi.org/10.1007/978-3-031-94219-8

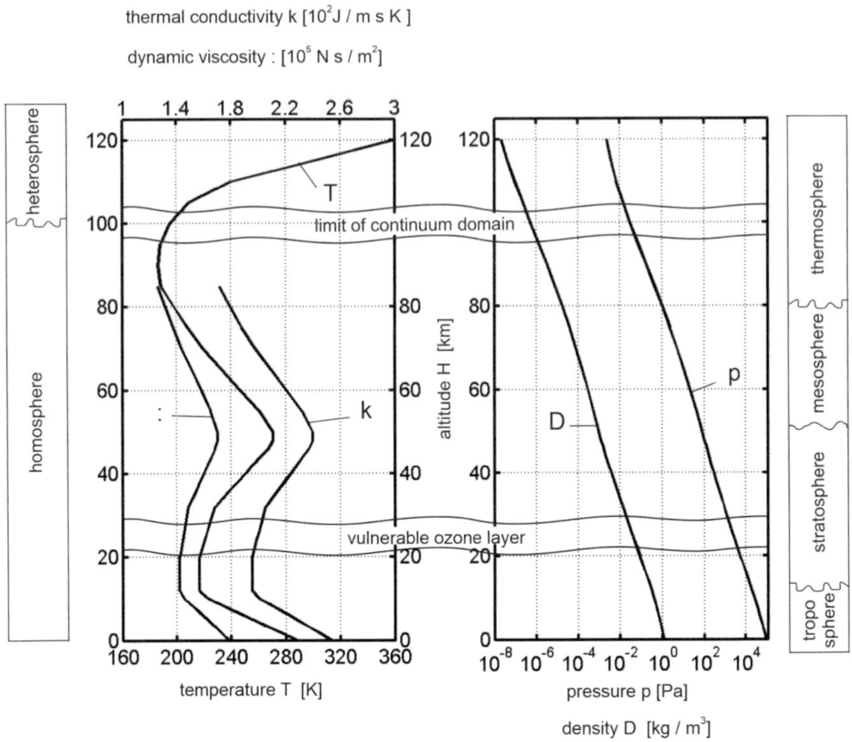

Fig. A.1 Atmospheric properties as function of the altitude [3]

The composition of the atmosphere can be considered as constant in the homosphere, up to approximately 100 km altitude. In the heterosphere, above 100 km altitude, it changes with altitude.

In aerothermodynamics we work usually with the U. S. standard atmosphere [4] in order to determine static pressure (p_∞), density (ρ_∞), static temperature (T_∞) et cetera as function of the altitude, Table A.1. The 15 °C standard atmosphere assumes a temperature of 15 °C at sea level.

Table A.1 Properties of the 15 °C U.S. standard atmosphere as function of the altitude H [4]

Altitude H [km]	Temperature T [K]	Pressure p [Pa]	Density ρ [kg/m^3]	Dynamic viscosity μ [$N \cdot s/\text{m}^2$]	Thermal conductivity k [$W/(m\ K)$]
0.0	288.150	$1.013 \cdot 10^5$	$1.225 \cdot 10^0$	$1.789 \cdot 10^{-5}$	$2.536 \cdot 10^{-2}$
1.0	281.651	$8.988 \cdot 10^4$	$1.112 \cdot 10^0$	$1.758 \cdot 10^{-5}$	$2.485 \cdot 10^{-2}$
2.0	275.154	$7.950 \cdot 10^4$	$1.007 \cdot 10^0$	$1.726 \cdot 10^{-5}$	$2.433 \cdot 10^{-2}$
3.0	268.659	$7.012 \cdot 10^4$	$9.092 \cdot 10^{-1}$	$1.694 \cdot 10^{-5}$	$2.381 \cdot 10^{-2}$
4.0	262.166	$6.166 \cdot 10^4$	$8.193 \cdot 10^{-1}$	$1.661 \cdot 10^{-5}$	$2.329 \cdot 10^{-2}$
5.0	255.676	$5.405 \cdot 10^4$	$7.364 \cdot 10^{-1}$	$1.628 \cdot 10^{-5}$	$2.276 \cdot 10^{-2}$
6.0	249.187	$4.722 \cdot 10^4$	$6.601 \cdot 10^{-1}$	$1.595 \cdot 10^{-5}$	$2.224 \cdot 10^{-2}$
7.0	242.700	$4.110 \cdot 10^4$	$5.900 \cdot 10^{-1}$	$1.561 \cdot 10^{-5}$	$2.170 \cdot 10^{-2}$
8.0	236.215	$3.565 \cdot 10^4$	$5.258 \cdot 10^{-1}$	$1.527 \cdot 10^{-5}$	$2.117 \cdot 10^{-2}$
9.0	229.733	$3.080 \cdot 10^4$	$4.671 \cdot 10^{-1}$	$1.493 \cdot 10^{-5}$	$2.063 \cdot 10^{-2}$
10.0	223.252	$2.650 \cdot 10^4$	$4.135 \cdot 10^{-1}$	$1.458 \cdot 10^{-5}$	$2.009 \cdot 10^{-2}$
12.0	216.650	$1.940 \cdot 10^4$	$3.119 \cdot 10^{-1}$	$1.421 \cdot 10^{-5}$	$1.953 \cdot 10^{-2}$
14.0	216.650	$1.417 \cdot 10^4$	$2.279 \cdot 10^{-1}$	$1.421 \cdot 10^{-5}$	$1.953 \cdot 10^{-2}$
16.0	216.650	$1.035 \cdot 10^4$	$1.665 \cdot 10^{-1}$	$1.421 \cdot 10^{-5}$	$1.953 \cdot 10^{-2}$
18.0	216.650	$7.565 \cdot 10^3$	$1.216 \cdot 10^{-1}$	$1.421 \cdot 10^{-5}$	$1.953 \cdot 10^{-2}$
20.0	216.650	$5.529 \cdot 10^3$	$8.891 \cdot 10^{-2}$	$1.421 \cdot 10^{-5}$	$1.953 \cdot 10^{-2}$
22.0	218.574	$4.047 \cdot 10^3$	$6.451 \cdot 10^{-2}$	$1.432 \cdot 10^{-5}$	$1.969 \cdot 10^{-2}$
24.0	220.560	$2.972 \cdot 10^3$	$4.694 \cdot 10^{-2}$	$1.443 \cdot 10^{-5}$	$1.986 \cdot 10^{-2}$
26.0	222.544	$2.188 \cdot 10^3$	$3.426 \cdot 10^{-2}$	$1.454 \cdot 10^{-5}$	$2.003 \cdot 10^{-2}$
28.0	224.527	$1.616 \cdot 10^3$	$2.508 \cdot 10^{-2}$	$1.465 \cdot 10^{-5}$	$2.019 \cdot 10^{-2}$
30.0	226.509	$1.197 \cdot 10^3$	$1.841 \cdot 10^{-2}$	$1.475 \cdot 10^{-5}$	$2.036 \cdot 10^{-2}$
32.0	228.490	$8.891 \cdot 10^2$	$1.355 \cdot 10^{-2}$	$1.486 \cdot 10^{-5}$	$2.053 \cdot 10^{-2}$
34.0	233.743	$6.634 \cdot 10^2$	$9.887 \cdot 10^{-3}$	$1.514 \cdot 10^{-5}$	$2.096 \cdot 10^{-2}$
36.0	239.282	$4.985 \cdot 10^2$	$7.258 \cdot 10^{-3}$	$1.543 \cdot 10^{-5}$	$2.142 \cdot 10^{-2}$
38.0	244.818	$3.771 \cdot 10^2$	$5.366 \cdot 10^{-3}$	$1.572 \cdot 10^{-5}$	$2.188 \cdot 10^{-2}$
40.0	250.350	$2.871 \cdot 10^2$	$3.996 \cdot 10^{-3}$	$1.601 \cdot 10^{-5}$	$2.233 \cdot 10^{-2}$
42.0	255.878	$2.200 \cdot 10^2$	$2.995 \cdot 10^{-3}$	$1.629 \cdot 10^{-5}$	$2.278 \cdot 10^{-2}$
44.0	261.403	$1.695 \cdot 10^2$	$2.259 \cdot 10^{-3}$	$1.657 \cdot 10^{-5}$	$2.323 \cdot 10^{-2}$
46.0	266.925	$1.313 \cdot 10^2$	$1.714 \cdot 10^{-3}$	$1.685 \cdot 10^{-5}$	$2.367 \cdot 10^{-2}$
48.0	270.650	$1.023 \cdot 10^2$	$1.317 \cdot 10^{-3}$	$1.704 \cdot 10^{-5}$	$2.397 \cdot 10^{-2}$
50.0	270.650	$7.978 \cdot 10^1$	$1.027 \cdot 10^{-3}$	$1.704 \cdot 10^{-5}$	$2.397 \cdot 10^{-2}$

(continued)

Table A.1 (continued)

Altitude H [km]	Temperature T [K]	Pressure p [Pa]	Density ρ [kg/m^3]	Dynamic viscosity μ [$N \cdot s/m^2$]	Thermal conductivity k [$J/(m \cdot s \cdot K)$]
55.0	260.771	$4.252 \cdot 10^1$	$5.681 \cdot 10^{-4}$	$1.654 \cdot 10^{-5}$	$2.318 \cdot 10^{-2}$
60.0	247.021	$2.196 \cdot 10^1$	$3.097 \cdot 10^{-4}$	$1.584 \cdot 10^{-5}$	$2.206 \cdot 10^{-2}$
65.0	233.292	$1.093 \cdot 10^1$	$1.632 \cdot 10^{-4}$	$1.512 \cdot 10^{-5}$	$2.093 \cdot 10^{-2}$
70.0	219.585	$5.221 \cdot 10^0$	$8.283 \cdot 10^{-5}$	$1.438 \cdot 10^{-5}$	$1.978 \cdot 10^{-2}$
75.0	208.399	$2.388 \cdot 10^0$	$3.992 \cdot 10^{-5}$	$1.376 \cdot 10^{-5}$	$1.883 \cdot 10^{-2}$
80.0	198.639	$1.052 \cdot 10^0$	$1.846 \cdot 10^{-5}$	$1.321 \cdot 10^{-5}$	$1.800 \cdot 10^{-2}$
85.0	188.893	$4.457 \cdot 10^{-1}$	$8.220 \cdot 10^{-6}$	$1.265 \cdot 10^{-5}$	$1.716 \cdot 10^{-2}$
90.0	186.870	$1.836 \cdot 10^{-1}$	$3.416 \cdot 10^{-6}$		
95.0	188.420	$7.597 \cdot 10^{-2}$	$1.393 \cdot 10^{-6}$		
100.0	195.080	$3.201 \cdot 10^{-2}$	$5.604 \cdot 10^{-7}$		
105.0	208.840	$1.448 \cdot 10^{-2}$	$2.325 \cdot 10^{-7}$		
110.0	240.000	$7.104 \cdot 10^{-3}$	$9.708 \cdot 10^{-8}$		
115.0	300.000	$4.010 \cdot 10^{-3}$	$4.289 \cdot 10^{-8}$		
120.0	360.000	$2.538 \cdot 10^{-3}$	$2.222 \cdot 10^{-8}$		

Appendix B
Constants, Functions, SI Dimensions and Conversions

We provide in Appendix B.1 constants and air properties. In Appendix B.2 the dimensions of the most important quantities are given. The dimensions are in general the SI units (Système International d'unités), see [5, 6], where also the constants given in Appendix C can be found. The basic units, the derived units, and conversions[1] to US customary units (English units) are given. Note that we write the dimensions as 'a/bc' instead of 'a/(bc)' or 'a b^{-1} c^{-1}'.

B.1 Constants and Air Properties

The molar universal gas constant R_0, the Stephan-Boltzmann constant σ, and standard gravitational acceleration g_0 are:

Molar universal gas constant R_0 $= 8.314472 \cdot 10^3$ kg m^2/s^2 kg-mol K
$= 4.97201 \cdot 10^4$ lb$_m$ ft^2/s^2 lb$_m$-mol °R,

Stephan-Boltzmann constant σ $= 5.670400 \cdot 10^{-8}$ W/m^2 K^4
$= 1.7123 \cdot 10^{-9}$ Btu/hr ft^2 °R^4.

Standard gravitational acceleration at sea level g_0 $= 9.80665$ m/s^2
$= 32.174$ ft/s^2.

[1] Details can be found, for instance, at http://physics.nist.gov/cuu/Units/units.html or at http://www.chemie.fu-berlin.de/chemistry/general/si_en.html.

E. H. Hirschel et al., *Elements of Hypersonic Airbreather Design and Development*,
https://doi.org/10.1007/978-3-031-94219-8

Table B.1 Molecular weights, gas constants, and intermolecular force parameters of air constituents for the low temperature domain [7, 8]. * is the U. S. standard atmosphere value, + the value from [8]

Gas	Molecular weight M [kg/kg-mol]	Specific gas constant R [$m^2/(s^2K)$]
Air	28.9644* (28.97^+)	287.06
N_2	28.02	296.73
O_2	32.00	259.83
NO	30.01	277.06
N	14.01	593.47
O	16.00	519.65
Ar	39.948	208.13
He	4.003	2077.06

In Table B.1 we list some properties of air and its major constituents. Listed are the molecular weights and the gas constants, all for the temperature domain of our interest.

B.2 Dimensions and Conversions

SI basic units and SI derived units are listed of the major flow, transport, and thermal quantities. In the left column name and symbol are given and in the right column the unit (dimension), with the symbol in → [ ] used in Appendix C, and in the line below its conversion to the U.S. customary units.

B.2.1 SI Basic Units

length, L

[m], → [L]
1 m = 100 cm = 3.28084 ft
(1,000 m = 1 km)

mass, m

[kg], → [M]
1 kg = 10^3 g = 2.20462 lb_m
1 to = 10^3 kg = 10^6 g

time, t	[s] (= [sec]), $\rightarrow$ [t]
temperature, T	[K], $\rightarrow$ [T] 1 K = 1.8 °R $\Rightarrow T_{Rankine} = T_{Fahrenheit} + 459.67$ $\Rightarrow T_{Kelvin} = (5/9)\, T_{Rankine}$ $\Rightarrow T_{Kelvin} = T_{Celsius} + 273.15$
amount of substance, *mole*	[kg-mol], $\rightarrow$ [mole] 1 kg-mol = 2.20462 lb_m-mol

B.2.2 SI Derived Units

area, A	[m^2], $\rightarrow$ [L^2] 1 m^2 = 10.76391 ft^2
volume, V	[m^3], $\rightarrow$ [L^3] 1.0 m^3 = 35.31467 ft^3
speed, velocity, v, u	[m/s], $\rightarrow$ [L/t] 1.0 m/s = 3.28084 ft/s
force, F	[N] = [kg m/s^2], $\rightarrow$ [M L/t^2] 1.0 N = 0.224809 lb_f
pressure, p	[Pa] = [N/m^2], $\rightarrow$ [M/L t^2] 1.0 Pa = 10^{-5} bar = $9.86923 \cdot 10^{-6}$ atm = = 0.020885 lb_f/ft^2
density, ρ	[kg/m^3], $\rightarrow$ [M/L^3] 1.0 kg/m^3 = 0.062428 lb_m/ft^3
(dynamic) viscosity, μ	[Pa s] = [N s/m^2], $\rightarrow$ [M/L t] 1.0 Pa s = 0.020885 lb_f s/ft^2
kinematic viscosity, ν	[m^2/s], $\rightarrow$ [L^2/t] 1.0 m^2/s = 10.76391 ft^2/s

shear stress, τ	[Pa] = [N/m^2], $\rightarrow$ [M/L t^2] 1.0 Pa = 0.020885 lb$_f$/ft^2
energy, enthalpy, work, quantity of heat	[J] = [N m], $\rightarrow$ [M L^2/t^2] 1.0 J = 9.47813$\cdot 10^{-4}$ BTU = = 23.73036 lb$_m$ft^2/s^2 = 0.737562 lb$_f$/s^2
(mass specific) internal energy, enthalpy, e, h	[J/kg] = [m^2/s^2], $\rightarrow$ [L^2/t^2] 1.0 m^2/s^2 = 10.76391 ft^2/s^2
(mass) specific heat, c_v,c_p specific gas constant, R	[J/kg K] = [m^2/s^2 K], $\rightarrow$ [L^2/t^{2}T] 1.0 m^2/s^{2}K = 5.97995 ft^2/s^2 °R
power, work per unit time	[W] = [J/s] = [N m/s], $\rightarrow$ [M L^2/t^3] 1.0 W = 9.47813$\cdot 10^{-4}$ BTU/s = = 23.73036 lb$_m$ft^2/s^3
thermal conductivity, k	[W/m K] = [N/s K], $\rightarrow$ [M L/t^{3}T] 1.0 W/m K = 1.60496$\cdot 10^{-4}$ BTU/s ft °R = = 4.018342 lb$_m$ ft/s^3 °R
heat flux, q	[W/m^2] = [J/m^{2}s], $\rightarrow$ [M/t^3] 1.0 W/m^2 = 0.88055$\cdot 10^{-4}$ BTU/s ft^2 = = 2.204623 lb$_m$/s^3
(binary) mass diffusivity, D_{AB}	[m^2/s], $\rightarrow$ [L^2/t] 1.0 m^2/s = 10.76391 ft^2/s
thermo diffusivity, D_A^T	[kg/m s], $\rightarrow$ [M/L t] 1.0 kg/m s = 0.67197 lb$_m$/ft s
diffusion mass flux, j	[kg/m^{2}s], $\rightarrow$ [M/L^{2}t] 1.0 kg/m^{2}s = 0.20482 lb$_m$/ft^{2}s

Appendix C
Symbols and Acronyms

Only the important symbols are listed. If a symbol appears only locally or infrequent, it is not included. In general the page number is indicated, where a symbol appears first or is defined. Dimensions are given in terms of the SI basic units: length [L], time [t], mass [M], and temperature [T], Appendix B. The dimensions are written as 'a/bc' instead of 'a/(bc)' or 'a b^{-1} c^{-1}'. For actual dimensions and their conversions see Appendix B.2.

C.1 Latin Letters

A	Surface, [L^2]
A_{ref}	Reference surface, [L^2]
a	Speed of sound, [L/t]
b	Width of combustor flow path, [L]
b	Wing span, [L]
C_D	Drag coefficient, [-]
C_L	Lift coefficient, [-]
C_m	Pitching moment coefficient [-]
C_X, C_Z	Axial and normal force coefficient, [-]
c_f	Skin-friction coefficient, [-]
c_p	Pressure coefficient, [-]
c_p	(Mass) specific heat at constant pressure, [L^2/t^2T]
c_v	(Mass) specific heat at constant volume, [L^2/t^2T]
D	Drag, [ML/t^2]
D_w	Wave drag, [ML/t^2]
F	Force, [ML/t^2]
g_0	Standard gravitational acceleration, [L/t^2]

E. H. Hirschel et al., *Elements of Hypersonic Airbreather Design and Development*,
https://doi.org/10.1007/978-3-031-94219-8

H	Altitude, [L]
I_{sp}	Specific impulse, [t]
K	Hypersonic similarity parameter, [-]
Kn	Knudsen number, [-]
k	Thermal conductivity, [ML/t^3T]
L	Length, [L]
L	Lift, [ML/t^2]
M	Mach number, [-]
M	Molecular weight, [M/mole]
M_∞	Flight Mach number, [-]
m, M	Mass, [M]
n	Load factor, [-]
p	Pressure, [M/Lt^2]
p_∞	Free-stream pressure, [M/Lt^2]
q	Heat flux, [M/t^3]
q_{gw}	Heat flux in the gas at the wall, [M/t^3]
q_w	Heat flux into the wall, [M/t^3]
q_{rad}	Thermal radiation heat flux, [M/t^3]
q_∞	Free-stream dynamic pressure, [M/Lt^2]
R	Gas constant, [L^2/t^2T]
R, r	Radius, [L]
R_0	Universal gas constant, [ML^2/t^2moleT]
Re	Reynolds number, [-]
Re^u	Unit Reynolds number, [1/L]
T	Thrust, [ML/t^2]
T	Temperature, [T]
T_{gw}	Temperature of the gas at the wall, [T]
T_{ra}	Radiation-adiabatic temperature, [T]
T_w	Wall temperature, [T]
T_r	Recovery temperature, [T]
T_0	Stagnation temperature, [T]
T_∞	Free-stream temperature, [T]
t	Time, [t]
u, v	Speed, [L/t]
u, v, w	Cartesian velocity components, [L/t]
u_∞, v_∞	Free-stream velocity, flight speed, [L/t]
x, y, z	Cartesian coordinates, [L]
x_N	Neutral-point location, [L]

C.2 Greek Letters

α Angle of attack, [°]
$\overline{\alpha}$ Surface inclination angle against the free stream, [°]
β Deflection angle, [°]
γ Flight-path angle, [°]
γ Ratio of specific heats, [-]
γ_{eff} Effective ratio of specific heats, [-]
Δ Characteristic boundary-layer thickness, [L]
Δ_0 Shock stand-off distance, [L]
δ Flow (ordinary) boundary-layer thickness, [L]
δ Ramp angle, [°]
δ Shock stand-off distance ($\delta \equiv \Delta_0$), [L]
ε Emissivity coefficient, [-]
ε_f Fictitious emissivity coefficient, [-]
$\eta_{ke,c}$ Kinetic energy recovery factor, [-]
$\eta_{pt,c}$ Total pressure recovery factor, [-]
θ Shock angle, [°]
Λ Aspect ratio, [-]
λ Mean free path, [L]
μ Viscosity, [M/Lt]
μ Mach angle, [°]
ν Kinematic viscosity, [L^2/t]
ν'_{ir}, ν''_{ir} Stoichiometric coefficients, [-]
ρ Density, [M/L^3]
τ_w Skin friction, wall shear stress, [M/t^2L]
Φ Angle, [°]
Φ Fuel-air equivalence ratio, [-]
φ, φ_0 Sweep angle of leading edge or cylinder, [°]

C.3 Indices

C.3.1 Upper Indices

^{u} Unit
* Reference-temperature/enthalpy value

C.3.2 *Lower Indices*

cog	Center of gravity
cc	Combustion chamber
$crit$	Critical
D	Drag
dd	Drag divergence
gw	Gas at the wall
L	Lift
N	Nose
n	Nozzle
r	Recovery
ra	Radiation adiabatic
rad	Radiation
ref	Reference
SL	Sea level
t	Total
$trim$	Trim, trimmed
vac	Vacuum
w	Wall
0	Reference
1	Ahead of the shock wave
2	Behind the shock wave
∞	Infinity

C.4 Acronyms

A/C	Aircraft
API	Airframe propulsion integration
ARV	Ascent and re-entry vehicle
ATR	Air turbo rocket
CAV	Cruise and acceleration vehicle
CFD	Computational fluid dynamics
CSM	Computational structure mechanics
DCR	Dual combustion ramjet
DMR	Dual mode ramjet
DNS	Direct numerical solution
EIS	Entry into service
JP	Jet propulsion fuel
LEO	Low Earth orbit

LH2	Liquid hydrogen
LCH4	Liquid methane
RAM	Ramjet
RANS	Reynolds-averaged Navier-Stokes
RV	Re-entry vehicle
SCRAM	Supersonic ramjet
SERN	Single expansion ramp nozzle
SSTO	Single stage to orbit
TRL	Technology readiness level
TSTO	Two stage to orbit
TPS	Thermal protection system
URANS	Unsteady Reynolds-averaged Navier-Stokes

Permissions

Figures reproduced with permission by

AGARD/RTO: Figs. 2.12, 3.40 to 3.42, 3.45, 5.22, 5.24, 5.28.
American Control Conference (ACC): Fig. 2.11,
American Institute of Aeronautics and Astronautics (AIAA): Figs. 1.1, 2.5, 2.12, 2.13, 2.16, 2.33 to 2.45, 3.32, 6.17 to 6.19, 6.22, 8.2, 8.3.
Aerospace America: Figs. 2.14, 2.15, 5.15.
ASME: Figs. 3.33, 4.2, 5.30, 5.31.
Astronautics and Aeronautics: Figs. 2.29, 2.30.
Birkhäuser: Figs. 3.2, 5.16.
CIMNE: Fig. 7.1.
DLR Braunschweig: Figs. 7.4 to 7.7.
IAF: Figs. 3.44, 6.9, 6.12, 6.21.
ICAS: Figs. 2.22, 2.23a.
Springer: Figs. 1.6b, 2.10, 2.20, 2.21, 3.3 to 3.9, 3.18, 3.19, 3.23, 4.3 to 4.5, 4.8, 5.4, 6.5, 6.6, 6.23, 6.24, 7.3.
Springer/Gorn: Figs. 1.4, 1.5, 1.6a, 6.13, 6.14.
J. Steelant: Figs. 1.2, 1.3, 5.9 to 5.12.
Verlag Dr. Hut: Figs. 2.2 to 2.4, 6.15, 6.16.
W. Schröder: Fig. 3.39.
ZFW: Figs. 6.7, 6.8.

The permission to reproduce Figs. 2.23b, 3.26, 3.43, 3.46, 5.6 to 5.8, 5.18, 5.19, 5.25, 5.32 to 5.35, 6.10, 6.11, which did originate in the German Hypersonics Technology Programme, was given by the programmes former director, Dr. H. Kuczera.

The permissions to reprint the figures provided directly from the authors—see the Acknowledgements at the beginning of the book—in all cases were given by the authors.

E. H. Hirschel et al., *Elements of Hypersonic Airbreather Design and Development*,
https://doi.org/10.1007/978-3-031-94219-8

References

1. N.N.: Guide to reference and standard atmosphere models. AIAA G-003 Revision C (2016)
2. Hirschel, E.H.: Basics of Aerothermodynamics: Second, Revised. Springer, Berlin (2015); Progress in Astronautics and Aeronautics, AIAA, Reston, Va, vol. 204. Springer, Heidelberg, New York (2004)
3. Hirschel, E.H., Weiland, C.: Selected Aerothermodynamic Design Problems of Hypersonic Flight Vehicles. Progress in Astronautics and Aeronautics, AIAA, Reston, Va, vol. 229. Springer, Berlin, Heidelberg (2009)
4. N.N.: U.S. standard atmosphere. Government Printing Office, Washington, D.C. (1976)
5. Taylor, B.N. (ed.): The international system of units (SI). US Department of Commerce, National Institute of Standards and Technology, NIST Special Publication 330, 2001. US Government Printing Office, Washington, D.C. (2001)
6. Taylor, B.N.: Guide for the use of the international system of units (SI). US Department of Commerce, National Institute of Standards and Technology, NIST Special Publication 811, p: US Government Printing Office, p. 1995. D.C, Washington (1995)
7. Hirschfelder, J.O., Curtiss, C.F., Bird, R.B.: Molecular Theory of Gases and Liquids. John Wiley & Sons, New York (1966)
8. Bird, R.B., Stewart, W.E., Lightfoot, E.N.: Transport Phenomena, 2nd edn. John Wiley & Sons, New York, London Sydney (2002)

Name Index

E. H. Hirschel et al., *Elements of Hypersonic Airbreather Design and Development*,
https://doi.org/10.1007/978-3-031-94219-8

Subject Index

E. H. Hirschel et al., *Elements of Hypersonic Airbreather Design and Development*,
https://doi.org/10.1007/978-3-031-94219-8

MIX
Papier aus verantwortungsvollen Quellen
Paper from responsible sources
FSC® C105338

If you have any concerns about our products,
you can contact us on
ProductSafety@springernature.com

In case Publisher is established outside the EU,
the EU authorized representative is:
Springer Nature Customer Service Center GmbH
Europaplatz 3, 69115 Heidelberg, Germany

Printed by Libri Plureos GmbH
in Hamburg, Germany